Classical Mechanics

Classical Mechanics

K. SANKARA RAO
Formerly Professor of Mathematics
Anna University, Chennai

PHI Learning Private Limited
Delhi-110092
2015

₹ 295.00

CLASSICAL MECHANICS
K. Sankara Rao

© 2005 by PHI Learning Private Limited, Delhi. All rights reserved. No part of this book may be reproduced in any form, by mimeograph or any other means, without permission in writing from the publisher.

ISBN-978-81-203-2676-7

The export rights of this book are vested solely with the publisher.

Fourth Printing ... ... **January, 2015**

Published by Asoke K. Ghosh, PHI Learning Private Limited, Rimjhim House, 111, Patparganj Industrial Estate, Delhi-110092 and Printed by Rajkamal Electric Press, Plot No. 2, Phase IV, HSIDC, Kundli-131028, Sonepat, Haryana.

To the memory of
Dr. VIKRAM SARABHAI,
the founding father of the Indian space programme, who inspired me to take interest in space related subjects as early as 1970

Contents

Preface *xi*

1. Kinematics and Dynamics of a Particle   **1–27**
    1.1 Introduction   1
    1.2 Newton's Laws of Motion   1
    1.3 Inertial Frames   3
    1.4 Units and Dimensions   4
    1.5 Conservation Laws   5
    1.6 Kinematics of a Particle   8
        1.6.1 Equations of Motion of a Particle   14
        1.6.2 Simple Pendulum   22
        1.6.3 The Harmonic Oscillator   26
    Exercises   27

2. Dynamics of a System of Particles and Rotating Coordinate Systems   **28–58**
    2.1 Motion of a System of Particles   28
    2.2 Principle of Angular Momentum   31
    2.3 Motion of a Rigid Body   34
        2.3.1 Angular Velocity   35
        2.3.2 Rigid Body Rotation about a Fixed Axis   36
        2.3.3 Rate of Change of a Vector in a Rotating Frame   38
    2.4 Moving Frames of Reference   40
        2.4.1 Frames of Reference with Translational Motion   40
        2.4.2 Motion of a Particle Relative to a Rotating Frame   41
        2.4.3 Uniformly Rotating Frames   43
        2.4.4 Absolute and Relative Velocity and Acceleration in Plane Motion   53
    2.5 Linear Impulse and Angular Impulse   56
    Exercises   57

3. Kinematics of a Rigid Body Motion   **59–112**
    3.1 Moments and Products of Inertia   59
        3.1.1 Moment of Inertia of a Body about Any Line Through the Origin of a Coordinate Frame   77

          3.1.2 The Momental Ellipsoid 80
          3.1.3 Rotation of Coordinate Axes 82
          3.1.4 Principal Axes and Principal Moments 84
    3.2 Kinetic Energy of a Rigid Body Rotating about
         a Fixed Point 87
    3.3 Angular Momentum of a Rigid Body 91
    3.4 Eulerian Angles 103
          3.4.1 Angular Velocities Expressed in Terms of
                Eulerian Angles 106
          3.4.2 Kinetic Energy and Angular Momentum in Terms of
                Eulerian Angles 108
    3.5 The Compound Pendulum 109
    Exercises 111

4. **Dynamics of a Rigid Body Motion in Space**   **113–146**
    4.1 Euler's Dynamical Equations for a Rigid Body Rotating
         about a Fixed Point 113
          4.1.1 Body-fixed Translational Equations 115
          4.1.2 Gyroscopic Motion 118
    4.2 Motion of a Symmetrical Spinning Top 119
    4.3 Torque-free Motion of a Symmetrical Rigid Body—
         Rotational Motion of the Earth 132
    4.4 Attitudinal Stability of Earth Satellites 139
    4.5 Gyroscope 143
    Exercises 145

5. **Orbital Motion**   **147–206**
    5.1 Kepler's Laws of Planetary Motion 147
    5.2 Newton's Law of Gravitation 149
    5.3 Some Definitions from Spherical Astronomy 150
    5.4 Central Force Motion 155
          5.4.1 Integrals of Energy 156
          5.4.2 Differential Equation of the Orbit 157
          5.4.3 The Inverse Square Force 159
          5.4.4 Geometry of an Orbit 163
    5.5 The Two-body Problem 180
          5.5.1 The Motion of the Centre of Mass 180
          5.5.2 Relative Motion 181
          5.5.3 Solution of the Two-body Problem 183
          5.5.4 Earth-bound Satellite Circular Orbits 185
    5.6 Classical Orbital Elements 186
    5.7 Kepler's Equation 189
    5.8 Position in the Elliptic Orbit 192
    5.9 Position in a Parabolic Orbit 201
    5.10 The Hyperbolic Orbit 203
    Exercises 205

6. Lagrange and Hamilton Equations — 207–247
6.1 Introduction 207
6.2 Classification of a Dynamical System 210
6.3 Lagrange's Equations for Simple Systems 211
6.4 Principle of Virtual Work—D'Alemberts Principle 225
6.5 Lagrange Equations for General Systems 226
6.6 Hamilton's Equations 231
6.7 Ignorable Coordinates 238
6.8 The Routhian Function 240
Exercises 245

7. Hamiltonian Methods — 248–294
7.1 Introduction 248
7.2 Hamilton's Principle 248
   7.2.1 Hamilton's Principle for a Conservative System 256
   7.2.2 Principle of Least Action 257
7.3 Characteristic Function and Hamilton–Jacobi Equation 258
7.4 Phase Space and Liouville's Theorem 267
7.5 Special Transformations 271
   7.5.1 Lagrange Brackets 274
   7.5.2 Poisson Brackets 281
7.6 Calculus of Variations 283
Exercises 293

8. Special Theory of Relativity — 295–318
8.1 Some Fundamental Concepts 295
8.2 The Lorentz Transformation 298
8.3 Immediate Consequences of Lorentz Transformations 302
8.4 The Mass of a Moving Particle 305
8.5 Equivalence of Mass and Energy 308
Exercises 317

9. Rocket Dynamics — 319–343
9.1 Introduction 319
9.2 Equation of Motion for Variable Mass 320
9.3 Performance of a Single-stage Rocket 321
   9.3.1 Exhaust Speed Parameters 324
   9.3.2 Effect of Gravity 326
9.4 Performance of a Two-stage Rocket 331
9.5 Optimization of a Multi-stage Rocket 335
Exercises 342

10. The Three- and n-Body Problems — 344–364
10.1 Introduction 344
10.2 Mathematical Formulation of n-Body Problem 345

10.3 Integrals of Motion   345
10.4 The Virial Theorem   349
10.5 The Equations of Relative Motion   351
10.6 The General Three-body Problem   353
    10.6.1 Mathematical Formulation of General Three-body Problem   353
    10.6.2 Equations of Relative Motion of Three Bodies   354
    10.6.3 Stationary Solutions of Three-body Problem   357
Exercises 364

Multiple-Choice Questions	*365–373*
Bibliography	*375–376*
Answers to Exercises	*377–386*
Answers to Multiple-Choice Questions	*387–396*
Index	*397–399*

Preface

We are living in a world which is technologically advanced. There is a need to motivate young minds to work for excellence. To achieve this objective we need quality textbooks on various subjects and one such subject is classical mechanics. The knowledge of classical mechanics is very vital to understand advanced technologies in several areas such as automation, industrial process control, aerospace, among others. Mathematicians, physicists, astronomers and engineers have shown keen interest during the last several centuries in the development of mechanics.

This book covers many advanced topics in classical mechanics, the knowledge of which will be invaluable for all postgraduate students of mathematics, physics, engineering, astronomy and celestial mechanics. In fact, this text is an outgrowth of my several years of experience of teaching classical mechanics and space mechanics to M.Sc. and M.Phil. students at Anna University, coupled with my work during 1970–80 as scientist/engineer in the satellite launch vehicle (SLV) program at Vikram Sarabhai Space Centre, Thiruvananthapuram. In my opinion, the topics covered in this book should inculcate excellence in basic sciences and mathematical modelling and create aptitude to convert concepts into real-life applications. The minimal prerequisite to understand the text is knowledge of advanced calculus and vector analysis. I have taken care to cover the standard postgraduate syllabi of many Indian universities and IITs. New topics such as rocket dynamics, three-body and n-body problems are also included to create interest among readers in the design concepts of multi-stage rockets and in the study of orbits of satellites.

The text starts with the Newton's laws of motion, units and dimensions. Chapter 2 presents the dynamics of a system of particles and rotating frames of reference. Kinematics of a rigid body motion and Eulerian angles are discussed in Chapter 3. Chapter 4 describes the dynamics of a rigid body motion in space. Orbital motion is discussed in Chapter 5. Lagrange and Hamilton methods are dealt with in detail in Chapters 6 and 7. The special theory of relativity is presented in Chapter 8. Chapter 9 discusses preliminary rocket design and performance characteristics of single-stage, two-stage and multi-stage rockets, along with vehicle optimization. Finally, three- and n-body problems are presented in the last chapter for the benefit of those students who wish to pursue advanced study in celestial/space mechanics.

Several typical worked examples are presented in each chapter to instill self-confidence in the students to solve many more difficult problems. Multiple-choice questions given at the end of the book are modelled on examination patterns of various Indian universities, GATE and other competitive examinations.

I wish to express my gratitude to various authors, whose books and research papers have been referred to, while writing this book; if not all, many of these have been listed in the bibliography.

It is a pleasure to thank Prof. A. Avudainayagam of Department of Mathematics, IIT Madras, for going through the original manuscript and giving constructive suggestions to bring it to the present form. I would also like to thank Prof. (Mrs.) Prabhamani Patil and Prof. P. Ganesan both of Department of Mathematics, Anna University, Prof. R. Parthasarathi of Department of Computer Science and Engineering, Pondicherry Engineering College, Pondicherry, Dr. Y.V.S.S.S. Raju of IIT Madras, and Tadikonda S. Krishna, Lecturer in Physics of Gupta College, Tenali, for their encouragement in bringing out this book.

I wish to record my deep gratitude to my wife Leela Sankar who has been the real source of inspiration while preparing the manuscript. My special thanks to my daughter Aruna and to my granddaughter P. Sangeetha (B.Tech.), for their understanding and constant encouragement while I was busy with the manuscript preparation.

Finally, I wish to thank the publishers, Prentice-Hall of India, for their careful processing of my hand-written manuscript, both at the editorial and production stages, besides their keen interest in publishing this book.

I am sure that both students and teaching fraternity will enjoy reading this book. Despite my best efforts, some errors are bound to exist and, I shall, therefore, be grateful to the readers who would be kind enough to bring these errors to my notice.

K. Sankara Rao

1

Kinematics and Dynamics of a Particle

You cannot teach a man anything. You can only help him find it within himself.
— GALILEO

1.1 INTRODUCTION

Our universe consists of material bodies which are in constant interaction and motion. The mechanical form of motion of a body is studied in mechanics. There are many fields of study in mechanics, such as classical mechanics, relativistic mechanics, quantum mechanics, continuum mechanics, space mechanics, and so on. This text concentrates on the study of classical mechanics. Mechanics is usually subdivided into three parts: kinematics, statics and dynamics. In *kinematics*, we consider the motion of bodies without considering the factors causing their motion. *Statics* deals with the laws of equilibrium of bodies, which topic is not discussed in the present text. In *dynamics*, we study the laws of motion of bodies and the causes producing their motion.

Historically, the developments in understanding the principles of mechanics started from the days of Aryabhatta (476 AD – 499 AD), Galileo, Huygens, Bhaskara, Newton (1642–1727), etc. The subject became popular with Newton publishing *Principia* in 1687, where he stated the laws of motion of bodies which today form the basis for classical mechanics. Newton was one of the earliest scientists to have introduced mathematical theory to observed data, thereby making theory and observation work in tandem to explain the world around us. After Newton, many developments in mechanics were contributed by Euler, D'Alembert, Lagrange, Laplace, Poinsot, Coriolis, Einstein and others. Newton modestly acknowledged that, if he had seen farther than others, it was by standing upon the shoulders of giants.

1.2 NEWTON'S LAWS OF MOTION

The three laws of motion which form the basis of classical mechanics or Newtonian mechanics are customarily stated as follows:
1. Every particle continues to move in a state of uniform motion in a straight line or remains at rest, unless acted upon by an external force.

2. The time rate of change of linear momentum of a particle is proportional to the force acting on it and is in the direction of this force.
3. The forces of action and reaction between two interacting bodies are equal in magnitude and opposite in direction and are collinear.

The Newton's laws of motion as stated above can be easily understood by applying them to the motion of a particle. What is a *particle?* A particle is the idealisation that an extended body's mass is concentrated at a particular point. A particle has no size, structure or orientation. For instance, to analyse the trajectory of a cricket ball, as it soars into the space, the particle approximation is adequate. At first, let us try to understand the laws of motion by applying them to the motion of a particle. Later, we can extend these ideas to treat the motion of a rigid body, which may be defined as a solid system of particles, wherein the distances between the particles remains essentially unchanged.

Suppose a force $\vec{F}$ acts on a particle of mass m, then the Newton's second law may be stated as

$$\vec{F} = k\frac{d}{dt}(m\vec{V}) = km\vec{a}. \qquad (1.1)$$

Here, the product $m\vec{V}$ is regarded as the linear momentum $\vec{p}$ and therefore, we may write

$$\vec{p} = m\vec{V}, \qquad (1.2)$$

Here $\vec{a}$ is the acceleration of the particle, and k is a universal positive constant, whose value depends upon the units chosen. In general, we take $k = 1$ by a special choice of the unit of force. Thus, Eq. (1.1) simplifies to

$$\vec{F} = m\vec{a}. \qquad (1.3)$$

In SI (International system of metric units) system, the unit of force is called newton or simply N. A newton (N) is defined as that force which will give to one kilogram of mass an acceleration of one metre per second per second. Thus,

$$1 \text{ newton} = (1 \text{ kg}) (1 \text{ m/s}^2).$$

Equivalently,

$$N = \text{kg-m/s}^2.$$

It may be noted that in SI units, the fundamental unit of mass (M) is kilogram (kg), length (L) is metre (m) and that of time (T) is seconds (s). If we consider the motion of a particle in the Cartesian coordinate system, the equation of motion (1.3), in vectorial form can be written as

$$F_x \hat{i} + F_y \hat{j} + F_z \hat{k} = m\left(\frac{d^2 x}{dt^2}\hat{i} + \frac{d^2 y}{dt^2}\hat{j} + \frac{d^2 z}{dt^2}\hat{k}\right)$$

and in component form as

$$F_x = m\ddot{x}, \qquad F_y = m\ddot{y}, \qquad F_z = m\ddot{z}. \qquad (1.4)$$

We may observe that the Newton's first law is simply a consequence of the second law. If the resultant force $\vec{F}$ acting on the particle is zero, then, we can see from Eq. (1.3), that there is no acceleration and therefore, the particle is either at rest or moving with a uniform velocity.

Newton's third law when applied to two particles, exerting forces on each other, the forces are equal in magnitude and opposite in direction, and act along the straight line joining the particles.

1.3 INERTIAL FRAMES

In Newtonian mechanics, it is assumed that both space and time are 'absolute' entities, which means that they are the same for everyone. This assumption immediately brings forth the notion or concept of a frame of reference. A reference frame can be thought of as an imaginary set of rectangular axes, whether fixed in a laboratory on the Earth or in the cabin of an aeroplane, with reference to which the location of the bodies may be determined. To complete the mechanical description, let us append a clock to each of the above imaginary set of rectangular axes, which pinpoints when a particle is at a given location in space. Thus, the trajectory of a particle through space can be determined as time varies, in a frame of reference. Newton's concept of 'absolute time', also means that the time measured in one frame is the same as the time measured in any other frame. This intuitive aspect of time fails dramatically when frames in relative motion are considered (see Chapter 8 on Special Theory of Relativity).

Now, the most important question is whether the Newton's laws apply in all frames or just in some special class of frames? The basic frame of reference in which the Newton's laws hold true is known as the *inertial* frame, which is an imaginary set of rectangular axes assumed to have no translation or rotation in space. For most Earthbound scientific and engineering problems, it is enough to consider the inertial frame as a set of rectangular axes attached to the Earth, wherein the distances considered are short relative to the radius of the Earth and the velocities are small compared to the escape velocity from the Earth. In this case, the inertial frame is also called *Newtonian* frame.

However, if one were to consider the computation of rocket flight trajectories or those of the satellite motion around the Earth, a non-rotating coordinate frame with origin on the Earth's axis of rotation is chosen, since the absolute motion of the Earth under the influence of gravitational forces exerted by the presence of Sun, Moon and other planets becomes an important parameter. For interplanetary travel, Newton assumed the existence of an inertial frame, whose origin is fixed at the centre of the Sun or more precisely at the centre of mass of the solar system and is non-rotating with respect to fixed stars. Thus, the choice of the inertial frame depends on the type of the problem under investigation.

1.4 UNITS AND DIMENSIONS

In SI system, also known as MKS system, the units of mass, length and time are taken as base units and they are symbolically represented as kilogram (kg), metre (m) and second (s), respectively.

Stating mere numbers has no meaning unless their units are specified too. That is, it is meaningless to say that the distance between Tenali and Delhi is 1600. This statement is meaningful only when followed by 'kilometres' or 'miles', i.e. whichever unit makes the statement true. In MKS system, the unit of force is that force which when acted on one kilogram mass, produces an acceleration of magnitude one metre per second per second. A force of this magnitude is called a *newton*. In other words, in SI system, the unit of force is newton ($= kg\text{-}m/s^2$). Another unit of measurement, which is also in common use is in CGS system, in which, centimetre, gram and second are fundamental units. In this system, the unit of force is called a *dyne*. For a scientist or engineer, it is frequently necessary to convert the units from one system to another. The units which are used in the measurement of a physical quantity such as length may be 'metres' or 'kilometres', which certainly differ in magnitude but of course, qualitatively, they are the same, in the sense that both are units of length. The qualitative aspect of a given unit is characterised by its dimension.

All units used in the study of mechanics can be expressed in terms of length, mass and time, denoted by the symbols L, M and T, respectively. Thus, a physical quantity such as velocity, v, has the dimension of length per unit time, written as $[LT^{-1}]$, irrespective of SI system or CGS system of units of measurement. Similarly, acceleration has the dimensions $[LT^{-2}]$. Since, area = length × width, volume = length × width × height, the dimensions of area and volume are respectively given as $[L^2]$ and $[L^3]$. The dimensions of v and $2v$ are same as $[LT^{-1}]$, as we ignore any numerical factor. Now, let us look at the dimensions of a force as defined in Eq. (1.1). Ignoring the numerical factor k, we may write

$$\text{Force} = \text{Mass} \times \text{Acceleration}.$$

Thus, force has the dimensions $[MLT^{-2}]$. Similarly, when we speak of an angle being measured in degrees or radians, it is really the ratio of arc length to the radius of an arc of a circle and thus has the dimensions $[L/L] = [L^0]$. An angle is therefore dimensionless. Therefore, we may observe that trigonometric functions, exponential functions, Bessel functions, etc. are all dimensionless. However, any physical quantity Q may have the dimensions

$$[Q] = [M^\alpha L^\beta T^\gamma],$$

where, α, β, and γ may be positive or negative powers and need not always be integers.

Finally, it may be noted that an important characteristic of the equations of physics is that they are always dimensionally homogeneous. That is, the dimensions of all terms in an equation must be same. The dimensionless quantities often serve as characteristic values, which signal the change from one qualitative behaviour or 'regime' to another. For example, the Reynolds number for a fluid flowing through a pipe signals the onset of turbulence when its value exceeds 2000.

Table 1.1 presents dimensions of certain physical quantities, which will be found to be useful in subsequent chapters.

Table 1.1 Dimensions of Some Physical Quantities

Physical quantity (Q)	Dimension
Velocity ($\vec{V}$)	[LT^{-1}]
Acceleration ($\vec{a}$)	[LT^{-2}]
Force ($\vec{F}$)	[MLT^{-2}]
Moment of a force or torque ($\vec{G}$)	[ML^{2}T^{-2}]
Linear momentum ($\vec{p}$)	[MLT^{-1}]
Angular momentum ($\vec{h}$ or $\vec{H}$)	[ML^{2}T^{-1}]
Energy	[ML^{2}T^{-2}]
Work	[ML^{2}T^{-2}]
Angular velocity	[T^{-1}]
Moment of inertia	[ML^2]
Power = Force × Velocity or rate of doing work	[ML^{2}T^{-3}]
Viscosity of the fluid (μ)	[ML^{-1}T^{-1}]

1.5 CONSERVATION LAWS

In this section, we shall discuss three conservation laws relating to the Newtonian motion of a particle, namely the conservation of linear momentum, angular momentum and energy.

Conservation of linear momentum

What is linear momentum? The linear momentum $\vec{p}$ is defined as the product of mass and velocity. That is,

$$\vec{p} = m\vec{V} = m\dot{\vec{r}}.$$

Theorem 1.1 If no force is acting on a particle, the linear momentum of the particle is conserved.

Proof As a consequence of Newton's second law, Eq. (1.1) can be rewritten as

$$\frac{d\vec{p}}{dt} = \vec{F}.$$

If $\vec{F} = 0$, then $d\vec{p}/dt = 0$. On integration, we get $\vec{p}$, a constant vector. It has the dimensions [MLT^{-1}].

The moment of a vector about a point (torque) The moment of a vector $\vec{F}$ about a point A, is defined as the product of the magnitude of the vector $\vec{F}$ and its shortest distance from A and is denoted by $\vec{G}$. Vectorially, let P be a point on the line of action of the force $\vec{F}$ (see Fig. 1.1), whose position vector is $\vec{r}$ with respect to the origin O. Let A be a point about which we have to find the moment, whose position vector is $\vec{a}$, and suppose, α is the angle between $\vec{F}$ and $(\vec{r} - \vec{a})$, then the moment of $\vec{F}$ about A is defined as

$$\vec{G}_A = |\vec{F}||\vec{r} - \vec{a}|\sin\alpha = (\vec{r} - \vec{a}) \times \vec{F}. \tag{1.5}$$

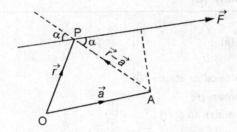

Figure 1.1 Moment of a vector.

Thus, if a particle with position vector $\vec{r}$ is in a force field $\vec{F}$, then we define the moment of a force $\vec{F}$ about O or torque as

$$\vec{G} = \vec{r} \times \vec{F}. \tag{1.6}$$

The magnitude of $\vec{G}$ indicates the turning effect produced on the particle by the applied force $\vec{F}$.

Angular momentum

A moment of momentum is angular momentum. Let P be a particle of mass m moving along a curve (see Fig. 1.2), then, the momentum of the particle is $m\dot{\vec{r}}$. The angular momentum denoted by the symbol h about A is defined as

$$\vec{h}_A = |m\dot{\vec{r}}| \times (\text{Shortest distance from A to the vector } \dot{\vec{r}}).$$

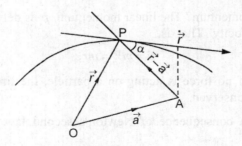

Figure 1.2 Angular momentum.

That is,

$$\vec{h}_A = |m\dot{\vec{r}}||\vec{r} - \vec{a}|\sin\alpha = m(\vec{r} - \vec{a}) \times \dot{\vec{r}}. \tag{1.7}$$

In case, A coincides with the origin O, then

$$\vec{h} = m\vec{r} \times \dot{\vec{r}} = m(\vec{r} \times \vec{V}) = \vec{r} \times \vec{p}. \qquad (1.8)$$

Now, angular momentum, as such has no meaning, unless we relate it to the equation of motion. Angular momentum has the dimensions $[ML^2T^{-1}]$.

Theorem 1.2 (*Angular momentum theorem*) The rate of change of angular momentum vector of a particle with respect to time is equal to the torque acting on the particle.

Proof From the definition of angular momentum, we have

$$\vec{h} = m(\vec{r} \times \vec{V}).$$

Therefore,

$$\frac{d\vec{h}}{dt} = \frac{d}{dt}[m(\vec{r} \times \vec{V})] = m\dot{\vec{r}} \times \vec{V} + m\vec{r} \times \dot{\vec{V}}.$$

That is,

$$\frac{d\vec{h}}{dt} = \vec{r} \times m\frac{d\vec{V}}{dt} = \vec{r} \times \vec{F} = \vec{G}. \qquad (1.9)$$

Hence the result. It may be noted that this result holds good even if the mass is variable and the force is non-conservative. This result is also known as the *principle of angular momentum*.

Theorem 1.3 If the torque about the origin O, of the forces acting on a particle is zero, then, the angular momentum of the particle about O is conserved.

Proof Given $\vec{G} = 0$. Therefore, from the angular momentum theorem, we have

$$\frac{d\vec{h}}{dt} = \vec{G} = 0,$$

which means $\vec{h}$ is constant. That is, if the total external torque acting on a particle is zero, its angular momentum vector is conserved.

Conservation of energy

Suppose, we assume that a particle P is moving on a curve C, whose parametric equation is

$$\vec{r} = \vec{r}(t), \qquad t_0 \leq t \leq t_1$$

and is acted upon by a conservative force $\vec{F}$, then the equation of motion of the particle is

$$m\ddot{\vec{r}} = \vec{F}. \qquad (1.10)$$

Taking the scalar product with $\dot{\vec{r}}$, we get

$$m\ddot{\vec{r}} \cdot \dot{\vec{r}} = \vec{F} \cdot \dot{\vec{r}},$$

which on rewriting becomes

$$\frac{d}{dt}\left(\frac{1}{2}m\dot{\vec{r}}\cdot\dot{\vec{r}}\right)=\vec{F}\cdot\dot{\vec{r}}.$$

If T stands for the kinetic energy of the particle, then the above equation assumes the form

$$\frac{dT}{dt}=\vec{F}\cdot\dot{\vec{r}}.$$

Since $\vec{F}$ is a conservative force, there exists a scalar function ϕ, such that $\vec{F}=\nabla\phi$. Thus,

$$\frac{dT}{dt}=\nabla\phi\cdot\dot{\vec{r}}=\left(\hat{i}\frac{\partial\phi}{\partial x}+\hat{j}\frac{\partial\phi}{\partial y}+\hat{k}\frac{\partial\phi}{\partial z}\right)\cdot\left(\hat{i}\frac{dx}{dt}+\hat{j}\frac{dy}{dt}+\hat{k}\frac{dz}{dt}\right)$$

$$=\frac{\partial\phi}{\partial x}\frac{dx}{dt}+\frac{\partial\phi}{\partial y}\frac{dy}{dt}+\frac{\partial\phi}{\partial z}\frac{dz}{dt}.$$

That is,

$$\frac{dT}{dt}=\frac{d\phi}{dt}. \tag{1.11}$$

Now, if we define the potential energy as

$$V=-\phi+\text{constant},$$

then

$$\frac{dV}{dt}=-\frac{d\phi}{dt}. \tag{1.12}$$

Adding Eqs. (1.11) and (1.12), we at once get

$$\frac{d}{dt}(T+V)=0.$$

On integration, we obtain

$$T+V=\text{Constant}=E, \tag{1.13}$$

which means that the total energy, that is, the sum of kinetic energy and potential energy is conserved, and has the dimensions $[ML^2T^{-2}]$. This is called the principle of conservation of energy. The implications of these concepts can be seen in subsequent chapters.

1.6 KINEMATICS OF A PARTICLE

Kinematics is the study of the motions of particles and rigid bodies disregarding the forces that cause their motion. In describing the motion of a particle, it is necessary to specify a frame of reference, because the motion will be different in general when viewed from different frames of reference. For instance, Newton's laws of motion hold in a particular frame of reference called the *inertial frame*. This frame, non-rotating but translating uniformly relative to another, constitutes a special or a preferred set for writing the dynamical equations of a particle. From the point of view of kinematics, there is no preferential frame of reference.

Let OXYZ be rectangular axes and $\hat{i}, \hat{j}$ and $\hat{k}$ be unit vectors along OX, OY and OZ respectively. For any particle located at (x, y, z), we define the following vectors:

$$\text{Position vector, } \vec{r} = x\hat{i} + y\hat{j} + z\hat{k}$$
$$\text{Velocity vector, } \vec{V} = \dot{x}\hat{i} + \dot{y}\hat{j} + \dot{z}\hat{k}$$
$$\text{Acceleration vector, } \vec{a} = \ddot{x}\hat{i} + \ddot{y}\hat{j} + \ddot{z}\hat{k}.$$

The simplest way of describing the motion of a particle is to give the vector function $\vec{r}(t)$ and trace the path of the particle and find its velocity and acceleration, at any instant of time. Very often, we require expressions for velocity and acceleration in directions other than $\hat{i}, \hat{j}$ and $\hat{k}$.

Velocity and acceleration along a curved path

The velocity and acceleration of a particle A as it moves along a curved path in space may be obtained as follows:

Let C be the path of a moving particle A (see Fig. 1.3). Let A_o be a fixed point on C, then let the arc length A_oA be denoted by s. As the particle moves an infinitesimal distance ds along the curve C, the corresponding change in the position vector $\vec{r}$ is given as

$$d\vec{r} = ds\hat{e}_t,$$

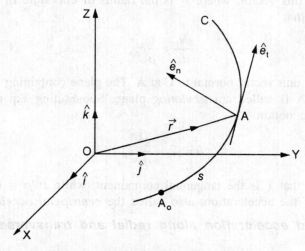

Figure 1.3 Motion of a particle in space.

that is,

$$\frac{d\vec{r}}{ds} = \hat{e}_t, \qquad (1.14)$$

where $\hat{e}_t$ is the unit vector tangential to the path C at A and points in the direction of s increasing. Therefore, the velocity of the particle at A is given as

$$\vec{V} = \frac{d\vec{r}}{dt} = \frac{d\vec{r}}{ds}\frac{ds}{dt} = \dot{s}\hat{e}_t. \qquad (1.15)$$

Hence, the velocity of the particle is directed along the tangent to the path, having $\dot{s}$ as its magnitude.

Now, for acceleration, we have

$$\begin{aligned}\vec{a} = \frac{d\vec{V}}{dt} &= \ddot{s}\hat{e}_t + \dot{s}\dot{\hat{e}}_t \\ &= \ddot{s}\hat{e}_t + \dot{s}\frac{d\hat{e}_t}{ds}\cdot\frac{ds}{dt} \\ &= \ddot{s}\hat{e}_t + \dot{s}^2\frac{d\hat{e}_t}{ds}. \end{aligned} \qquad (1.16)$$

The unit vector $\hat{e}_t$, which is tangential to the curve at A, is clearly a vector function of s. Since $\hat{e}_t \cdot \hat{e}_t = 1$, we have

$$\hat{e}_t \cdot \frac{d\hat{e}_t}{ds} + \hat{e}_t \cdot \frac{d\hat{e}_t}{ds} = 0, \qquad \text{implying } \hat{e}_t \cdot \frac{d\hat{e}_t}{ds} = 0,$$

which means that the vector $d\hat{e}_t/ds$ is normal to C at A. Let $1/\rho$ denote the magnitude of this vector, where ρ is the radius of curvature of C at A. It then follows that

$$\frac{d\hat{e}_t}{ds} = \frac{1}{\rho}\hat{e}_n, \qquad (1.17)$$

where $\hat{e}_n$ is a unit vector normal to C at A. The plane containing $\hat{e}_t$ and $\hat{e}_n$ at any point A is called an *osculating* plane. Substituting Eq. (1.17) into Eq. (1.16), we obtain

$$\vec{a} = \ddot{s}\hat{e}_t + \frac{\dot{s}^2}{\rho}\hat{e}_n. \qquad (1.18)$$

Thus, we see that $\ddot{s}$ is the tangential component, while $\dot{s}^2/\rho$ is the normal component of the acceleration, also called the *centripetal acceleration*.

Velocity and acceleration along radial and transverse directions

Consider a particle A moving in a plane, its position being described in polar coordinates (r, θ) as in Fig. 1.4.

Suppose $\hat{e}_r$ be the unit vector along OA and $\hat{e}_\theta$ the unit vector perpendicular to $\hat{e}_r$. Then, we can see from Fig. 1.4 that

$$\left.\begin{array}{l}\hat{e}_r = \hat{i}\cos\theta + \hat{j}\sin\theta \\ \hat{e}_\theta = -\hat{i}\sin\theta + \hat{j}\cos\theta.\end{array}\right\} \quad (1.19)$$

Since $\hat{e}_r$ and $\hat{e}_\theta$ are time dependent as the particle moves along C, we have

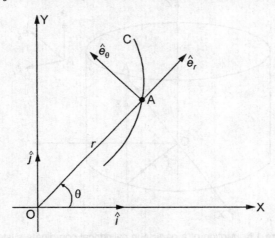

Figure 1.4 Motion of a particle in radial and transverse coordinate system.

$$\frac{d\hat{e}_r}{dt} = (-\hat{i}\sin\theta + \hat{j}\cos\theta)\frac{d\theta}{dt} = \dot{\theta}\hat{e}_\theta \quad (1.20)$$

$$\frac{d\hat{e}_\theta}{dt} = (-\hat{i}\cos\theta - \hat{j}\sin\theta)\frac{d\theta}{dt} = -\dot{\theta}\hat{e}_r. \quad (1.21)$$

Since $\vec{r} = r\hat{e}_r$, the velocity vector is

$$\vec{V} = \dot{\vec{r}} = \dot{r}\hat{e}_r + r\dot{\hat{e}}_r = \dot{r}\hat{e}_r + r\dot{\theta}\hat{e}_\theta. \quad (1.22)$$

Again differentiating with respect to time, we get acceleration as

$$\vec{a} = \dot{\vec{V}} = \ddot{r}\hat{e}_r + \dot{r}\dot{\hat{e}}_r + \dot{r}\dot{\theta}\hat{e}_\theta + r\ddot{\theta}\hat{e}_\theta + r\dot{\theta}\dot{\hat{e}}_\theta$$

$$= \ddot{r}\hat{e}_r + \dot{r}\dot{\theta}\hat{e}_\theta + \dot{r}\dot{\theta}\hat{e}_\theta + r\ddot{\theta}\hat{e}_\theta + r\dot{\theta}(-\dot{\theta}\hat{e}_r)$$

$$= (\ddot{r} - r\dot{\theta}^2)\hat{e}_r + (r\ddot{\theta} + 2\dot{r}\dot{\theta})\hat{e}_\theta.$$

Therefore,

$$\vec{a} = (\ddot{r} - r\dot{\theta}^2)\hat{e}_r + \frac{1}{r}\frac{d}{dt}(r^2\dot{\theta})\hat{e}_\theta. \quad (1.23)$$

Velocity and acceleration in cylindrical coordinate system

Consider the case, where the position of the particle A at an instant relative to OXYZ system, as shown in Fig. 1.5, is expressed in cylindrical coordinates r, ϕ, z. In this case, the position vector of A is

$$\vec{R} = \overrightarrow{OA} = r\hat{e}_r + z\hat{e}_z. \tag{1.24}$$

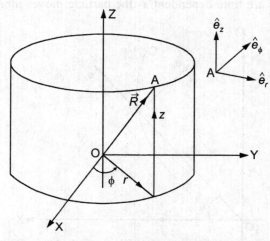

Figure 1.5 Motion of a particle in cylindrical coordinate system.

It may be noted that the unit vectors $\hat{e}_r, \hat{e}_\phi$ and $\hat{e}_z$ form a mutually orthogonal triad, whose directions are given by the directions in which A moves for small increases in r, ϕ and z respectively. Differentiating Eq. (1.24) with respect to time, we get velocity as

$$\vec{V} = \dot{\vec{R}} = \dot{r}\hat{e}_r + r\dot{\hat{e}}_r + \dot{z}\hat{e}_z. \tag{1.25}$$

since $\hat{e}_r$ and $\hat{e}_\phi$ change their directions in space as A moves, while $\hat{e}_z$ remains always parallel to Z-axis. It may further be noted that changes in the directions of $\hat{e}_r$ and $\hat{e}_\phi$ are solely due to changes in ϕ, corresponding to rotations about Z-axis.

Here, (r, θ) of Fig. 1.4 corresponds to (r, ϕ) in Fig. 1.5. We therefore have

$$\frac{d\hat{e}_r}{dt} = \dot{\phi}\hat{e}_\phi \tag{1.26}$$

$$\frac{d\hat{e}_\phi}{dt} = -\dot{\phi}\hat{e}_r. \tag{1.27}$$

Equation (1.25) now becomes

$$\vec{V} = \dot{r}\hat{e}_r + r\dot{\phi}\hat{e}_\phi + \dot{z}\hat{e}_z. \tag{1.28}$$

Differentiating Eq. (1.28) once again with respect to time, we get an expression for acceleraton as

$$\vec{a} = \ddot{\vec{R}} = \dot{\vec{V}} = \ddot{r}\hat{e}_r + \dot{r}\dot{\hat{e}}_r + (\dot{r}\dot{\phi} + r\ddot{\phi})\hat{e}_\phi + r\dot{\phi}\dot{\hat{e}}_\phi + \dot{z}\dot{\hat{e}}_z + \ddot{z}\hat{e}_z.$$

Using Eqs. (1.26) and (1.27) and noting that $\dot{\hat{e}}_z = 0$, we have

$$\vec{a} = \ddot{r}\hat{e}_r + \dot{r}\dot{\phi}\hat{e}_\phi + (\dot{r}\dot{\phi} + r\ddot{\phi})\hat{e}_\phi + r\dot{\phi}(-\dot{\phi}\hat{e}_r) + \ddot{z}\hat{e}_z,$$

that is,

$$\vec{a} = [\ddot{r} - r\dot{\phi}^2]\hat{e}_r + [r\ddot{\phi} + 2\dot{r}\dot{\phi}]\hat{e}_\phi + \ddot{z}\hat{e}_z$$

$$= [\ddot{r} - r\dot{\phi}^2]\hat{e}_r + \left[\frac{1}{r}\frac{d}{dt}(r^2\dot{\phi})\right]\hat{e}_\phi + \ddot{z}\hat{e}_z. \quad (1.29)$$

Similarly, in spherical coordinate system, as shown in Fig. 1.6, we can obtain expressions for velocity and acceleration as

$$\vec{V} = \dot{r}\hat{e}_r + r\dot{\theta}\hat{e}_\theta + r\dot{\phi}\sin\theta\hat{e}_\phi \quad (1.30)$$

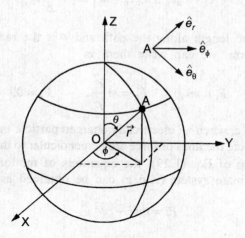

Figure 1.6 Motion of a particle in spherical coordinate system.

and

$$\vec{a} = (\ddot{r} - r\dot{\theta}^2 - r\dot{\phi}^2\sin^2\theta)\hat{e}_r + (r\ddot{\theta} + 2\dot{r}\dot{\theta} - r\dot{\phi}^2\sin\theta\cos\theta)\hat{e}_\theta$$
$$+ (r\ddot{\phi}\sin\theta + 2\dot{r}\dot{\phi}\sin\theta + 2r\dot{\theta}\dot{\phi}\cos\theta)\hat{e}_\phi. \quad (1.31)$$

Here, θ is the angle between the radius vector and +Z-axis which is called *zenith* angle, while ϕ is the angle between the projection of radius vector $\overrightarrow{OA}$ in the XY-plane and the +X-axis measured in the direction shown in Fig. 1.6, and is called *azimuth* angle.

1.6.1 Equations of Motion of a Particle

Suppose a force $\vec{F}$ acts on a particle of mass m, then we have from Newton's second law

$$\vec{F} = \frac{d}{dt}(m\vec{V}) = m\vec{a}, \tag{1.32}$$

where $\vec{V}$ is the velocity of the particle and $\vec{a}$ is the acceleration relative to a Newtonian frame of reference. This is the equation of motion in a rectangular coordinate system. In component form, we write as

$$F_x = m\ddot{x}, \qquad F_y = m\ddot{y}, \qquad F_z = m\ddot{z}. \tag{1.33}$$

The other forms of equations of motion, in various coordinate systems can be written as follows:

Let $\hat{e}_t, \hat{e}_n$ and $\hat{e}_b$ be the unit vectors along the tangent, normal and binormal to the path of the particle. Then, using Eq. (1.18), the equations of motion of the particle can be written as

$$F_t \hat{e}_t + F_n \hat{e}_n + F_b \hat{e}_b = m\left(\ddot{s}\hat{e}_t + \frac{\dot{s}^2}{\rho}\hat{e}_n\right), \tag{1.34}$$

where s is the arc length along the path and ρ is the radius of curvature. In component form, we can write them as

$$F_t = m\ddot{s}, \qquad F_n = m\frac{\dot{s}^2}{\rho}, \qquad F_b = 0. \tag{1.35}$$

This situation arise, when an electrically charged particle moves in a uniform magnetic field where the lines of force are perpendicular to the plane of motion.

With the help of Eq. (1.29), the equations of motion of a particle in cylindrical coordinate system (r, ϕ, z) can be obtained as

$$\left. \begin{array}{l} F_r = m(\ddot{r} - r\dot{\phi}^2) \\[6pt] F_\phi = m\dfrac{1}{r}\dfrac{d}{dt}(r^2\dot{\phi}) \\[6pt] F_z = m\ddot{z}. \end{array} \right\} \tag{1.36}$$

Similarly Eq. (1.31) gives us the equations of motion of a particle in spherical coordinate system (r, θ, ϕ) as

$$\left. \begin{array}{l} F_r = m(\ddot{r} - r\dot{\theta}^2 - r\dot{\phi}^2 \sin^2\theta) \\[4pt] F_\theta = m(r\ddot{\theta} + 2\dot{r}\dot{\theta} - r\dot{\phi}^2 \sin\theta \cos\theta) \\[4pt] F_\phi = m(r\ddot{\phi}\sin\theta + 2\dot{r}\dot{\phi}\sin\theta + 2r\dot{\theta}\dot{\phi}\cos\theta). \end{array} \right\} \tag{1.37}$$

If the path of the particle is known before hand, it is advised to use Eqs. (1.35). However, if the path of the particle is to be found, it is convenient to use equations such as (1.33), (1.36) or (1.37) depending upon the coordinate system at hand. It may be noted that these equations of motion can be solved provided the initial position and velocity of the particle along with the forces that are acting are known.

If we regard our universe as composed of several particles, and if their initial position and velocity are given, in principle, we should be able to solve these equations of motion and be able to predict the future of the universe or we may uncover the past history of the universe. It will be interesting if it is true.

Example 1.1 A particle moves along a straight line with constant acceleration, a. Show that $v = v_0 + at$, $v^2 = v_0^2 + 2as$, where s is the distance travelled by the particle from the instant $t = 0$, v_0 is the initial velocity and v is the final velocity. Hence explain the concepts of work, energy, impulse and power.

Solution Assuming that a force acts on a particle, which does not depend on where the particle is located, the equation of motion of the particle is

$$\vec{F} = m\ddot{x} = m\frac{dv}{dt}. \tag{1}$$

Since both m and $\vec{F}$ are constants, we have

$$\ddot{x} = \frac{d^2x}{dt^2} = a. \tag{2}$$

Now, integrating Eq. (2) twice with respect to time, we at once get velocity and position of the particle as

$$v = v_0 + at \tag{3}$$

$$x = x_0 + v_0 t + \frac{1}{2}at^2, \tag{4}$$

where x_0 and v_0 are the initial position and velocity of the particle. Eliminating t between Eqs. (3) and (4), we get

$$x - x_0 = v_0\frac{v - v_0}{a} + \frac{1}{2}a\left(\frac{v - v_0}{a}\right)^2.$$

That is,

$$v^2 = v_0^2 + 2a(x - x_0) = v_0^2 + 2as. \tag{5}$$

Integrating Eq. (1) with respect to x, we have

$$\int_{x_1}^{x_2} F\, dx = m\int_{x_1}^{x_2} \frac{dv}{dt} dx = m\int_{v_1}^{v_2} \frac{dx}{dt} dv = m\int_{v_1}^{v_2} v\, dv. \tag{6}$$

Since F does not depend on location of the particle, it can be taken outside the integral sign, to obtain

$$F(x_2 - x_1) = \frac{1}{2}mv_2^2 - \frac{1}{2}mv_1^2 \tag{7}$$

The left-hand side of Eq. (7) is defined as the work done by a constant force on a particle as it moves from x_1 to x_2. If $T = (1/2)mv^2$ denotes the kinetic energy of the particle of mass m, moving with speed v, then, in terms of linear momentum p, $T = p^2/(2m)$. Thus, Eq. (7) can be interpreted as

Work done by the force = Increase in kinetic energy.

Symbolically we write this as

$$\Delta W = \Delta T. \tag{8}$$

This is known as work–energy theorem/principle.

Now, consider a particle in free fall under gravity, which provides an approximately constant force close to the Earth's surface. Taking the z-direction vertically upwards, the z-component of the force acting on the particle is

$$F_z = -mg.$$

Now, the work done by the gravity is

$$W = \int_{z_0}^{z} F_z \, dz = \int_{z_0}^{z} -mg \, dz = -mg(z - z_0).$$

Using the work–energy theorem, we obtain

$$W = -mgz + mgz_0 = \frac{1}{2}(mv^2 - mv_0^2),$$

so that

$$\frac{mv^2}{2} + mgz = \frac{mv_0^2}{2} + mgz_0. \tag{9}$$

Here, the term mgz is referred to as the *potential energy* and also represents the work done against gravity in raising the particle to a particular height z. This together with the kinetic energy of the particle constitutes the total mechanical energy. Thus, Eq. (9) shows that the total mechanical energy is the same at some general point of the motion at the beginning and is therefore, constant in time, which means that the total energy is conserved.

Impulse and power: Integrating Eq. (1) with respect to time from 0 to t, we have

$$\int_0^t m \frac{dv}{dt} dt = \int_0^t F \, dt \tag{10}$$

that is,

$$\int_{v_0}^{v} m \, dv = mv - mv_0 = \int_0^t F \, dt. \tag{11}$$

By defining the right-hand side of (11) as impulse, the above expression shows that the impulse of a force on a body is the change in linear momentum experienced by the body. Impulse is used in contexts where a force acts on a body for a short duration, for example, when a ball is struck by a bat. The interaction time in such a case is so short that the force may be considered as constant over that period. Thus,

$$\text{Impulse} = Ft.$$

Another quantity of interest is the power P, which is defined as the rate of doing work. Thus,

$$P = \frac{dW}{dt} = F\frac{dx}{dt} = \text{Force} \times \text{Velocity}. \tag{12}$$

To understand the concept of power, consider the situation of driving a car along the road with the accelerator depressed. The car moves, because, a constant force is applied to the wheels through the burning of fuel. However, resistance from the road and air, opposes the car's motion so that, in a steady state no net force is applied. According to Newton's first law, the car then moves with a constant velocity and the power developed by the engine is that required to overcome friction and to maintain the car's speed.

Example 1.2 A particle of mass m is subjected to two forces, a central force f_1 and frictional force f_2 given by

$$f_1 = f(r), \qquad f_2 = -\lambda \vec{V} \quad (\lambda > 0)$$

where $\vec{V}$ is the velocity of the particle. If the particle initially has angular momentum h_0 about $r=0$, find its angular momentum at subsequent times.

Solution The equations of motion of the particle in polar coordinates, using Eq. (1.23), can be written as

$$m(\ddot{r} - r\dot{\theta}^2) = f(r) \tag{1}$$

and

$$m(\dot{r}\dot{\theta} + r\ddot{\theta}) = -\lambda \vec{V} = -\lambda r\dot{\theta}. \tag{2}$$

Now, Eq. (2) may be rewritten as

$$\frac{1}{r}\frac{d}{dt}(mr^2\dot{\theta}) = -\lambda r\dot{\theta}. \tag{3}$$

It may be noted that the momentum vector (see Fig. 1.4) has components $m\dot{r}$ and $mr\dot{\theta}$ along and perpendicular to the radius vector respectively. The former component has no moment about the origin, while the moment due to the latter component, also called angular momentum, is given by

$$\vec{h} = mr^2\dot{\theta}. \tag{4}$$

Thus, Eq. (3) can be written as

$$\frac{d\vec{h}}{dt} = -\frac{\lambda \vec{h}}{m}. \tag{5}$$

Integrating and using the initial condition, we get

$$\log \vec{h} = -\frac{\lambda}{m}t + \log \vec{h}_0.$$

That is,

$$\vec{h} = \vec{h}_0 e^{-(\lambda/m)t} \tag{6}$$

is the required solution.

Example 1.3 · A particle of mass m moves along a trajectory given by $x = x_0 \cos \omega_1 t$, $y = y_0 \sin \omega_2 t$
 (i) Find the x, y components of the force acting on the particle.
 (ii) Find the potential energy as functions of x and y.
 (iii) Determine the kinetic energy of the particle and hence show that the total energy is conserved.

Solution (i) From the given data, we have

$$\dot{x} = -x_0 \omega_1 \sin \omega_1 t, \qquad \ddot{x} = -x_0 \omega_1^2 \cos \omega_1 t$$

$$\dot{y} = y_0 \omega_2 \cos \omega_2 t, \qquad \ddot{y} = -y_0 \omega_2^2 \sin \omega_2 t.$$

But Newton's second law states

$$\vec{F} = m(\ddot{x}\hat{i} + \ddot{y}\hat{j})$$

$$= -m[x_0 \omega_1^2 \cos(\omega_1 t)\hat{i} + y_0 \omega_2^2 \sin(\omega_2 t)\hat{j}]$$

$$= -m(\omega_1^2 x\hat{i} + \omega_2^2 y\hat{j}).$$

Thus, the x, y components of the force acting on the particle are given by

$$F_x = -m\omega_1^2 x, \qquad F_y = -m\omega_2^2 y. \tag{1}$$

(ii) For a conservative force field, we know that

$$\vec{F} = -\nabla V.$$

That is,

$$F_x = -\frac{\partial V}{\partial x}, \qquad F_y = -\frac{\partial V}{\partial y}.$$

Integrating these equations with respect to x and y respectively and using the result (1), we get

$$V = \frac{1}{2}m\omega_1^2 x^2 + f(y) = \frac{1}{2}m\omega_2^2 y^2 + f(x).$$

Hence, the potential energy is given by

$$V = \frac{1}{2} m (\omega_1^2 x^2 + \omega_2^2 y^2). \qquad (2)$$

(iii) Finally, the kinetic energy denoted in general by T is

$$T = \frac{1}{2} m (\dot{x}^2 + \dot{y}^2)$$

$$= \frac{1}{2} m [x_0^2 \omega_1^2 \sin^2(\omega_1 t) + y_0^2 \omega_2^2 \cos^2(\omega_2 t)]. \qquad (3)$$

We can also observe from Eqs. (2) and (3) that the total energy E is

$$E = T + V = \frac{1}{2} m \{[x_0^2 \omega_1^2 \sin^2(\omega_1 t) + y_0^2 \omega_2^2 \cos^2(\omega_2 t)]$$

$$+ [\omega_1^2 x_0^2 \cos^2(\omega_1 t) + \omega_2^2 y_0^2 \sin^2(\omega_2 t)]\}$$

$$= \frac{1}{2} m (x_0^2 \omega_1^2 + y_0^2 \omega_2^2)$$

$$= \text{constant}.$$

Hence, the total energy is conserved.

Example 1.4 An insect moves along the diameter of a revolving round table that is rotating at a uniform rate of p radians per second. Assuming that the insect has a constant speed s, relative to the table, find its acceleration as it passes through the centre of the table.

Solution Let P be the typical position of the insect at an instant t. Suppose that the insect started initially at Q on the rim of the table, which is at a distance of R, from its centre (see Fig. 1.7).

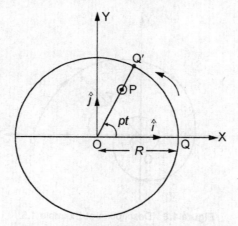

Figure 1.7 Description of Example 1.4.

Then, after time t, let OQ occupy the position OQ', so that
$$PQ' = st, \qquad OP = R - st.$$
It may be observed that the radius along which, the insect is moving, will have turned through an angle pt relative to the fixed x-axis. Thus, the position vector of P is given as

$$\vec{r} = [(R-st)\cos pt]\hat{i} + [(R-st)\sin pt]\hat{j}. \tag{1}$$

Differentiating Eq. (1) twice with respect to time t, we get
$$\dot{\vec{r}} = [-s\cos pt - (R-st)p\sin pt]\hat{i} + [-s\sin pt + (R-st)p\cos pt]\hat{j}$$
and
$$\ddot{\vec{r}} = [2sp\sin pt - (R-st)p^2\cos pt]\hat{i} + [-2sp\cos pt + (R-st)p^2\sin pt]\hat{j}.$$
That is,
$$\ddot{\vec{r}} = 2sp[(\sin pt)\hat{i} - (\cos pt)\hat{j}] - p^2\vec{r}. \tag{2}$$

Therefore, Eq. (2) indicates that, when the insect passes through the centre of the table, that is, as and when $\vec{r} = 0$, we find that the acceleration of the insect has a magnitude $2sp$.

Example 1.5 A particle travels on a path of a spiral described by the equation $r = 2e^{-0.3\phi}$ metres, such that $\dot{\phi}$ is a constant and equal to 1.5 rad/s. Find the velocity and acceleration of the particle when $\phi = 120°$.

Solution Refer to Fig. 1.8. From the given data, we have $\dot{\phi} = 1.5$ rad/s and also $\ddot{\phi} = 0$. Thus,

$$r = 2e^{-0.3\phi} \text{ m}$$
$$\dot{r} = -0.6\dot{\phi}e^{-0.3\phi} = -0.9e^{-0.3\phi} \text{ m/s}$$
$$\ddot{r} = -0.27\dot{\phi}e^{-0.3\phi} = 0.405e^{-0.3\phi} \text{ m/s}^2.$$

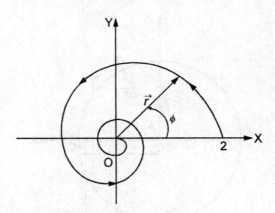

Figure 1.8 Description of Example 1.5.

In polar coordinates, that is, in radial and transverse coordinate system, recalling Eqs. (1.22) and (1.23), we have velocity and acceleration respectively given by

$$\vec{V} = \dot{r}\hat{e}_r + r\dot{\phi}\hat{e}_\phi = -0.9 e^{-0.3\phi}\hat{e}_r + 3.0 e^{-0.3\phi}\hat{e}_\phi. \tag{1}$$

and

$$\begin{aligned}\vec{a} &= (\ddot{r} - r\dot{\phi}^2)\hat{e}_r + (r\ddot{\phi} + 2\dot{r}\dot{\phi})\hat{e}_\phi \\ &= (0.405 - 4.5)e^{-0.3\phi}\hat{e}_r - (2.7 e^{-0.3\phi})\hat{e}_\phi \\ &= -4.095 e^{-0.3\phi}\hat{e}_r - 2.7 e^{-0.3\phi}\hat{e}_\phi. \end{aligned} \tag{2}$$

When $\phi = 210° = 7\pi/6$ radians

$$e^{-0.3\phi} = e^{-7\pi/20} = e^{-1.10} = 0.333.$$

Finally Eqs. (1) and (2) give us the required velocity and acceleration as

$$\vec{V} = -0.3 \hat{e}_r + 0.99 \hat{e}_\phi \text{ m/s}$$

and

$$\vec{a} = -1.36 \hat{e}_r - 0.899 \hat{e}_\phi \text{ m/s}^2.$$

Example 1.6 The meteorologist normally expresses atmospheric pressure in bars (1 bar = 10^6 dynes/cm^2). Similarly, laboratory pressures are in general expressed in mm of mercury. Obtain the conversion between these two units of measurement. It is known that the density of mercury is 13.59 g/cm^3.

Solution The density of mercury is given by

$$13.59 \frac{g}{cm^3} = 13.59 \frac{g}{cm^3}\left(\frac{1 \text{ cm}}{10 \text{ mm}}\right)^3 = 13.59 \times 10^{-3} \frac{g}{mm^3}.$$

It may be noted that the term in parentheses has a numerator and a denominator that are equivalent. The above equation means that a cubic mm of mercury will have a mass equal to 13.59×10^{-3} g. Then its weight is found to be

$$13.59 \times 10^{-3} \text{ g} \times \frac{980.6 \text{ cm}}{s^2} = 13.33 \text{ dynes}.$$

It means that a pressure of 1 mm of mercury is equivalent to 13.33 dynes/mm^2. Therefore,

$$1 \text{ mm of Hg (or mercury)} = 13.33 \frac{\text{dynes}}{\text{mm}^2}\left(\frac{1 \text{ bar}}{10^6 \text{ dynes/cm}^2}\right)\left(\frac{10 \text{ mm}}{1 \text{ cm}}\right)^2$$

$$= 13.33 \times 10^{-4} \text{ bars}.$$

It is known that, an atmospheric pressure corresponds to 760 mm of mercury. Hence

$$1 \text{ atm} = 760 \text{ mm Hg} = 760 \times 13.33 \times 10^{-4} \text{ bars} = 1.013 \text{ bars}.$$

1.6.2 Simple Pendulum

A simple pendulum consists of a heavy particle of mass m called the bob, attached to one end of a weightless string of length l, which oscillates back and forth in a vertical plane, while the other end of the string is attached to a fixed point O, as shown in Fig. 1.9. This is one example of a dynamical system of frequent occurrence.

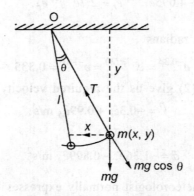

Figure 1.9 Simple pendulum.

Taking the fixed point O as the origin of the coordinate system, let the typical position of the particle at time t be given by (x, y), or (r, θ) in polar coordinates. Here, the particle moves under the action of two forces, namely the gravity and the tension T in the string. Now, using Eq. (1.23), the equation of motion of the particle in polar coordinates with $r = l$ (constant), $\dot{r} = \ddot{r} = 0$, can be written as

$$F_r = -ml\dot{\theta}^2 = mg \cos\theta - T \tag{1.38}$$

and

$$F_\theta = ml\ddot{\theta} = -mg \sin\theta, \tag{1.39}$$

where θ is the inclination of the string to the downward vertical. Since, we are interested only in $\theta(t)$, Eq. (1.39) can be recast as

$$\ddot{\theta} + \omega^2 \sin\theta = 0, \tag{1.40}$$

where $\omega = \sqrt{g/l}$, is called the *angular* frequency. To discuss its motion, we present below two simple cases:

Case 1: For small oscillations, that is for θ small, $\sin\theta \approx \theta$, then Eq. (1.40) can be written as

$$\ddot{\theta} + \omega^2 \theta = 0. \tag{1.41}$$

It may be noted that, both the functions $\cos \omega t$ and $\sin \omega t$ satisfy Eq. (1.41) and therefore, using the principle of superposition, the general solution of Eq. (1.41) is

$$\theta(t) = A \cos \omega t + B \sin \omega t, \quad (1.42)$$

where A and B are constants to be determined from the initial conditions. A motion described by Eq. (1.41) is called *simple harmonic*. The solution given by Eq. (1.42) can also be written in equivalent form as

$$\theta(t) = a \cos(\omega t + \varepsilon), \quad (1.43)$$

where constants are related by

$$A = a \cos \varepsilon, \qquad B = -a \sin \varepsilon. \quad (1.44)$$

Thus,

$$a = \sqrt{A^2 + B^2}, \qquad \cos \varepsilon = \frac{A}{\sqrt{A^2 + B^2}}, \qquad \sin \varepsilon = -\frac{B}{\sqrt{A^2 + B^2}}. \quad (1.45)$$

Here, a is called the *amplitude* of the motion, $(\omega t + \varepsilon)$ is called the *phase*. It can also be observed that the motion is periodic and the period of the small oscillation is given by

$$\tau = \frac{2\pi}{\omega} = 2\pi \sqrt{\frac{l}{g}}, \quad (1.46)$$

which is independent of the amplitude and the initial phase ε.

Case 2: To solve for the motion exactly, we integrate Eq. (1.40) after rewriting as

$$\ddot{\theta} = \frac{d}{dt}(\dot{\theta}) = \frac{d\dot{\theta}}{d\theta}\frac{d\theta}{dt} = \dot{\theta}\frac{d\dot{\theta}}{d\theta} = \frac{d}{d\theta}\left(\frac{\dot{\theta}^2}{2}\right) \quad (1.47)$$

to get

$$\int_0^{\dot{\theta}} d\left(\frac{\dot{\theta}^2}{2}\right) = -\frac{g}{l} \int_\alpha^\theta \sin \theta \, d\theta, \quad (1.48)$$

where α is the maximum angle of the motion obtained when $\dot{\theta} = 0$, that is, $\dot{\theta} = 0$ for $\theta = \pm \alpha$. Thus, Eq. (1.48) now simplifies to

$$\dot{\theta}^2 = \frac{2g}{l}(\cos \theta - \cos \alpha). \quad (1.49)$$

At this point, we define a new variable ϕ by

$$\sin \phi = \frac{\sin(\theta/2)}{\sin(\alpha/2)} \quad (1.50)$$

such that

$$\frac{1}{2}\cos\frac{\theta}{2}\dot\theta = \sin\frac{\alpha}{2}\cos\phi\,\dot\phi. \qquad (1.51)$$

Multiplying Eq. (1.49) by $(1/4)\cos^2(\theta/2)$, we get after using Eq. (1.51),

$$\frac{1}{4}\cos^2\frac{\theta}{2}\dot\theta^2 = \sin^2\frac{\alpha}{2}\cos^2\phi\,\dot\phi^2$$

$$= \frac{2g}{l}\frac{1}{4}\cos^2\frac{\theta}{2}(\cos\theta - \cos\alpha)$$

$$= \frac{g}{l}\cos^2\frac{\theta}{2}\left(\sin^2\frac{\alpha}{2} - \sin^2\frac{\theta}{2}\right)$$

$$= \frac{g}{l}\cos^2\frac{\theta}{2}\sin^2\frac{\alpha}{2}\left[1 - \frac{\sin^2(\theta/2)}{\sin^2(\alpha/2)}\right]$$

$$= \frac{g}{l}\cos^2\frac{\theta}{2}\sin^2\frac{\alpha}{2}\cos^2\phi \qquad \text{[using Eq. (1.50)]}.$$

Therefore,

$$\dot\phi^2 = \frac{g}{l}\cos^2\frac{\theta}{2}$$

$$= \frac{g}{l}\left(1 - \sin^2\frac{\theta}{2}\right)$$

$$= \frac{g}{l}\left(1 - \sin^2\frac{\alpha}{2}\sin^2\phi\right), \qquad (1.52)$$

which can be rewritten as

$$\sqrt{\frac{g}{l}}\,dt = \frac{d\phi}{\sqrt{1 - \sin^2(\alpha/2)\sin^2\phi}}.$$

On integration, we get

$$\sqrt{\frac{g}{l}}\int_0^t dt = \int_\phi^{\pi/2} \frac{d\phi}{\sqrt{1 - \sin^2(\alpha/2)\sin^2\phi}}.$$

That is,

$$\sqrt{\frac{g}{l}}\,t = \int_\phi^{\pi/2} \frac{d\phi}{\sqrt{1 - \sin^2(\alpha/2)\sin^2\phi}}. \qquad (1.53)$$

When $\theta = 0$, hence $\phi = 0$, we will have a quarter period and hence the period is given by

$$\tau = 4\sqrt{\frac{l}{g}} \int_0^{\pi/2} \frac{d\phi}{\sqrt{1-\sin^2(\alpha/2)\sin^2\phi}}. \qquad (1.54)$$

Suppose, we define

$$\sin\frac{\alpha}{2} = k. \qquad (1.55)$$

Equation (1.54) gives the periodic time of the pendulum as

$$\tau = 4\sqrt{\frac{l}{g}} \int_0^{\pi/2} \frac{d\phi}{\sqrt{1-k^2\sin^2\phi}}. \qquad (1.56)$$

This integral is known as *elliptic integral of the first kind* which cannot be evaluated in a closed form. Thus, to find an approximate period, we expand Eq. (1.56) in terms of power series as

$$\tau = 4\sqrt{\frac{l}{g}} \int_0^{\pi/2} \left(1 + \frac{1}{2}k^2\sin^2\phi + \frac{1\times 3}{2\times 4}k^4\sin^4\phi + \cdots\right) d\phi. \qquad (1.57)$$

But, we know that

$$\int_0^{\pi/2} d\phi = \frac{\pi}{2},$$

$$\int_0^{\pi/2} \sin^2\phi \, d\phi = \frac{\pi}{2}\frac{1}{2},$$

$$\int_0^{\pi/2} \sin^4\phi \, d\phi = \frac{\pi}{2}\frac{3}{4}\frac{1}{2}, \text{ and so on.}$$

Substituting these values into Eq. (1.57), we obtain

$$\tau = 4\sqrt{\frac{l}{g}}\frac{\pi}{2}\left[1 + \left(\frac{1}{2}\right)^2 k^2 + \left(\frac{1\times 3}{2\times 4}\right)^2 k^4 + \cdots\right], \qquad (1.58)$$

where $k = \sin(\alpha/2)$. When the amplitude α is very small, we get the first approximation as

$$\tau_0 = 2\pi\sqrt{\frac{l}{g}}. \qquad (1.59)$$

However, the second approximation is

$$\tau = 2\pi \sqrt{\frac{l}{g}} \left[1 + \frac{1}{4} \sin^2\left(\frac{\alpha}{2}\right) \right]$$

$$\approx 2\pi \sqrt{\frac{l}{g}} \left(1 + \frac{\alpha^2}{16} \right)$$

$$= \tau_0 \left(1 + \frac{\alpha^2}{16} \right). \tag{1.60}$$

Evidently, Eq. (1.58) indicates that the periodic time increases steadily with the amplitude. From Eq. (1.60) we may also note that

$$\frac{\tau - \tau_0}{\tau_0} = \frac{\alpha^2}{16}. \tag{1.61}$$

Suppose, we assume that the maximum amplitude $\alpha = 30°$, then, the fractional lengthening of the period is

$$\frac{\tau - \tau_0}{\tau_0} = \frac{1}{16} \left(\frac{30 \times \pi}{180} \right)^2 = \frac{(0.5236)^2}{16} = 0.017,$$

which is less than 2 per cent.

1.6.3 The Harmonic Oscillator

Many problems of oscillations can be discussed through a single mathematical model called harmonic oscillator. A system undergoing periodic steady-state motion under the action of a spring is called a *harmonic oscillator*. The harmonic oscillator consists of a spring and a particle which can move on a straight line, which we shall take for convenience as x-axis (see Fig. 1.10).

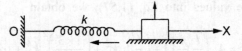

Figure 1.10 The harmonic oscillator.

The particle is attracted towards the origin by a controlling force $-kx\hat{i}$ (varying as the distance), where $\hat{i}$ is the unit vector in the positive direction of the x-axis and k is the spring constant. The equation of motion for this simple harmonic oscillator is given by

$$m\ddot{x} = -kx, \qquad k > 0. \tag{1.62}$$

Its general solution is found to be

$$x(t) = A \cos \omega t + B \sin \omega t, \tag{1.63}$$

where $\omega = \sqrt{(k/m)}$. An equivalent form of the solution is
$$x(t) = a \cos(\omega t + \varepsilon), \tag{1.64}$$
where a is the amplitude of the motion and ε is called the initial phase.

The period of the oscillator is
$$\tau = \frac{2\pi}{\omega}, \tag{1.65}$$
while the frequency of the oscillator is
$$\nu = \frac{1}{\tau} = \frac{\omega}{2\pi}. \tag{1.66}$$

EXERCISES

1.1 Newton's universal law of gravitation states
$$\vec{F} = \frac{GMm}{r^2}$$
between two masses M and m separated by a distance r. Find the dimensions of GM.

1.2 Determine the value of GM in Newton's universal gravitational law when the attracting force is Earth.

1.3 Under what conditions will both the linear and angular momentum of a moving particle are constant?

1.4 The velocity of a particle is given by
$$\dot{\vec{r}} = 12t^2 \hat{i} + 16t^3 \hat{j} + \sin(\pi t)\hat{k} \text{ m/s}.$$
The displacement vector $\vec{r}_0$ at time $t = 0$ is known to be $\vec{r}_0 = (4\hat{j} + 3\hat{k})m$. Find the position vector as a function of time.

1.5 Experiments are conducted to measure the drag force encountered by a sphere moving at a constant speed through a viscous fluid. Assuming that the drag force is given by
$$F_D = C\rho^\alpha v^\beta R^\gamma \mu^\delta$$
where C is a dimensionless coefficient, ρ is fluid density, v the speed, R the radius of the sphere, and μ the viscosity of the fluid. Using dimensional analysis, show that
$$F_D = C\rho v^2 R^2 \left(\frac{\rho v R}{\mu}\right)^{-\delta},$$
where δ may have any value.

2

Dynamics of a System of Particles and Rotating Coordinate Systems

What we know is not much, what we do not know is immense.
—LAPLACE

2.1 MOTION OF A SYSTEM OF PARTICLES

Consider a system of n particles, with masses $m_1, m_2, \ldots, m_n$, moving with velocities $\vec{V}_1, \vec{V}_2, \ldots, \vec{V}_n$, in a Newtonian frame. Then, the linear momentum $\vec{p}$ of the system is defined as a vector

$$\vec{p} = \sum_{i=1}^{n} m_i \vec{V}_i. \tag{2.1}$$

Differentiating with respect to time t, we get

$$\dot{\vec{p}} = \sum_{i=1}^{n} m_i \dot{\vec{V}}_i = \sum_{i=1}^{n} m_i \vec{a}_i, \tag{2.2}$$

where, $\vec{a}_i$ is the acceleration of the ith particle. Ignoring the inertial forces, let $\vec{F}_i$ be the external force acting on the ith particle, then, we can write Eq. (2.2) as

$$\dot{\vec{p}} = \sum_{i=1}^{n} \vec{F}_i = \vec{F}, \tag{2.3}$$

where $\vec{F}$ is the vector sum of all the forces. Physically, Eq. (2.3) means that the rate of change of linear momentum of a system is equal to the vector sum of all external forces acting on the system. This is the *principle of linear momentum* of a system of n particles.

Centre of mass and centre of gravity Consider a system of n particles of masses $m_1, m_2, \ldots, m_n$ whose position vectors are denoted by $\vec{r}_1, \vec{r}_2, \ldots, \vec{r}_n$, with respect to the origin O, as shown in Fig. 2.1.

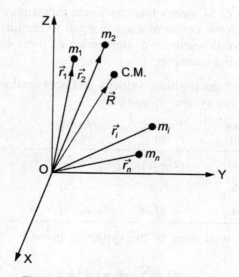

Figure 2.1 Centre of mass of a system.

Then, the position vector of the centre of mass of the system, denoted by C.M. is defined as

$$\vec{R} = \frac{1}{M} \sum_{i=1}^{n} m_i \vec{r}_i, \quad (2.4)$$

where M is the total mass of the system. Physically, the C.M. of the system can be thought of as a weighted average position of the system of particles.

In the special case of a uniform gravitational field the centre of mass coincides with the centre of gravity.

Now, we shall establish the law of centre of mass of a system of particles in the following steps. Without loss of generality, by dropping arrows in Eq. (2.4), its differentiation with respect to time t yields an expression for the velocity of the centre of mass of the system as

$$M\vec{V} = \sum_{i=1}^{n} m_i \vec{v}_i. \quad (2.5)$$

Differentiating once again, the above equation gives us

$$M\dot{\vec{V}} = M\vec{a} = \sum_{i=1}^{n} m_i \vec{a}_i = \vec{F} = \dot{\vec{p}} \quad (2.6)$$

showing that the centre of mass of the system moves like a particle whose mass is equal to the total mass of the system, acted upon by an external force equal to the vector sum of all the external forces acting on the system. This is the law of motion of the centre of mass of the system.

Physically, Eq. (2.5) means that the linear momentum of the fictitious particle moving with the centre of mass is equal to the linear momentum of the system under consideration. For illustration of these concepts, we shall consider the following examples:

Example 2.1 Find the position, velocity and acceleration of the centre of mass of the following system of particles:

Particle	Mass (kg)	Position (m)	Velocity (m/sec)	Acceleration (m/sec^2)
1	0.5	$2\hat{i} - 3\hat{j}$	$-4\hat{k}$	$-3\hat{j} + 5\hat{k}$
2	0.3	$-2\hat{i} + 4\hat{j}$	$-\hat{i} + 3\hat{j}$	$4\hat{i} - 3\hat{j}$
3	0.4	$-\hat{i} - 2\hat{j} - \hat{k}$	$4\hat{i} - 3\hat{j}$	$\hat{i} + \hat{j} + \hat{k}$

Solution The total mass of the system is given by

$$M = \sum_{i=1}^{3} m_i = 1.2 \text{ kg}$$

Let $\vec{R}$ be the position vector of the centre of mass of the given system, then

$$\sum_{i=1}^{3} m_i \vec{r}_i = (0.5)(2\hat{i} - 3\hat{j}) + (0.3)(-2\hat{i} + 4\hat{j}) + (0.4)(-\hat{i} - 2\hat{j} - \hat{k}) = -1.1\hat{j} - 0.4\hat{k}.$$

Therefore, the position of the centre of mass of the given system is obtained by using the formula given by Eq. (2.4) as

$$\vec{R} = \frac{1}{M}\sum_{i=1}^{3} m_i \vec{r}_i = \frac{1}{1.2}(-1.1\hat{j} - 0.4\hat{k}) = -0.92\hat{j} - 0.33\hat{k}. \tag{1}$$

Similarly, the velocity of the centre of mass of the system can be computed from the formula given by Eq. (2.5). Thus,

$$\vec{V} = \frac{1}{M}\sum m_i \vec{v}_i$$

$$= \frac{1}{1.2}[(0.5)(-4\hat{k}) + (0.3)(-\hat{i} + 3\hat{j}) + (0.4)(4\hat{i} - 3\hat{j})]$$

$$= \frac{1}{1.2}(1.3\hat{i} - 0.3\hat{j} - 2.0\hat{k})$$

$$= 1.08\hat{i} - 0.25\hat{j} - 1.67\hat{k} \text{ m/s}. \tag{2}$$

Finally, the acceleraton of the centre of mass of the system is given by Eq. (2.6) and is computed as

$$\vec{a} = \frac{1}{M} \sum_{i=1}^{3} m_i \vec{a}_i$$

$$= \frac{1}{1.2}[(0.5)(-3\hat{j}+5\hat{k}) + (0.3)(4\hat{i}-3\hat{j}) + (0.4)(\hat{i}+\hat{j}+\hat{k})]$$

$$= \frac{1}{1.2}(1.6\hat{i} - 2.0\hat{j} + 2.9\hat{k})$$

$$= (1.33\hat{i} - 1.67\hat{j} + 2.42\hat{k}) \text{ m/s}^2. \tag{3}$$

Example 2.2 Find the resultant external force acting on the system of particles as given in Example 2.1. Also find the angular momentum vector with respect to the origin of the given system.

Solution (i) The resultant external force acting on the system can be computed as

$$\vec{F} = \sum_{i=1}^{3} m_i \vec{a}_i = 1.6\hat{i} - 2.0\hat{j} + 2.9\hat{k} \text{ kg-m/s}^2.$$

(ii) The total angular momentum vector of the given system can be obtained, using the formula given by Eq. (2.8). That is,

$$\vec{H} = \sum_{i=1}^{3} \vec{r}_i \times m_i \vec{v}_i$$

$$= \begin{vmatrix} \hat{i} & \hat{j} & \hat{k} \\ 2 & -3 & 0 \\ 0 & 0 & -2 \end{vmatrix} + \begin{vmatrix} \hat{i} & \hat{j} & \hat{k} \\ -2 & 4 & 0 \\ -0.3 & 0.9 & 0 \end{vmatrix} + \begin{vmatrix} \hat{i} & \hat{j} & \hat{k} \\ -1 & -2 & -1 \\ 1.6 & -1.2 & 0 \end{vmatrix}$$

$$= (6\hat{i} + 4\hat{j}) + (-0.7\hat{k}) + (-1.2\hat{i} - 1.6\hat{j} + 4.4\hat{k})$$

$$= (4.8\hat{i} + 2.4\hat{j} + 3.7\hat{k}) \text{ kg-m}^2/\text{s}.$$

2.2 PRINCIPLE OF ANGULAR MOMENTUM

The generalization of the principle of angular momentum of a system is explained in the following theorem:

Theorem 2.1 The angular momentum of a moving system about a point O is the sum of

(a) the angular momentum about O of a particle moving with the centre of mass, whose mass is equal to that of the total mass of the system, and

(b) the angular momentum of the system about the centre of mass.

Proof Consider a system of n particles with masses $m_1, m_2, ..., m_n$, whose position vectors are $\vec{r}_1, \vec{r}_2, ..., \vec{r}_n$, with reference to a fixed point O, in a Newtonian frame at a given instant as shown in Fig. 2.2. Then, the angular momentum of the particle m_i about O is

$$\vec{h}_i = \vec{r}_i \times m_i \dot{\vec{r}}_i. \qquad (2.7)$$

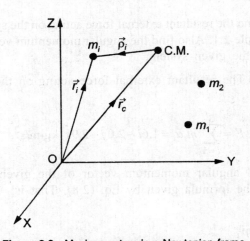

Figure 2.2 Moving system in a Newtonian frame.

Hence, the total angular momentum of the system is found by summing the angular momenta of individual particles. Thus,

$$\vec{H} = \sum_{i=1}^{n} \vec{h}_i = \sum_{i=1}^{n} \vec{r}_i \times m_i \dot{\vec{r}}_i. \qquad (2.8)$$

From Fig. 2.2, we have

$$\vec{r}_i = \vec{r}_c + \vec{\rho}_i.$$

Using this relation, we can shift the reference point 'O' to the centre of mass. Therefore,

$$\vec{H} = \sum_{i=1}^{n} (\vec{r}_c + \vec{\rho}_i) \times m_i (\dot{\vec{r}}_c + \dot{\vec{\rho}}_i)$$

$$= \sum_{i=1}^{n} (\vec{r}_c \times m_i \dot{\vec{r}}_c + \vec{r}_c \times m_i \dot{\vec{\rho}}_i + \vec{\rho}_i \times m_i \dot{\vec{r}}_c + \vec{\rho}_i \times m_i \dot{\vec{\rho}}_i). \qquad (2.9)$$

Recalling the definition of centre of mass of a system

$$\vec{r}_c = \frac{1}{M} \sum_{i=1}^{n} m_i \vec{r}_i$$

where M is the total mass of the system. On rewriting, we have

$$M\vec{r}_c = \sum_{i=1}^{n} m_i \vec{r}_i$$

$$= \sum_{i=1}^{n} m_i (\vec{\rho}_i + \vec{r}_c)$$

$$= M\vec{r}_c + \sum_{i=1}^{n} m_i \vec{\rho}_i,$$

which gives

$$\sum_{i=1}^{n} m_i \vec{\rho}_i = 0 \quad \text{and} \quad \sum m_i \dot{\vec{\rho}}_i = 0.$$

Thus, the second and fourth terms on the right-hand side of Eq. (2.9) vanish and this equation reduces to

$$\vec{H} = \sum \vec{r}_c \times m_i \dot{\vec{r}}_c + \sum \vec{\rho}_i \times m_i \dot{\vec{r}}_c.$$

That is,

$$\vec{H} = \vec{r}_c \times M\dot{\vec{r}}_c + \vec{H}_c. \tag{2.10}$$

Hence, the theorem is proved.

Now, to establish the principle of angular momentum, we differentiate Eq. (2.8) with respect to time to get

$$\dot{\vec{H}} = \sum_{i=1}^{n} \vec{r}_i \times m_i \ddot{\vec{r}}_i + \sum_{i=1}^{n} \dot{\vec{r}}_i \times m_i \dot{\vec{r}}_i$$

$$= \sum_{i=1}^{n} (\vec{r}_i \times m_i \ddot{\vec{r}}_i)$$

$$= \sum_{i=1}^{n} (\vec{r}_i \times m_i a_i).$$

Therefore,

$$\dot{\vec{H}}_O = \sum \vec{r}_i \times \vec{F}_i = \vec{G}_O, \tag{2.11}$$

where $\vec{F}_i$ are the external forces and $\vec{G}$ is the total moment about O due to external forces acting on the system. Thus, the principle of angular momentum can be summed up as follows:

For a system of particles, the time rate of change of angular momentum about a fixed point O, is equal to the total moment about O of the external forces acting on the system.

This fundamental relation (2.11) still holds when the frame of reference OXYZ is replaced by another frame with origin at the centre of mass, C, of the system of particles. Thus,

$$\dot{\vec{H}}_c = \vec{G}_c. \qquad (2.11a)$$

Relative motion of two particles

Consider two particles A and B moving along a straight line, as shown in Fig. 2.3.

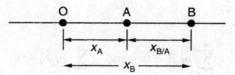

Figure 2.3 A simple system.

With respect to O, let the coordinates of A and B be denoted by x_A and x_B. Then, the difference, $x_B - x_A$, defines the relative position of B with respect to A and is denoted by $x_{B/A}$. Thus, we may write

$$x_{B/A} = x_B - x_A \quad \text{or} \quad x_B = x_A + x_{B/A}.$$

The sign for $x_{B/A}$ is positive if B is to the right of A and negative if B is to the left of A. If we designate the relative velocity of B with respect to A by $v_{B/A}$, then, on differentiating the above equation, we may write

$$v_{B/A} = v_B - v_A \quad \text{or} \quad v_B = v_A + v_{B/A}.$$

Similarly, the rate of change of $v_{B/A}$ is called *relative acceleration* of B with respect to A and is denoted by $a_{B/A}$. Thus, we arrive at

$$a_{B/A} = a_B - a_A \quad \text{or} \quad a_B = a_A + a_{B/A}.$$

2.3 MOTION OF A RIGID BODY

A *rigid body* is defined as the one in which the distance between any two particles is fixed. Such a body in general has a motion which is a superposition of both translation and rotation.

Translational motion A body is said to have *translational motion*, if the position of any line inscribed on the body remains parallel to its original position during its motion. It may be noted that during translational motion, all particles of a rigid body will have the same velocity with respect to some reference frame, at any instant of time. In fact, the velocity vector may change from time to time. For example, consider a situation as shown in Fig. 2.4. In either case, the motion is translational, as the line PQ inscribed on the body remains parallel to its original position at all instants of time.

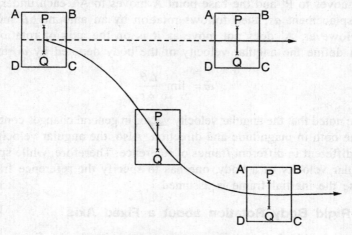

Figure 2.4 Translational motion.

Rotation A body is said to have *rotational* motion, if the body rotates such that along some straight line, all the particles of the body have zero velocity with respect to some reference frame. Such a line of stationary particles is called the *axis of rotation*. During rotation, the distance of any point of the body from the axis of rotation remains unaltered.

2.3.1 Angular Velocity

Consider the motion of a rigid body during an infinitesimal interval Δt, as shown in Fig. 2.5. During this interval, suppose, we have a translational displacement ds of all points in the body together with a rotational displacement

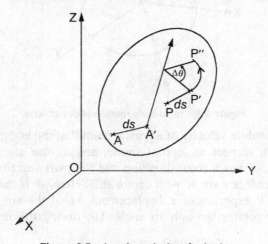

Figure 2.5 Angular velocity of a body.

$\Delta\theta$ about an axis through the base point A' fixed in the body. The order of performing the translation and rotation is immaterial. In Fig. 2.5, we have shown that translation occurs first, followed by rotation. Thus a typical point P moves to P' and the base point A moves to A', each undergoes the same displacement ds, then follows rotation by an amount $\Delta\theta$ moving P' to P''. However, A' does not move as it is on the axis of rotation. Now, we shall define the angular velocity of the body denoted by $\vec{\omega}$, as

$$\vec{\omega} = \lim_{\Delta t \to 0} \frac{\Delta\theta}{\Delta t}. \qquad (2.12)$$

It may be noted that the angular velocity vector in general changes continuously with time both in magnitude and direction. Also, the angular velocity of the body is different in different frames of reference. Therefore, while specifying the angular velocity of a body, one has to specify the reference frame too, otherwise the inertial frame is assumed.

2.3.2 Rigid Body Rotation about a Fixed Axis

Consider a rigid body which is constrained to rotate about a fixed axis AA', that is fixed in an inertial frame OXYZ, as shown in Fig. 2.6. For convenince, we shall assume that the origin is on AA', and Z-axis coincides with AA'.

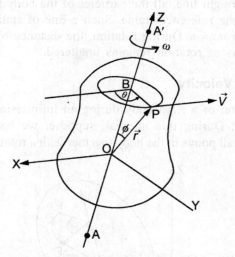

Figure 2.6 Rigid body rotation about an axis.

Let $\vec{V}$ be the absolute velocity of a typical point P of the body, whose position vector is $\vec{r}$ with respect to inertial frame, and $\vec{\omega}$, the angular velocity of the body. Then, $\vec{\omega}$ has a fixed direction and the path described by the point P is a circle of radius $r \sin\phi$, with centre at B. Here, ϕ is the angle between AA' and $\vec{r}$. If P experiences a displacement Δs of the arc described by P when the body rotates through an angle $\Delta\theta$, during the interval Δt, then,

$$\Delta s = (BP)\Delta\theta = (r\sin\phi)\Delta\theta$$

and the magnitude of the velocity of P is

$$|\vec{V}| = \lim_{\Delta t \to 0}\frac{\Delta s}{\Delta t} = \lim_{\Delta t \to 0}(r\sin\phi)\frac{\Delta\theta}{\Delta t} = (r\sin\phi)\omega.$$

Thus, we write

$$\vec{V} = \vec{\omega}\times\vec{r}. \tag{2.13}$$

Here, the vector

$$\vec{\omega} = \omega\hat{k} = \dot{\theta}\hat{k}$$

is called the *angular velocity* of the body.

The acceleration $\vec{a}$ of P can be obtained by differentiating Eq. (2.13) with respect to time. Thus, we write

$$\vec{a} = \frac{d\vec{V}}{dt} = \frac{d}{dt}(\vec{\omega}\times\vec{r}) = \frac{d\vec{\omega}}{dt}\times\vec{r} + \vec{\omega}\times\frac{d\vec{r}}{dt}.$$

If we define the vector $d\vec{\omega}/dt$ by $\vec{\alpha}$, called *angular acceleration* of the body, the above equation becomes

$$\vec{a} = \vec{\alpha}\times\vec{r} + \vec{\omega}\times(\vec{\omega}\times\vec{r}). \tag{2.14}$$

We also have

$$\vec{\alpha} = \alpha\hat{k} = \dot{\omega}\hat{k} = \ddot{\theta}\hat{k}.$$

In Eq. (2.14), the term $\vec{\omega}\times(\vec{\omega}\times\vec{r})$, a vector that is directed radially inwards from point P towards the instantaneous axis of rotation and perpendicular to it, is called *centripetal acceleration*, while the term $\vec{\alpha}\times\vec{r} = \dot{\vec{\omega}}\times\vec{r}$ is called the *tangential acceleration*.

Rotation of a representative plane slab

The rotation of a rigid body about a fixed axis can also be visualized as the rotation of a representative slab in a reference plane perpendicular to the axis of rotation. Thus, we can choose XY-plane as a reference plane that coincides with the plane of the figure and Z-axis pointing away from the reference plane but perpendicular to it (see Fig. 2.7).

Noting that $\vec{\omega} = \omega\hat{k}$ and a positive value of ω corresponds to a counter-clockwise rotation and a negative value to a clockwise rotation, we can express the velocity of any point P of the slab by the relation

$$\vec{V} = \vec{\omega}\times\vec{r} = \omega\hat{k}\times\vec{r}.$$

Since the vectors $\hat{k}$ and $\vec{r}$ are mutually orthogonal, the magnitude of the velocity can be written as

$$v = r\omega.$$

Also, substituting $\vec{\omega} = \omega\hat{k}$ and $\vec{\alpha} = \alpha\hat{k}$ into Eq. (2.14), we can express the acceleration of a typical point P of the slab as

$$\vec{a} = \alpha \hat{k} \times \vec{r} + \omega \hat{k} \times (\omega \hat{k} \times \vec{r})$$
$$= \alpha \hat{k} \times \vec{r} + [(\omega \hat{k} \cdot \vec{r})\omega \hat{k} - (\omega \hat{k} \cdot \omega \hat{k})\vec{r}]$$

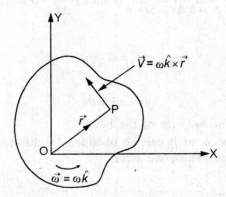

Figure 2.7 Rotation of a slab.

or
$$\vec{a} = \alpha \hat{k} \times \vec{r} + 0 - \omega^2 \vec{r} = \alpha \hat{k} \times \vec{r} - \omega^2 \vec{r}. \qquad (2.15)$$

Thus, we notice that the acceleration has two components: tangential and normal components. We write them as

$$\left.\begin{array}{ll} \vec{a}_t = \alpha \hat{k} \times \vec{r} & \text{or} \quad a_t = r\alpha \\ \vec{a}_n = -\omega^2 \vec{r} & \text{or} \quad a_n = r\omega^2. \end{array}\right\} \qquad (2.15a)$$

The normal component points in the direction opposite to that of $\vec{r}$, that is, towards O.

Alternative notations used in the literature for $\vec{\omega}$ and $\vec{\alpha}$ are given below:

$$\vec{\omega} = \frac{d\theta}{dt} \quad \text{and} \quad \vec{\alpha} = \frac{d\vec{\omega}}{dt} = \frac{d^2\theta}{dt^2} = \vec{\omega}\frac{d\vec{\omega}}{d\theta}.$$

2.3.3 Rate of Change of a Vector in a Rotating Frame

Suppose O'X'Y'Z' is a rotating frame S' which rotates with an angular velocity $\vec{\omega}$ relative to the Newtonian or inertial frame OXYZ denoted by S. Let $\hat{i}, \hat{j}, \hat{k}$ be the unit vectors along the coordinate axes in S' frame. If P is a fixed point in S' frame, its position vector may be expressed in the form

$$\vec{P} = P_1 \hat{i} + P_2 \hat{j} + P_3 \hat{k} \qquad (2.16)$$

with respect to the origin O' of S' frame as shown in Fig. 2.8.

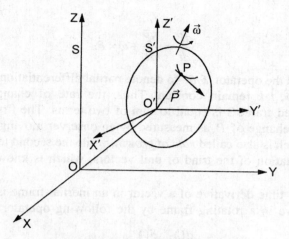

Figure 2.8 Vector in a rotating frame.

Since S′ frame is rotating, the point P which is not on the axes describes a circle lying in a plane perpendicular to the axis of rotation and has a velocity. That is, the vector $\vec{P}$ is changing with time. To an observer fixed relative to S′ frame, the time rate of change of $\vec{P}$ is $\delta \vec{P}/\delta t$. Now, what is the time rate of change of $\vec{P}$, for an observer who is fixed relative to S frame? That is, we have to compute $d\vec{P}/dt$. Noting that, not only P_1, P_2, P_3 change with time, but also $\hat{i}, \hat{j}, \hat{k}$ for the observer in S, and therefore we have

$$\frac{d\vec{P}}{dt} = \frac{dP_1}{dt}\hat{i} + \frac{dP_2}{dt}\hat{j} + \frac{dP_3}{dt}\hat{k} + P_1\frac{d\hat{i}}{dt} + P_2\frac{d\hat{j}}{dt} + P_3\frac{d\hat{k}}{dt}. \quad (2.17)$$

Since i is a fixed vector in S′ frame, which rotates with angular velocity $\vec{\omega}$, we may imagine $\hat{i}$ as the position vector of a particle P of a rigid body (S′) relative to the base point O′, the origin of $\hat{i}$. Then, $d\hat{i}/dt$ is the velocity of the particle P relative to O′, and therefore, utilizing Eq. (2.13), we can at once write

$$\frac{d\hat{i}}{dt} = \vec{\omega} \times \hat{i}. \quad (2.18)$$

On similar reasoning, we have

$$\frac{d\hat{j}}{dt} = \vec{\omega} \times \hat{j} \quad \text{and} \quad \frac{d\hat{k}}{dt} = \vec{\omega} \times \hat{k}. \quad (2.18a)$$

Substituting these results of the time derivatives of unit vectors, into Eq. (2.17), we get

$$\frac{d\vec{P}}{dt} = \frac{\delta \vec{P}}{\delta t} + P_1(\vec{\omega} \times \hat{i}) + P_2(\vec{\omega} \times \hat{j}) + P_3(\vec{\omega} \times \hat{k}).$$

Briefly,

$$\frac{d\vec{P}}{dt} = \frac{\delta \vec{P}}{\delta t} + \vec{\omega} \times \vec{P}. \qquad (2.19)$$

Here, we used the operator $\delta/\delta t$ to denote partial differentiation in which, the unit vectors $\hat{i}, \hat{j}, \hat{k}$ remain constant. Thus, the rate of change of $\vec{P}$ with respect to fixed frame S is equal to sum of two terms. The first term $\delta \vec{P}/\delta t$ is the rate of change of $\vec{P}$ as measured by an observer moving with rotating frame S', which is also called *rate of growth*, while the second term $(\vec{\omega} \times \vec{P})$ is due to the rotation of the triad of unit vectors, which is known as the *rate of transport*.

Thus, the time derivative of a vector in an inertial frame is related to its time derivative in a rotating frame by the following operator relation

$$\frac{d(\,)}{dt} = \frac{\delta (\,)}{\delta t} + \vec{\omega} \times (\,). \qquad (2.20)$$

2.4 MOVING FRAMES OF REFERENCE

Very often we come across many practical problems whose solutions could be easily obtained by referring to moving frames of reference. Here, we consider two frames of reference, one fixed in the laboratory and the other fixed on the moving system. Studying the motion of a rocket or a missile from a ground station is one of the specific examples. The laboratory frame is called a *Newtonian* frame or *inertial* frame or a *fixed* frame. A moving frame in general possesses both translational and rotational velocities relative to the inertial frame. A frame of reference moving with a uniform velocity relative to an inertial frame is also called an inertial frame. However, if a frame of reference is accelerated relative to an inertial frame, it is known as *non-inertial* frame.

2.4.1 Frames of Reference with Translational Motion

Let S(OXYZ) be an inertial frame and suppose S'(O'X'Y'Z') is a frame fixed on a certain body which is moving with translational velocity relative to S-frame. Let $\vec{R}$ be the position vector of the origin O' of S'-frame, with respect to the origin O of S-frame. Also, let $\vec{r}$ and $\vec{r}'$ be the position vectors of a typical particle P with mass m relative to the origins of S and S', respectively. Then from Fig. 2.9, we observe

$$\vec{r} = \vec{R} + \vec{r}'. \qquad (2.21)$$

Differentiating twice with respect to time t, we get

$$\ddot{\vec{r}} = \ddot{\vec{R}} + \ddot{\vec{r}}'.$$

In other words,

$$\vec{a} = \vec{a}_0 + \vec{a}', \qquad (2.22)$$

where $\vec{a}_0$ is the acceleration of S'-frame relative to S-frame.

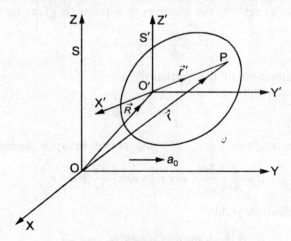

Figure 2.9 Translational motion.

In Newtonian or inertial frame, the equation of motion of the particle P is

$$m\vec{a} = \vec{F}. \tag{2.23}$$

Here, m is the mass of the particle and $\vec{F}$ is the force acting on it. In the moving frame S', its equation of motion using Eqs. (2.22) and (2.23) can be written as

$$m\vec{a}' = m\vec{a} - m\vec{a}_0 = \vec{F} - m\vec{a}_0 = \vec{F}_{\text{effective}}. \tag{2.24}$$

If we define the force on the particle in an accelerated system by $\vec{F}_n$ and $\vec{F}$ is the real force acting on the particle, while $\vec{F}_0 (= -m a_0)$ is the fictitious force, then we may also write Eq. (2.24) as

$$\vec{F}_n = \vec{F} + \vec{F}_0 = \vec{F}_{\text{effective}}. \tag{2.24a}$$

Thus, in a moving frame S' with acceleration a_0, the effective force acting on the particle P is reduced from the actual force $\vec{F}$ by an amount $m\vec{a}_0$. Therefore, we may regard S'- frame as Newtonian or inertial by adding to the actual force, a fictitious force—ma_0 to each particle in the system. Suppose $a_0 = 0$, then the equations of motion of the particle will be identical in both S- and S'- frames. In other words, Newton's laws of motion hold in both the frames moving with a uniform relative velocity. This principle is known as *principle of Newtonian relativity*. The fact that the form of the equations of motion remains identical in both the systems, S and S', can also be expressed by saying that the equations of motion are invariant when the coordinate systems S and S' are in uniform translation.

2.4.2 Motion of a Particle Relative to a Rotating Frame

Suppose S' is a rotating frame which rotates with an angular velocity $\vec{\omega}$ about a point O', and S be an inertial frame. Let $\vec{r}$ be the position vector of a moving particle P relative to S'- frame, as shown in Fig. 2.8. The velocity of the moving particle, as seen by the observer in S-frame, is given by Eq. (2.19) as

$$\vec{V} = \frac{d\vec{r}}{dt} = \frac{\delta \vec{r}}{\delta t} + \vec{\omega} \times \vec{r}. \qquad (2.25)$$

Then, the acceleration of the particle P is

$$\vec{a} = \frac{d\vec{V}}{dt} = \frac{\delta \vec{V}}{\delta t} + \vec{\omega} \times \vec{V}. \qquad (2.26)$$

Substituting for $\vec{V}$ from Eq. (2.25), the above equation becomes

$$\vec{a} = \frac{\delta}{\delta t}\left(\frac{\delta \vec{r}}{\delta t} + \vec{\omega} \times \vec{r}\right) + \vec{\omega} \times \left(\frac{\delta \vec{r}}{\delta t} + \vec{\omega} \times \vec{r}\right).$$

Further simplification yields

$$\vec{a} = \frac{\delta^2 \vec{r}}{\delta t^2} + \frac{\delta \vec{\omega}}{\delta t} \times \vec{r} + 2\vec{\omega} \times \frac{\delta \vec{r}}{\delta t} + \vec{\omega} \times (\vec{\omega} \times \vec{r}).$$

Now, if $\vec{V}'$ and $\vec{a}'$ denote velocity and acceleration of the particle P relative to S'- frame, then the above equation can be rewritten as

$$\vec{a} = \vec{a}' + \left[\frac{\delta \vec{\omega}}{\delta t} \times \vec{r} + \vec{\omega} \times (\vec{\omega} \times \vec{r})\right] + 2\vec{\omega} \times \frac{\delta \vec{r}}{\delta t}. \qquad (2.27)$$

Since

$$\frac{d\vec{\omega}}{dt} = \frac{\delta \vec{\omega}}{\delta t} + \vec{\omega} \times \vec{\omega} = \frac{\delta \vec{\omega}}{\delta t}$$

which means that the angular acceleration is same in both the frames of reference. Thus, Eq. (2.27) reduces to the form

$$\vec{a} = \vec{a}' + \left[\frac{d\vec{\omega}}{dt} \times \vec{r} + \vec{\omega} \times (\vec{\omega} \times \vec{r})\right] + 2\vec{\omega} \times \frac{\delta \vec{r}}{\delta t}, \qquad (2.27a)$$

which may be written as

$$\vec{a} = \vec{a}' + \vec{a}_t + \vec{a}_{cor}, \qquad (2.28)$$

where

$$\vec{a}_t = \frac{d\vec{\omega}}{dt} \times \vec{r} + \vec{\omega} \times (\vec{\omega} \times \vec{r}) \qquad (2.29)$$

is called the *accleration of the transport*, while

$$a_{cor} = 2\vec{\omega} \times \frac{\delta \vec{r}}{\delta t} \qquad (2.30)$$

is called the *acceleration of the Coriolis*. It, therefore follows, that if a particle of mass m is moving under the action of an external force $\vec{F}$, then the equation of motion of the particle in S-frame is

$$m\vec{a} = \vec{F}.$$

However, the equation of motion of the particle in S'-frame is

$$m\vec{a}' = \vec{F} - m\vec{a}_t - m\vec{a}_{cor}. \qquad (2.31)$$

Thus, the rotation of S'-frame gives rise to two fictitious forces—$m\vec{a}_t$, centrifugal force and $m\vec{a}_{cor}$, Coriolis force.

2.4.3 Uniformly Rotating Frames

It is a known fact that Earth rotates about its axis and completes one revolution in 24 hours. Therefore, any frame that is fixed with the Earth will also rotate with it and hence will be a non-inertial frame.

Let O'X'Y'Z' be a rotating frame denoted by S'. Suppose, it is rotating with an angular velocity $\vec{\omega}$ relative to the inertial frame S(OXYZ). For simplicity, let us assume that both the frames have common origin O and also common Z-axis, as shown in Fig. 2.10.

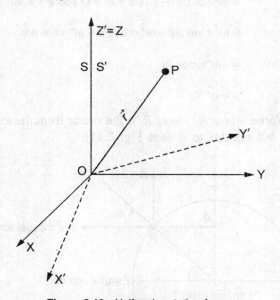

Figure 2.10 Uniformly rotating frame.

In the case of Earth, the common origin O may be taken as the centre of the Earth, while the Z-axis may be considered as its rotational axis. S' may be taken as a rotating frame with Earth relative to non-rotating frame S.

Let P be the position of a typical particle with mass m, then, its position vector is $\vec{r}$ in both S- and S'-frames. However, the velocity and acceleration of the moving particle as seen by the observer in S-frame are as given in Eqs. (2.25) and (2.27). Since $\vec{\omega}$ is constant in the case of the Earth, we have $(d\omega/dt) = 0$ and therefore, Eq. (2.27) becomes

$$\vec{a} = \vec{a}' + \vec{\omega} \times (\vec{\omega} \times \vec{r}) + 2\vec{\omega} \times \frac{\delta \vec{r}}{\delta t}. \qquad (2.32)$$

Noting that m is the mass of the particle considered, the force in the rotating frame is seen to be

$$m\vec{a}' = m\vec{a} - m\vec{\omega} \times (\vec{\omega} \times \vec{r}) - 2m\vec{\omega} \times \frac{\delta \vec{r}}{\delta t}. \qquad (2.33)$$

Here, $-m\vec{\omega} \times (\vec{\omega} \times \vec{r})$ is called *centrifugal* force, while $2m\vec{\omega} \times (\delta \vec{r}/\delta t)$ is the *Coriolis* force, usually denoted by $\vec{F}_{cor}$. The centrifugal force is one of the fictitious forces acting on a particle which is at rest in the rotating frame. Suppose, the particle P is situated at a place whose latitude is ϕ as depicted in Fig. 2.11, then, since

$$\vec{\omega} \times (\vec{\omega} \times \vec{r}) = (\vec{\omega} \cdot \vec{r})\vec{\omega} - (\vec{\omega} \cdot \vec{\omega})\vec{r}$$

$$= \omega r \cos(90 - \phi)\omega \hat{k} - \omega^2 r(\hat{i} \cos \phi + \hat{k} \sin \phi)$$

$$= \omega^2 r \sin \phi \hat{k} - \omega^2 r \hat{i} \cos \phi - \omega^2 r \sin \phi \hat{k}$$

$$= -\omega^2 r \cos \phi \hat{i}$$

$$= -\omega^2 \vec{R},$$

the centrifugal force $= m\omega^2 \vec{R}$. Here, $\vec{R}$ is the vector from the axis of the Earth to the particle but normal to it (see Fig. 2.11).

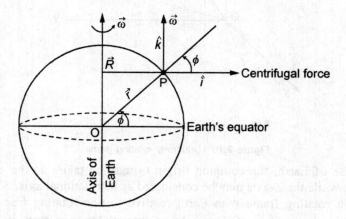

Figure 2.11 Motion of a particle on the Earth.

We may note that the earth rotates counterclockwise about the North Pole with an angular velocity relative to the fixed stars given by

$$\omega = \frac{2\pi}{24 \times 3600} = 7.272 \times 10^{-5} \text{ rad/s}.$$

With this value of ω and taking the equatorial radius of the Earth $r = 6378$ km, the maximum centripetal acceleration is found to be

$$\omega^2 r = 0.034 \text{ m/s}^2$$

or about 0.3% of the acceleration due to gravity.

Now the Coriolis force, $\vec{F}_{cor}$, is given by

$$\vec{F}_{cor} = -2m\vec{\omega} \times \frac{\delta \vec{r}}{\delta t} = -2m\vec{\omega} \times \vec{V}. \qquad (2.34)$$

This force acts on a particle only if it is moving with respect to the rotating frame. Thus, in the rotating frame, if a particle moves with a velocity $\vec{V}$, then, it always experiences a force equal to $-2m\vec{\omega} \times \vec{V}$, perpendicular to its path, opposite to the direction of the vector product $\vec{\omega} \times \vec{V}$. Due to Coriolis force, a moving particle is always found to be deflected to the right of its path in the Northern Hemisphere and to the left of its path in the Southern Hemisphere.

This important effect of the Coriolis force on the true vertical path of the freely falling body is studied in the following example:

Example 2.3 A body is dropped from rest at a height h above the surface of the Earth and at a latitude ϕ. Discuss the lateral displacement of the point of impact due to Coriolis force.

Solution Assume that the Earth is rotating about its axis, with an angular velocity $\vec{\omega}$. If the body has a mass m, in the rotating frame of the Earth, a Coriolis force, $2m\vec{\omega} \times \vec{V}$, is seen to act on the body. We choose a frame with origin at the point P on the Earth's surface below the starting point of the body with X-axis vertically, Y-axis along East and Z-axis along North. Let $\hat{i}, \hat{j}, \hat{k}$ be the unit vectors along these axes (see Fig. 2.12). Then, the angular velocity $\vec{\omega}$ is given by

$$\vec{\omega} = \omega \cos\left(\frac{\pi}{2} - \phi\right)\hat{i} + \omega \cos\phi \hat{k}$$

or

$$\vec{\omega} = \omega \sin\phi \hat{i} + \omega \cos\phi \hat{k}. \qquad (1)$$

As the effective value of the acceleration due to gravity g is the combined effect of the centrifugal acceleration and the acceleration in the inertial frame, we may write

$$-g\hat{i} = \vec{a} - \vec{\omega} \times (\vec{\omega} \times \vec{r}). \qquad (2)$$

The equation of motion of the body in the rotating Earth's frame is

$$m\vec{a} = m\vec{a}' + m\vec{\omega} \times (\vec{\omega} \times \vec{r}) + 2m(\vec{\omega} \times \vec{V}). \qquad (3)$$

Substituting Eq. (2) into (3), we get

$$-mg\hat{i} = m\vec{a}' + 2m\vec{\omega} \times \vec{V}. \qquad (4)$$

Here, the velocity $\vec{V}$ of the body is almost along the X-axis with negligible Y and Z-components and therefore, we take

$$\vec{V} = \frac{dx}{dt}\hat{i}. \qquad (5)$$

Writing $\vec{a}'$ in component form and substituting Eqs. (1) and (5) into Eq. (4), we have

$$-mg\hat{i} = m\left(\frac{d^2x}{dt^2}\hat{i} + \frac{d^2y}{dt^2}\hat{j} + \frac{d^2z}{dt^2}\hat{k}\right) + 2\omega(\sin\phi\,\hat{i} + \cos\phi\,\hat{k}) \times \hat{i}\frac{dx}{dt}.$$

That is,

$$-g\hat{i} = \frac{d^2x}{dt^2}\hat{i} + \left(\frac{d^2y}{dt^2} + 2\omega\frac{dx}{dt}\cos\phi\right)\hat{j} + \frac{d^2z}{dt^2}\hat{k}.$$

Thus, the equations of motion in component form can be written as

$$\frac{d^2x}{dt^2} = -g, \qquad (6)$$

$$\frac{d^2y}{dt^2} = -2\omega\frac{dx}{dt}\cos\phi, \qquad (7)$$

$$\frac{d^2z}{dt^2} = 0. \qquad (8)$$

Integrating Eq. (6) with respect to t, we get

$$\frac{dx}{dt} = -gt + c.$$

Using the initial condition: at $t = 0$, $(dx/dt) = 0$, we get $c = 0$. Thus, we obtain after integrating once again

$$x = \frac{-gt^2}{2} + c'.$$

Now, using the initial condition: at $t = 0$, the distance of the body from the Earth is $x = h$, thus giving $c' = h$. Hence,

$$x = \frac{-gt^2}{2} + h. \qquad (9)$$

Finally, when $t = T$, let the body touch the Earth, that is, when $x = 0$, we have

$$T = \sqrt{\frac{2h}{g}}. \qquad (10)$$

Now, Eq. (7) gives

$$\frac{d^2y}{dt^2} = -2\omega \frac{dx}{dt} \cos\phi = 2\omega gt \cos\phi.$$

Integrating twice with respect to t, we get

$$y = \frac{\omega gt^3}{3} \cos\phi. \tag{11}$$

Here, we have taken the constant of integration to be zero as, initially, the body has no displacement and the velocity in the Y-direction is zero (Fig. 2.12). Finally, when $t = T$, the horizontal displacement or deflection at impact after using Eq. (10) is found to be

$$y = \frac{\omega g}{3} \left(\frac{2h}{g}\right)^{3/2} \cos\phi. \tag{12}$$

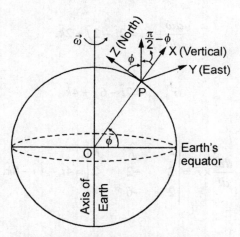

Figure 2.12 Description of Example 2.3.

An order of magnitude of the deflection can be obtained by assuming $\phi = 0°$, which corresponds to the equator and $h = 200$ m, for which we find the deflection as

$$\frac{\omega g}{3}\left(\frac{2h}{g}\right)^{3/2} = (2.378 \times 10^{-4})(40.775)^{3/2} \approx 0.0619 \text{ m}.$$

Example 2.4 Let an S'-frame be rotating with respect to a fixed frame S having the same origin. Assume that the angular velocity vector $\vec{\omega}$ of the S'-frame is given by

$$\vec{\omega} = 2t\hat{i} - t^2\hat{j} + (2t+4)\hat{k},$$

where t is the time, and the position vector of a typical particle at an instant t as observed in S'-frame is given by

$$\vec{r} = (t^2 + 1)\hat{i} - 6t\hat{j} + 4t^3\hat{k}.$$

Find (a) the acceleration of transport and (b) the Coriolis acceleration at time $t = 1$.

Solution From Eq. (2.29), we have a formula to compute the acceleration of the transport as

$$\vec{a}_t = \frac{d\vec{\omega}}{dt} \times \vec{r} + \vec{\omega} \times (\vec{\omega} \times \vec{r}). \tag{1}$$

From the given data, we have

$$\vec{\omega} = 2t\hat{i} - t^2\hat{j} + (2t + 4)\hat{k}.$$

Therefore

$$\frac{d\vec{\omega}}{dt} = 2\hat{i} - 2t\hat{j} + 2\hat{k}.$$

Then,

$$\left.\frac{d\vec{\omega}}{dt}\right|_{t=1} = 2\hat{i} - 2\hat{j} + 2\hat{k}.$$

Also

$$\vec{r}\,|_{t=1} = 2\hat{i} - 6\hat{j} + 4\hat{k}.$$

Thus,

$$\frac{d\vec{\omega}}{dt} \times \vec{r} = \begin{vmatrix} \hat{i} & \hat{j} & \hat{k} \\ 2 & -2 & 2 \\ 2 & -6 & 4 \end{vmatrix} = 4\hat{i} - 4\hat{j} - 8\hat{k}. \tag{2}$$

Now,

$$(\vec{\omega} \times \vec{r})|_{t=1} = \begin{vmatrix} \hat{i} & \hat{j} & \hat{k} \\ 2 & -1 & 6 \\ 2 & -6 & 4 \end{vmatrix} = 32\hat{i} + 4\hat{j} - 10\hat{k}$$

and

$$\vec{\omega} \times (\vec{\omega} \times \vec{r})|_{t=1} = \begin{vmatrix} \hat{i} & \hat{j} & \hat{k} \\ 2 & -1 & 6 \\ 32 & 4 & -10 \end{vmatrix} = -14\hat{i} + 212\hat{j} + 40\hat{k}. \tag{3}$$

Hence, the acceleration of transport is obtained by substituting equations (3) and (2) into (1). Thus,

$$\vec{a}_t = (4\hat{i} - 4\hat{j} - 8\hat{k}) + (-14\hat{i} + 212\hat{j} + 40\hat{k}) = -10\hat{i} + 208\hat{j} + 32\hat{k}. \tag{4}$$

Similarly, the Coriolis acceleration $\vec{a}_c$ is obtained using the formula

$$\vec{a}_{\text{cor}} = 2\vec{\omega} \times \frac{\delta \vec{r}}{\delta t}.$$

In this problem

$$\frac{\delta \vec{r}}{\delta t} = 2t\hat{i} - 6\hat{j} + 12t^2\hat{k}.$$

Therefore,

$$\left.\frac{\delta \vec{r}}{\delta t}\right|_{t=1} = 2\hat{i} - 6\hat{j} + 12\hat{k}.$$

Hence,

$$\vec{a}_c = (4\hat{i} - 2\hat{j} + 12\hat{k}) \times (2\hat{i} - 6\hat{j} + 12\hat{k})$$

$$= \begin{vmatrix} \hat{i} & \hat{j} & \hat{k} \\ 4 & -2 & 12 \\ 2 & -6 & 12 \end{vmatrix}$$

$$= 48\hat{i} - 24\hat{j} - 20\hat{k}. \tag{5}$$

Hence, Eqs. (4) and (5) constitute the required solution.

Example 2.5 A circular wheel is rolling with constant speed along a straight level track. If the angular velocity of the wheel is $\vec{\omega}$, find the acceleration of every particle of the wheel.

Solution Consider a circular wheel rolling along a level track, as shown in Fig. 2.13.

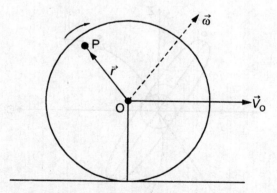

Figure 2.13 A wheel rolling on a level track.

Let us take the centre O of the wheel as the base point and denote its velocity by $\vec{V}_O$, which is given as a constant vector and it lies in the plane of the wheel and parallel to the straight track. Of course, the angular velocity of the wheel $\vec{\omega}$ is perpendicular to the plane of the wheel and is also a constant vector. Let P be a typical particle of the wheel. Its velocity is given as the sum of the velocity of the base point and the velocity of P relative to O. Thus, $\vec{V} = \vec{V}_O + \vec{V}_{P/O}$.

Since the centre of the wheel is moving

$$\vec{V} = \vec{V}_O + \vec{\omega} \times \vec{r} \qquad \text{where} \quad \vec{r} = \overline{OP}.$$

Since $\vec{\omega} \times \vec{r}$ is a vector perpendicular to $\vec{\omega}$, it lies in the plane of the wheel. Since $\vec{\omega}$ and $\vec{V}_O$ are constant vectors, the acceleration of P is given by

$$\vec{a} = \vec{a}_O + \frac{d\vec{\omega}}{dt} \times \vec{r} + \vec{\omega} \times (\vec{\omega} \times \vec{r}).$$

That is,

$$\vec{a} = \vec{\omega} \times (\vec{\omega} \times \vec{r}) \qquad \text{(since } d\vec{\omega}/dt = 0, \; V_O = \text{constant}, \; a_O = 0\text{)}$$

$$= \vec{\omega}(\vec{\omega} \cdot \vec{r}) - \vec{r}(\vec{\omega} \cdot \vec{\omega}) \qquad \text{(Since } \vec{\omega} \text{ and } \vec{r} \text{ are mutually perpendicular)}$$

$$= -\vec{r}\omega^2.$$

Hence, each particle of the wheel has an acceleration of magnitude $r\omega^2$ which is directed towards O, the centre of the wheel.

Example 2.6 Assume that the propeller of an aeroplane when turning, describes a horizontal circle of radius b with centre C and with constant speed v_0. Find the velocity and acceleration of the tip of the propeller.

Solution At any instant, let the position of the propeller be as shown in Fig. 2.14, such that the line from the centre to the tip P of the propeller makes an angle θ with the Z-axis (the vertical).

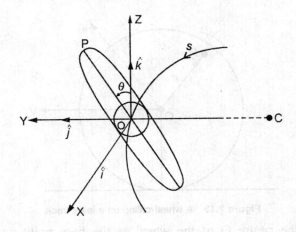

Figure 2.14 Motion of an aeroplane propeller.

For simplicity, we assume that the centre O of the propeller describes a horizontal circle s with constant speed v_0 of radius b with centre at C. Suppose, $\hat{i}, \hat{j}, \hat{k}$ be a triad of unit orthogonal vectors at O, $\hat{j}$ pointing along CO or OY, $\hat{k}$ pointing vertically upwards along OZ, and $\hat{i}$ completing the triad. The vector $\hat{i}$ is clearly the unit tangent vector to s at O. Let us consider O as the base point which itself is moving with velocity $v_0\hat{i}$. Now, the angular velocity of the propeller $\vec{\omega}$ consists of two parts, given by

   (i) the spin $(d\theta/dt)\hat{i}$ imparted by the engine,
   (ii) the angular velocity $(v_0/b)\hat{k}$, due to the turning of aeroplane.

Thus,

$$\vec{\omega} = \frac{d\theta}{dt}\hat{i} + \frac{v_0}{b}\hat{k}. \qquad (1)$$

Now, any point of the propeller will have the velocity

$$\vec{v} = v_0\hat{i} + \vec{\omega} \times \vec{r},$$

where $\vec{r}$ is the position vector of any point of the propeller relative to O. In particular, the velocity of P is

$$\vec{v}_P = v_0\hat{i} + \left(\frac{d\theta}{dt}\hat{i} + \frac{v_0}{b}\hat{k}\right) \times (l\sin\theta\,\hat{j} + l\cos\theta\,\hat{k}),$$

where $OP = l$. Simplifying further, we have

$$\vec{v}_P = v_0\hat{i} + \frac{d\theta}{dt}l\sin\theta\,\hat{k} - \frac{d\theta}{dt}l\cos\theta\,\hat{j} - \frac{v_0}{b}l\sin\theta\,\hat{i}. \qquad (2)$$

If

$$v = \frac{d\theta}{dt}l \qquad (3)$$

then

$$\vec{v}_P = \left(1 - \frac{l}{b}\sin\theta\right)v_0\hat{i} - v\cos\theta\,\hat{j} + v\sin\theta\,\hat{k}. \qquad (4)$$

Hence, the absolute speed of the tip of the propeller of the aeroplane is

$$v_P^2 = v^2 + v_0^2\left(1 - \frac{l}{b}\sin\theta\right)^2. \qquad (5)$$

In practice l/b will be actually small and, therefore,

$$v_P^2 = v^2 + v_0^2 \quad \text{(approximately)}.$$

Now, the acceleration of P can be obtained by differentiating Eq. (2) with respect to time t. We assume that $d\theta/dt$ is constant, in that case, v is constant and v_0, b and the vector $\hat{k}$ are also constants, while θ and the vectors $\hat{i}$

and $\hat{j}$ are variables. We have already noted that the angular velocity of any point of the propeller rotating about O is given by

$$\omega \hat{k} = \frac{v_0}{b} \hat{k}.$$

Hence, using Eq. (2.18), we have

$$\frac{d\hat{i}}{dt} = \frac{v_0}{b} \hat{k} \times \hat{i} = \frac{v_0}{b} \hat{j} \quad \text{and} \quad \frac{d\hat{j}}{dt} = \frac{v_0}{b} \hat{k} \times \hat{j} = -\frac{v_0}{b} \hat{i}. \tag{6}$$

Now differentiating Eq. (2) with respect to t and at the same time using results of Eqs. (3) and (6), we get

$$\vec{a}_P = v_0^2 \hat{j} + v \cos\theta \frac{d\theta}{dt} \hat{k} + v \sin\theta \frac{d\theta}{dt} \hat{j} + \frac{v v_0}{b} \cos\theta \hat{i} - \frac{v_0}{b} l \left(\cos\theta \frac{d\theta}{dt} \hat{i} + \sin\theta \frac{v_0}{b} \hat{j} \right)$$

$$= \left(\frac{v v_0}{b} \cos\theta - \frac{v v_0}{b} \cos\theta \right) \hat{i} + \left(\frac{v_0^2}{b} + \frac{v^2}{l} \sin\theta - \frac{v_0^2}{b^2} l \sin\theta \right) \hat{j} + v \cos\theta \frac{d\theta}{dt} \hat{k}.$$

That is,

$$\vec{a}_P = \left(\frac{v_0^2}{b} + \frac{v^2}{l} \sin\theta - \frac{v_0^2 l}{b^2} \sin\theta \right) \hat{j} + \frac{v^2}{l} \cos\theta \, \hat{k}. \tag{7}$$

It may be noted that the term v^2/l normally far exceeds the first bracketed term in Eq. (7) in magnitude, which indicates that the acceleration is more or less entirely due to the spin of the propeller.

Example 2.7 A rod of length l rotates at a constant angular velocity Ω in the XY-plane with one end O fixed. A circular disc of radius r is pinned at the other end P of the rod. The disc can rotate freely about its centre at P with constant angular velocity ω. Find the acceleration of any point Q on the rim of the disc.

Solution In this example, we can consider the motion of the system in two parts: firstly, the motion of P relative to O and secondly, the motion of Q relative to P.

At some instant, let θ be the angle which the rod makes with the X-axis (see Fig. 2.15). Since the rod rotates with constant angular velocity, we have

$$\dot{\theta} = \Omega = \text{constant} \quad \text{and} \quad \ddot{\theta} = \dot{\Omega} = 0. \tag{1}$$

Similarly, let ϕ be the angle from a line through P parallel to X-axis to the radius of the disc PQ. Since the disc also rotates with constant angular velocity, we have

$$\dot{\phi} = \omega = \text{constant}, \quad \ddot{\phi} = \dot{\omega} = 0. \tag{2}$$

Let $I_{xx}^*, I_{yy}^*, I_{zz}^*$ be the roots of the characteristic equation $|H - \lambda I| = 0$, where I is a unit matrix, which on expansion can be seen as

$$\begin{vmatrix} I_{xx} - \lambda & -I_{xy} & -I_{xz} \\ -I_{yx} & I_{yy} - \lambda & -I_{yz} \\ -I_{zx} & -I_{zy} & I_{zz} - \lambda \end{vmatrix} = 0. \quad (3.35)$$

This is a cubic equation in λ, the solution of which yields three roots, namely $I_{xx}^*, I_{yy}^*, I_{zz}^*$ which are called *principal moments of inertia* at a point in question and the corresponding directions are the principal directions, such as OX*, OY*, OZ*. The important observation here is that all P.Is about the principal axes at a given point vanish. In other words,

$$I_{yz} = I_{zx} = I_{xy} = 0.$$

Corresponding to the principal M.I., that is, I_{xx}^*, the direction cosines l, m, n of the principal direction can be obtained by solving the system of equations

$$\left. \begin{array}{l} (I_{xx} - I_{xx}^*)l - I_{xy}m - I_{xz}n = 0 \\ -I_{yx}l + (I_{yy} - I_{yy}^*)m - I_{yz}n = 0 \\ -I_{zx}l - I_{zy}m + (I_{zz} - I_{zz}^*)n = 0 \end{array} \right\} \quad (3.36)$$

subject to the condition

$$l^2 + m^2 + n^2 = 1.$$

Similarly, the directions corresponding to I_{yy}^* and I_{zz}^* can be obtained by replacing I_{xx}^* by I_{yy}^* and I_{zz}^* respectively in Eqs. (3.36). To conclude, it can be stated that for any point on the body, there exists three mutually perpendicular principal directions about which the M.Is are $I_{xx}^*, I_{yy}^*, I_{zz}^*$, while the corresponding P.Is are zero. To illustrate this important concept, we shall consider the following example.

Example 3.11 Find the principal M.Is and the principal directions at a corner of a uniform cube of length $2a$.

Solution In Example 3.6, we obtained the M.Is of the cube about OX, OY, OZ as

$$I_{xx} = I_{yy} = I_{zz} = \frac{8}{3} Ma^2,$$

and the corresponding P.Is as

$$I_{yz} = I_{zx} = I_{xy} = Ma^2,$$

where M is the mass of the cube. Let the M.Is at the corner O be $I_{xx}^*, I_{yy}^*, I_{zz}^*$, which can be obtained as the roots of the characteristic equation

$$\begin{vmatrix} \dfrac{8}{3}Ma^2 - \lambda & -Ma^2 & -Ma^2 \\ -Ma^2 & \dfrac{8}{3}Ma^2 - \lambda & -Ma^2 \\ -Ma^2 & -Ma^2 & \dfrac{8}{3}Ma^2 - \lambda \end{vmatrix} = 0.$$

Its expansion gives

$$\left(\dfrac{8}{3}Ma^2 - \lambda\right)^3 - 3(Ma^2)^2\left(\dfrac{8}{3}Ma^2 - \lambda\right) - 2(Ma^2)^3 = 0,$$

which, when factorized, yields

$$\left(\dfrac{11}{3}Ma^2 - \lambda\right)^2 \left(\dfrac{2}{3}Ma^2 - \lambda\right) = 0.$$

Hence the required principal M.Is are

$$I_{xx}^* = \dfrac{11}{3}Ma^2, \qquad I_{yy}^* = \dfrac{11}{3}Ma^2, \qquad I_{zz}^* = \dfrac{2}{3}Ma^2. \tag{1}$$

The principal direction corresponding to, say, I_{zz}^* is obtained from Eqs. (3.36), that is, from equations

$$\left.\begin{aligned} \left(\dfrac{8}{3}Ma^2 - \dfrac{2}{3}Ma^2\right)l - Ma^2 m - Ma^2 n &= 0, \\ -Ma^2 l + \left(\dfrac{8}{3}Ma^2 - \dfrac{2}{3}Ma^2\right)m - Ma^2 n &= 0, \\ -Ma^2 l - Ma^2 m + \left(\dfrac{8}{3}Ma^2 - \dfrac{2}{3}Ma^2\right)n &= 0 \end{aligned}\right\} \tag{2}$$

which simplify to

$$2l - m - n = 0, \qquad -l + 2m - n = 0, \qquad -l - m + 2n = 0. \tag{3}$$

and whose solution is found to be

$$l = m = n.$$

Since the direction cosines have to satisfy the constraint

$$l^2 + m^2 + n^2 = 1$$

we get,

$$l = m = n = \dfrac{1}{\sqrt{3}}$$

which indicates that the principal direction is that of the diagonal of a cube. Similarly, the other principal directions associated with I_{xx}^*, I_{yy}^* can be found.

3.2 KINETIC ENERGY OF A RIGID BODY ROTATING ABOUT A FIXED POINT

Let OXYZ be any rectangular system of coordinate axes. Imagine a rigid body rotating about a fixed point O with angular velocity $\vec{\omega}$ at a given instant, as shown in Fig. 3.17.

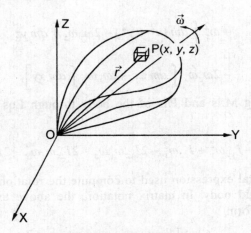

Fig. 3.17 A rotating rigid body.

Now, consider a typical particle P(x, y, z) of mass dm. We note that the speed v of this particle P is equal to the magnitude of the velocity given by the expression

$$\vec{V} = \vec{\omega} \times \vec{r}. \qquad (3.37)$$

Then, the kinetic energy of the particle P is

$$T = \frac{1}{2} dm\, v^2.$$

Hence, the total rotational kinetic energy T_{rot} of the body at a given instant is

$$T_{\text{rot}} = \frac{1}{2} \int dm\, v^2 = \frac{1}{2} \int dm\, |\vec{\omega} \times \vec{r}|^2. \qquad (3.38)$$

Now, let us seek an alternate expression for rotational kinetic energy, involving angular velocity $\vec{\omega}$ and the M.Is of the rotating body.

Let $\hat{i}, \hat{j}, \hat{k}$ be the unit vectors along OX, OY, OZ so that $\overrightarrow{OP} = \vec{r}$. Thus,

$$\vec{r} = \hat{i}x + \hat{j}y + \hat{k}z \quad \text{and} \quad \vec{\omega} = \hat{i}\omega_x + \hat{j}\omega_y + \hat{k}\omega_z.$$

Then,

$$\vec{\omega} \times \vec{r} = \begin{vmatrix} \hat{i} & \hat{j} & \hat{k} \\ \omega_x & \omega_y & \omega_z \\ x & y & z \end{vmatrix} = (\omega_y z - \omega_z y)\hat{i} + (\omega_z x - \omega_x z)\hat{j} + (\omega_x y - \omega_y x)\hat{k}.$$

Substituting the preceding expression into Eq. (3.38), we get

$$T_{rot} = \frac{1}{2} \int dm \left[(\omega_y z - \omega_z y) + (\omega_z x - \omega_x z)^2 + (\omega_x y - \omega_y x)^2 \right]$$

$$= \frac{1}{2} \left[\omega_x^2 \int dm (y^2 + z^2) + \omega_y^2 \int dm (z^2 + x^2) \right.$$

$$+ \omega_z^2 \int dm (x^2 + y^2) - 2\omega_y \omega_z \int dm \, yz$$

$$\left. - 2\omega_z \omega_x \int dm \, zx - 2\omega_x \omega_y \int dm \, xy \right].$$

Now, introducing M.Is and P.Is of the body through Eqs. (3.4) and (3.5), we get

$$T_{rot} = \frac{1}{2} \left(I_{xx} \omega_x^2 + I_{yy} \omega_y^2 + I_{zz} \omega_z^2 - 2 I_{yz} \omega_y \omega_z - 2 I_{zx} \omega_z \omega_x - 2 I_{xy} \omega_x \omega_y \right) \quad (3.39)$$

This is the general expression used to compute the rotational kinetic energy of a rotating rigid body. In matrix notation, the above expression can be recast into the form

$$T_{rot} = \frac{1}{2} \begin{pmatrix} \omega_x \\ \omega_y \\ \omega_z \end{pmatrix}^T \begin{bmatrix} I_{xx} & -I_{xy} & -I_{xz} \\ -I_{yx} & I_{yy} & -I_{yz} \\ -I_{zx} & -I_{zy} & I_{zz} \end{bmatrix} \begin{pmatrix} \omega_x \\ \omega_y \\ \omega_z \end{pmatrix}. \quad (3.40)$$

In case the coordinate axes coincide with the principal axes, all P.Is vanish and the corresponding expression for the rotational kinetic energy reduces to

$$T_{rot} = \frac{1}{2} (I_{xx} \omega_x^2 + I_{yy} \omega_y^2 + I_{zz} \omega_z^2). \quad (3.41)$$

Suppose, the rigid body is constrained to rotate about a fixed axis through O, then Eq. (3.41) further simplifies to

$$T_{rot} = \frac{1}{2} I \omega^2, \quad (3.42)$$

where I is the M.I. of the rigid body about the instantaneous axis of rotation and ω is the instantaneous angular velocity of the rigid body. In practice, such a situation can be visualized in the case of motion of a rocket spinning about its axis, during its initial lift off.

It may be noted that the variables ω_x, ω_y and ω_z in Eq. (3.40) do not correspond to the time derivatives of any set of coordinates that specify the orientation of the rigid body. Thus, we are led to search for a new set of three coordinates by which the orientation of the rigid body can be specified as required in many practical applications. To meet this requirement, we have introduced a set of generalized coordinates known as Eulerian angles, as described in Section 3.4.

Kinetic energy of a rigid body in general

To find the kinetic energy of a rigid body moving generally in space which has both translational and rotational motion we use the theorem of Konig which is stated below:

Theorem 3.3 (*Theorem of Konig*) The kinetic energy of a moving body quite generally in space is equal to the sum of

(a) the kinetic energy of a fictitious particle whose mass is equal to the total mass of the body moving with its centre of mass and
(b) the kinetic energy of the body relative to its centre of mass.

Proof Let OXYZ be any rectangular frame. Suppose $\vec{R}$ be the position vector of the centre of mass G of the rigid body. Let dm be the mass of the typical element located at P, whose position vector with respect to G is $\vec{s}$. Also, let $\vec{r}$ be the position vector of P with respect to O. Then, from Fig. 3.18, we have

$$\vec{r} = \vec{R} + \vec{s}.$$

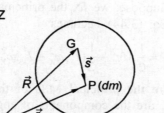

Fig. 3.18 General motion of a rigid body.

Hence, the kinetic energy of the rigid body is

$$T = \frac{1}{2} \int dm \, v^2$$

$$= \frac{1}{2} \int dm \, (\dot{\vec{r}} \cdot \dot{\vec{r}})$$

$$= \frac{1}{2} \int (\dot{\vec{R}} + \dot{\vec{s}}) \cdot (\dot{\vec{R}} + \dot{\vec{s}}) \, dm$$

$$= \frac{1}{2} \int (\dot{\vec{R}} \cdot \dot{\vec{R}}) \, dm + \frac{1}{2} \int (\dot{\vec{s}} \cdot \dot{\vec{s}}) \, dm + \dot{\vec{R}} \cdot \int \dot{\vec{s}} \, dm.$$

That is,

$$T = \frac{1}{2} M v_c^2 + \frac{1}{2} \int \dot{s}^2 \, dm + \dot{\vec{R}} \cdot \int \dot{\vec{s}} \, dm,$$

where v_c is the speed of the centre of mass, $\dot{\vec{s}}$ is the relative velocity of dm as viewed by a non-rotating observer moving with the centre of mass. Now, as a consequence of the definition of centre of mass

$$\int \dot{\vec{s}}\, dm = 0.$$

Hence,

$$T = \frac{1}{2} M v_c^2 + \frac{1}{2} \int \dot{s}^2 dm. \tag{3.43}$$

Here, the first term on the right-hand side corresponds to the translational kinetic energy, while the second term corresponds to the rotational kinetic energy of the rigid body. Thus, the kinetic energy of the rigid body moving generally in space is given by

$$T = \frac{1}{2} M v_c^2 + T_{\text{rot}}, \tag{3.44}$$

where T_{rot} is the rotational kinetic energy due to rotation of the body about its centre of mass. Suppose, we fix the principal axis at the centre of mass of the body, then Eq. (3.44) becomes

$$T = \frac{1}{2} M v_c^2 + \frac{1}{2}(I_{xx}^* \omega_x^2 + I_{yy}^* \omega_y^2 + I_{zz}^* \omega_z^2), \tag{3.45}$$

where $I_{xx}^*, I_{yy}^*, I_{zz}^*$ are the principal M.Is of the body about its centre of mass and $\omega_x, \omega_y, \omega_z$ are the components of angular velocity $\vec{\omega}$. If the rigid body is constrained to rotate about a fixed line through O, then

$$T = \frac{1}{2} M v_c^2 + \frac{1}{2} I \omega^2. \tag{3.46}$$

Definition 3.1 If the three components of the M.I. about the principal axis are equal, that is, if

$$I_{xx}^* = I_{yy}^* = I_{zz}^*,$$

then the body is called a *spherical top*. If two of them are equal, that is, if

$$I_{xx}^* = I_{yy}^* \neq I_{zz}^*,$$

then the body is called a *symmetrical top*. If $I_{xx}^*, I_{yy}^*, I_{zz}^*$ are unequal, the body is called an *asymmetrical top*.

In order to illustrate the ideas developed thus far, we shall consider the following examples:

Example 3.12 Find the kinetic energy of a homogeneous circular cylinder of mass m and radius a, rolling on a plane with linear velocity v.

Solution We know that the kinetic energy of a rigid body moving quite generally in space can be obtained from the Eq. (3.44), that is,

$$T = \frac{1}{2}mv_c^2 + T_{\text{rot}},$$

where T_{rot} is the kinetic energy of the rotating body about its centre of mass. Since the cylinder is constrained to roll about its axis, we have

$$T = \frac{1}{2}mv_c^2 + \frac{1}{2}I\omega^2,$$

where I is the M.I. of the cylinder of mass m and radius a about its axis, which is known to be

$$I = \frac{1}{2}ma^2.$$

Hence, the required kinetic energy is

$$T = \frac{1}{2}mv^2 + \frac{1}{2}\frac{ma^2}{2}\frac{v^2}{a^2} = \frac{3}{4}mv^2.$$

3.3 ANGULAR MOMENTUM OF A RIGID BODY

It is known that the general motion of a rigid body possesses both translational and rotational motion. In rotational motion, angular momentum and torque will come into play. Before we study them in detail, let us recall our understanding of angular momentum of a particle. Consider the motion of a particle of mass m, moving with velocity $\vec{V}$, relative to some fixed frame of reference OXYZ. Then, the linear momentum of the particle is defined as $m\vec{V}$. Now, we can define the angular momentum $\vec{h}$ of the particle about any point, say O, as the moment of $m\vec{V}$ about O. Thus,

$$\vec{h} = \vec{r} \times m\vec{V},$$

where $\vec{r}$ is the position vector of the particle with respect to O. To extend this concept to a rigid body—a rigid body in rotation about a fixed point possesses momentum. Let us consider a typical mass dm situated at P, whose position vector is $\vec{r}$ relative to the origin O (see Fig. 3.17). Then, the angular momentum of the typical mass dm at P about O is

$$d\vec{h} = \vec{r} \times dm\vec{V} = \vec{r} \times (\vec{\omega} \times \vec{r})\, dm.$$

Hence, the total angular momentum of the rigid body about O can be found by integrating the momenta of each differential element. That is,

$$\vec{H} = \int \vec{r} \times (\vec{\omega} \times \vec{r})\, dm = \int (\vec{r} \cdot \vec{r})\vec{\omega}\, dm - \int (\vec{r} \cdot \vec{\omega})\vec{r}\, dm.$$

In other words,

$$\vec{H} = \int \vec{\omega} r^2\, dm - \int \vec{r}(\vec{\omega} \cdot \vec{r})\, dm. \qquad (3.47)$$

Here, the integral is over the volume of the body.

Let $\hat{i}, \hat{j}, \hat{k}$ be the unit vectors along the coordinate axes, then,

$$\vec{r} = \hat{i}x + \hat{j}y + \hat{k}z, \quad \text{(therefore, } r^2 = x^2 + y^2 + z^2\text{)}$$

$$\vec{\omega} = \hat{i}\omega_x + \hat{j}\omega_y + \hat{k}\omega_z,$$

$$\vec{H} = \hat{i}H_x + \hat{j}H_y + \hat{k}H_z.$$

Now, Eq. (3.47) gives the component in the direction of $\hat{i}$ as

$$H_x = \int (x^2 + y^2 + z^2)\omega_x \, dm - \int (x\omega_x + y\omega_y + z\omega_z)x \, dm$$

$$= \omega_x \int (y^2 + z^2) \, dm - \omega_y \int xy \, dm - \omega_z \int xz \, dm$$

$$= I_{xx}\omega_x - I_{xy}\omega_y - I_{xz}\omega_z.$$

Similarly, the components in the $\hat{j}$ and $\hat{k}$ directions yield

$$H_y = -I_{yx}\omega_x + I_{yy}\omega_y - I_{yz}\omega_z,$$

$$H_z = -I_{zx}\omega_x - I_{zy}\omega_y + I_{zz}\omega_z.$$

Thus, the complete set of components of angular momentum vector of a rotating rigid body can be written in matrix notation as

$$\begin{pmatrix} H_x \\ H_y \\ H_z \end{pmatrix} = \begin{bmatrix} I_{xx} & -I_{xy} & -I_{xz} \\ -I_{yx} & I_{yy} & -I_{yz} \\ -I_{zx} & -I_{zy} & I_{zz} \end{bmatrix} \begin{pmatrix} \omega_x \\ \omega_y \\ \omega_z \end{pmatrix} \quad (3.48)$$

or simply

$$\vec{H} = [I]\vec{\omega}. \quad (3.49)$$

Here, the matrix $[I]$ is called *inertia tensor* of a rigid body and is a dyadic operator which changes angular velocity into angular momentum. Thus, we observe that the angular momentum vector of a rotating rigid body does not coincide with its angular velocity vector. It may be noted that the elements of the dyadic or inertia tensor depend on the shape and mass distribution of the body and the coordinate system to which they are referred. In fact, when a body fixed coordinate system is employed in the study of angular motions of a rigid body such as a missile or a rocket, the elements of inertia matrix would be a function of time.

In case $\hat{i}, \hat{j}, \hat{k}$ coincides with the principal axes, then,

$$H_x = I_{xx}^* \omega_x, \qquad H_y = I_{yy}^* \omega_y, \qquad H_z = I_{zz}^* \omega_z. \quad (3.50)$$

where $I_{xx}^*, I_{yy}^*, I_{zz}^*$ are the principal M.Is of the body. Suppose, the rigid body is constrained to rotate about a fixed axis, say Z-axis, with angular velocity ω, then Eq. (3.48) simply reduces to

$$H_x = -I_{xz}\,\omega, \qquad H_y = -I_{yz}\omega, \qquad H_z = I_{zz}\,\omega. \qquad (3.51)$$

and then the angular momentum vector does not lie along the axis of rotation, unless the latter is a principal axis of inertia. However, the component H_z along the axis of rotation is equal to the product of the M.I. about that axis and the angular velocity. This idea is of fundamental importance to understand the phenomenon of gyroscopic motion.

Example 3.13 If a rigid body rotates about a fixed point with an angular velocity $\vec{\omega}$ and has an angular momentum $\vec{H}$, show that the kinetic energy T is given by

$$T = \frac{1}{2}\vec{\omega}\cdot\vec{H}.$$

Solution We know that the kinetic energy in the given situation is given by

$$T = \frac{1}{2}\int dm\, v^2,$$

where dm is the mass of the typical element of the body which can be rewritten as

$$T = \frac{1}{2}\int dm\,(\vec{V}\cdot\vec{V}) = \frac{1}{2}\int dm\,(\vec{\omega}\times\vec{r})\cdot(\vec{\omega}\times\vec{r}),$$

where $\vec{r}$ is the position vector of the typical particle. Since '·' and '×' can be interchanged, we have

$$T = \frac{1}{2}\int dm\,\{\vec{\omega}\cdot[\vec{r}\times(\vec{\omega}\times\vec{r})]\}.$$

Hence,

$$T = \frac{1}{2}\vec{\omega}\cdot\vec{H}.$$

Example 3.14 A body turns about a fixed point. Show that the angle between its angular velocity vector and its angular momentum vector about the fixed point is always acute.

Solution Let $\vec{\omega} = \omega_x\hat{i} + \omega_y\hat{j} + \omega_z\hat{k}$ be the angular velocity vector with which the given body is turning about a fixed point, then the angular momentum vector about the fixed point is given by Eq. (3.49) as

$$\vec{H} = [I]\,\vec{\omega}, \qquad (1)$$

94 | *Kinematics of a Rigid Body Motion (Ch. 3)*

where $[I]$ is an inertia tensor. Thus,

$$\vec{\omega} \cdot \vec{H} = (\omega_x \hat{i} + \omega_y \hat{j} + \omega_z \hat{k}) \cdot [I](\omega_x \hat{i} + \omega_y \hat{j} + \omega_z \hat{k})$$

$$= [I](\omega_x^2 + \omega_y^2 + \omega_z^2).$$

We have seen in Example 3.13, that

$$\vec{\omega} \cdot \vec{H} = 2T. \tag{2}$$

But,

$$\vec{\omega} \cdot \vec{H} = |\vec{\omega}||\vec{H}|\cos\theta, \tag{3}$$

where θ is the angle between $\vec{\omega}$ and $\vec{H}$. Therefore, from Eqs. (2) and (3), we find that

$$\cos\theta = \frac{2T}{|\vec{\omega}||\vec{H}|} \quad \text{(always positive)}$$

implying $\theta \leq 90°$, which means θ is always acute.

Example 3.15 A uniform sphere of mass M and radius a is released from rest on a plane inclined at an angle α to the horizontal. If the sphere rolls down without slipping, show that the acceleration of the centre of the sphere is constant and equal to $(5/7)g \sin \alpha$.

Solution Let O be the point of contact of the sphere with the plane, initially. Since rolling is without slipping, imagine a sphere of radius a, rolling down a plane inclined at an angle α to the horizontal (Fig. 3.19).

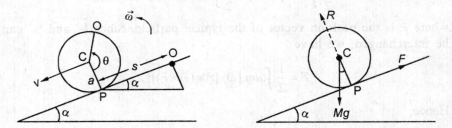

Figure 3.19 Illustration for Example 3.15.

When the body has descended a distance s, let P be the point of contact of the sphere with the plane. Let $\angle \text{PCO} = \theta$, then $\dot{s} = a\omega$, where $\dot{s}$ is the instantaneous velocity of C, the centre of the sphere, and ω is the instantaneous angular velocity about C.

At any time, the forces acting on the sphere are

    (i) its weight Mg acting vertically downwards through its centre of mass C,

   (ii) reaction $\vec{R}$ of the plane perpendicular to the plane itself, and

  (iii) the frictional force $\vec{F}$ acting upwards on the plane.

Since $\vec{R}$ and $\vec{F}$ both act through P which is instantaneously at rest and they do no work during the motion and hence the energy equation holds true, the kinetic energy of the system T is given by Eq. (3.46) as

$$T = \frac{1}{2} M v_c^2 + \frac{1}{2} I \omega^2,$$

that is,

$$T = \frac{1}{2} M \dot{s}^2 + \frac{1}{2} I \left(\frac{\dot{s}}{a} \right)^2. \tag{1}$$

When the sphere is at P, there is a loss of potential energy given by

$$V = -Mgs \sin \alpha. \tag{2}$$

For energy conservation, we must have from Eqs. (1) and (2)

$$\frac{1}{2} M \dot{s}^2 + \frac{1}{2} I \frac{\dot{s}^2}{a^2} - Mgs \sin \alpha = E \text{ (constant)}.$$

Using the initial conditions, $s = \dot{s} = 0$ when $t = 0$ implying $E = 0$ and thus, we have

$$\frac{1}{2} M \dot{s}^2 + \frac{1}{2} I \frac{\dot{s}^2}{a^2} = Mgs \sin \alpha$$

or

$$\left(M + \frac{I}{a^2} \right) \dot{s}^2 = 2 Mgs \sin \alpha. \tag{3}$$

Now, differentiating Eq. (3) with respect to t, we get after cancelling $2\dot{s}$ on both sides

$$\ddot{s} = \frac{Mg \sin \alpha}{M + I/a^2}.$$

But, for sphere, the moment of inertia $I = (2/5) Ma^2$. Hence,

$$\ddot{s} = \frac{Mg \sin \alpha}{M + (2/5)M} = \frac{5}{7} g \sin \alpha \tag{4}$$

from which, we can infer that the acceleration of the centre of the sphere is constant.

Example 3.16 A uniform circular disc of radius R and mass M is rigidly mounted on one end of a thin light shaft CD of length L. The shaft is normal to the disc at the centre C. The disc rolls on a rough horizontal plane, D, being fixed in this plane by a smooth universal joint. If the centre of the disc rotates without slipping about the vertical through D with constant angular velocity Ω, find the angular velocity, the kinetic energy and the angular momentum of the disc about D.

Solution The description of the problem can be seen as in Fig. 3.20.

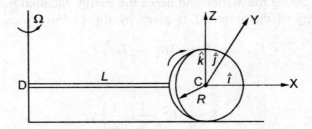

Figure 3.20 Illustration for Example 3.16.

The linear velocity of the centre of the wheel is

$$\vec{V} = L\Omega \hat{j}. \quad (1)$$

Therefore,

$$\text{Linear momentum} = M\vec{V} = ML\Omega \hat{j}, \quad (2)$$

where M is the mass of the wheel. Since the wheel is rolling without slipping, its angular velocity $\vec{\omega}$ in component form is

$$\vec{\omega} = \left(-\frac{L\Omega}{R}\hat{i},\ 0,\ \Omega\hat{k}\right). \quad (3)$$

If the wheel has principal moments of inertia I_{xx}, $I_{yy} = I_{zz}$, I_{zz}, then the angular momentum about the point C is found to be

$$H_x = I_{xx}\omega_x = -\frac{I_{xx}L\Omega}{R}\hat{i}$$

$$H_y = I_{yy}\omega_y = 0$$

$$H_z = I_{zz}\omega_z = I_{zz}\omega_z.$$

Therefore,

$$\vec{H}_C = -\frac{I_{xx}L\Omega}{R}\hat{i} + I_{zz}\Omega\hat{k}. \quad (4)$$

The angular momentum about D can then be obtained as

$$\vec{H}_D = \vec{H}_C + \vec{r}_{DC} \times M\vec{V} = \vec{H}_C + \vec{r}_{DC} ML\Omega\hat{j}$$

Now, using Eq. (4), we obtain

$$\vec{H}_D = -I_{xx}\frac{L}{R}\Omega\hat{i} + I_{zz}\Omega\hat{k} + L\hat{i} \times ML\Omega\hat{j}$$

$$= -\frac{I_{xx}L\Omega}{R}\hat{i} + (I_{zz} + ML^2)\Omega\hat{k}. \quad (5)$$

But, we know that the M.I. of the wheel about CX and that about CZ are given by

$$I_{xx} = \frac{MR^2}{4}, \qquad I_{zz} = \frac{MR^2}{2}.$$

Hence Eq. (5) becomes

$$\vec{H}_D = -\frac{MR^2 L\Omega}{4R} \hat{i} + \left(\frac{MR^2}{2} + ML^2\right) \Omega \hat{k}. \tag{6}$$

Finally, the kinetic energy of the wheel using Konig's theorem is

$$T = \frac{1}{2} M v_c^2 + \frac{1}{2} I \omega^2$$

$$= \frac{M}{2} L^2 \Omega^2 + \frac{1}{2} \frac{MR^2}{2} \left(\frac{L^2}{R^2} \Omega^2 + \Omega^2\right)$$

$$= \frac{M}{2} L^2 \Omega^2 + \frac{M\Omega^2}{4} (L^2 + R^2). \tag{7}$$

Example 3.17 A uniform circular disc of mass M and radius R is so mounted that it can turn freely about its centre, which is fixed. It is spinning with angular velocity $\vec{\omega}$ about the perpendicular to its plane at the centre, the plane being horizontal. A particle of mass m, falling vertically, hits the disc near the edge and adheres to it. Prove that immediately afterwards the particle is moving in a direction inclined to the horizontal at an angle α given by

$$\tan \alpha = 4 \frac{m(M+2m)}{M(M+4m)} \frac{v}{R\omega},$$

where v is the speed of the particle just before impact.

Solution Let $\hat{i}, \hat{j}, \hat{k}$ be the unit vectors along the fixed coordinate axes OX, OY, OZ fixed at O, the centre of the circular disc, as shown in Fig. 3.21. Since the impulsive force acting on the particle and the disc are

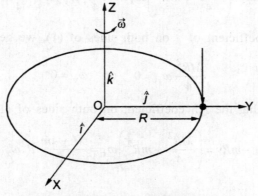

Figure 3.21 Illustration for Example 3.17.

equal and opposite, the resultant moment turns out to be zero. Similarly, the moment of the impulsive reaction at O is also zero. Therefore, the angular momentum of the system about O will be equal to that after the impact about O.

Before the disc is struck, its angular momentum about O is $I\omega\hat{k}$, where I is the moment of inertia of the disc about its axis of rotation. Hence the angular momentum of the disc, before the impact, about O is found to be $(MR^2/2)\,\omega\hat{k}$. Immediately after the particle strikes the disc, let $\omega_x, \omega_y, \omega_z$ be the components of its angular velocity, then the angular momentum of the disc about O, after impact, is given by

$$\vec{H}_{\text{disc}} = I_{xx}\omega_x\hat{i} + I_{yy}\omega_y\hat{j} + I_{zz}\omega_z\hat{k},$$

that is,

$$\vec{H}_{\text{disc}} = \frac{MR^2}{4}\omega_x\hat{i} + \frac{MR^2}{4}\omega_y\hat{j} + \frac{MR^2}{2}\omega_z\hat{k}.$$

Now, the angular momentum of the particle, before impact, about O is

$$R\hat{j} \times m(-v)\,\hat{k} = -mRv\hat{i}$$

Similarly, the angular momentum of the particle about O, after impact, is

$$\vec{H}_{\text{particle}} = R\hat{j} \times [m\,(\omega_x\hat{i} + \omega_y\hat{j} + \omega_z\hat{k}) \times R\hat{j}]$$
$$= R^2 m\,(\omega_x\hat{i} + \omega_y\hat{j} + \omega_z\hat{k}) - mR\omega_y R\hat{j},$$

that is,

$$\vec{H}_{\text{particle}} = R^2 m\,(\omega_x\hat{i} + \omega_z\hat{k}).$$

Using the fact that the angular momentum of the system before impact is equal to that after impact, we obtain

$$-mRv\hat{i} + \frac{MR^2}{2}\omega\hat{k} = \left(\frac{MR^2}{4} + mR^2\right)\omega_x\hat{i} + \frac{MR^2}{4}\omega_y\hat{j} + \left(\frac{MR^2}{4} + mR^2\right)\omega_z\hat{k}. \quad (1)$$

Equating the coefficient of $\hat{j}$ on both sides of (1), we get

$$\frac{MR^2}{4}\omega_y = 0 \quad \text{or} \quad \omega_y = 0. \qquad (2)$$

Similarly, equating the $\hat{i}$th coefficient on both sides of (1), we have

$$-mRv = \left(\frac{MR^2}{4} + mR^2\right)\omega_x = \frac{M + 4m}{4}R^2\omega_x$$

which gives

$$\omega_x = -\frac{4mv}{(M+4m)R}. \tag{3}$$

Equating the $\hat{k}$ th coefficient on both sides of (1), we get

$$\frac{MR^2}{2}\omega = \left(\frac{MR^2}{2} + mR^2\right)\omega_z$$

that is,

$$\omega_z = \frac{M\omega}{M+2m}. \tag{4}$$

Thus, using Eqs. (2)–(4), the velocity of the particle immediately after impact is found to be

$$\vec{v} = \vec{R}\times\vec{\omega} = R\hat{j}\times(\omega_x\hat{i} + \omega_y\hat{j} + \omega_z\hat{k})$$
$$= -R\omega_x\hat{k} + R\omega_z\hat{i}$$
$$= \frac{4mv}{M+4m}\hat{k} + \frac{M\omega R}{M+2m}\hat{i}. \tag{5}$$

Hence, the required angle is obtained as

$$\tan\alpha = \frac{4mv}{M+4m}\frac{M+2m}{M\omega R},$$

that is,

$$\tan\alpha = \frac{4m}{M}\frac{M+2m}{M+4m}\frac{v}{R\omega}.$$

Example 3.18 A slender bar of length l and mass m slides on the smooth floor and wall and has counterclockwise angular velocity $\vec{\omega}$ at the instant shown in Fig. 3.22(a). What is the bar's acceleration.

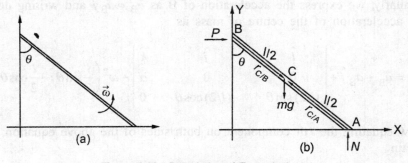

Figure 3.22 Description of Example 3.18.

Solution Let P and N be the normal forces acting on the rod exerted by the wall and the floor respectively, as shown in Fig. 3.22(b). Let C be the point of centre of mass of the rod. Writing the acceleration of the point C (centre of mass) as

$$\vec{a}_C = a_x \hat{i} + a_y \hat{j}.$$

Newton's second law gives

$$\sum F_x = P = ma_x \qquad (1)$$

$$\sum F_y = N - mg = ma_y. \qquad (2)$$

Let $\vec{\alpha}$ denote the bar's counterclockwise angular acceleration, then the principle of angular momentum, that is

$$\vec{G} = \dot{\vec{H}} = I\vec{\alpha}$$

gives

$$G_C = N \frac{l}{2} \sin\theta - P \frac{l}{2} \cos\theta = I\alpha, \qquad (3)$$

where I is the M.I. of the bar about its centre of mass. Now, we have only three equations in five unknowns P, N, a_x, a_y and α. To have a solution, we must relate the acceleration of the centre of mass C of the bar to its angular acceleration. Expressing acceleration of A as $\vec{a}_A = a_A \hat{i}$ and using Eq. (2.36), the acceleration of the centre of mass C is given by

$$\vec{a}_C = \vec{a}_A + \vec{\alpha} r_{C/A} - \omega^2 r_{C/A}$$

Therefore,

$$a_x \hat{i} + a_y \hat{j} = a_A \hat{i} + \begin{vmatrix} \hat{i} & \hat{j} & \hat{k} \\ 0 & 0 & \alpha \\ (-l/2)\sin\theta & (l/2)\cos\theta & 0 \end{vmatrix} - \omega^2 \left(-\frac{l}{2} \sin\theta \hat{i} + \frac{l}{2} \cos\theta \hat{j} \right).$$

Equating the $\hat{j}$ th component on both sides of the above equation, we get

$$a_y = -\frac{l}{2} (\alpha \sin\theta + \omega^2 \cos\theta). \qquad (4)$$

Similarly, we express the acceleration of B as $\vec{a}_B = a_B \hat{j}$ and writing down the acceleration of the centre of mass as

$$\vec{a}_C = \vec{a}_B + a_B \hat{j} + \begin{vmatrix} \hat{i} & \hat{j} & \hat{k} \\ 0 & 0 & \alpha \\ (l/2)\sin\theta & -(l/2)\cos\theta & 0 \end{vmatrix} - \omega^2 \left(\frac{l}{2} \sin\theta \hat{i} - \frac{l}{2} \cos\theta \hat{j} \right).$$

Now equating the $\hat{i}$ th component on both sides of the above equation, we obtain

$$a_x = \frac{l}{2}\left(\alpha \cos \theta - \omega^2 \sin \theta\right). \tag{5}$$

Further, we know that the M.I. of the rod about its C.M. is given as

$$I = \frac{1}{12} ml^2.$$

With the two kinematical Eqs. (4) and (5) we have five equations in five unknowns. Equations (1), (2), (4) and (5) give us

$$P = ma_x = \frac{mL}{2}(\alpha \cos \theta - \omega^2 \sin \theta) \tag{6}$$

$$N = mg - \frac{ml}{2}(\alpha \sin \theta + \omega^2 \cos \theta). \tag{7}$$

Now Eq. (3) gives

$$I\alpha = \frac{mgl}{2}\sin\theta - \frac{ml^2\alpha}{4}\sin^2\theta - \frac{ml^2}{4}\omega^2\sin\theta\cos\theta$$

$$-\frac{ml^2}{4}\alpha\cos^2\theta + \frac{ml^2}{4}\sin\theta\cos\theta,$$

which simplifies to

$$\alpha\left(\frac{ml^2}{12} + \frac{ml^2}{4}\right) = \frac{mgl}{2}\sin\theta \quad \left[\text{since } I = \frac{ml^2}{12}\right]$$

or

$$\alpha = \frac{3}{2}\frac{g}{l}\sin\theta, \tag{8}$$

which is the required angular acceleration of the bar.

Example 3.19 A slender bar of mass m is released from rest in the horizontal position, as shown in Fig. 3.23(a), at an instant. Determine the bar's angular acceleration and the force exerted on the bar by the support A.

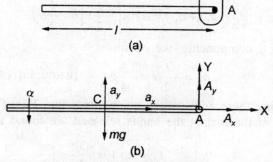

Figure 3.23 Description of Example 3.19.

Kinematics of a Rigid Body Motion (Ch. 3)

Solution Suppose A_x and A_y are the reactions at A when the bar is released. Let the acceleration of the point C (centre of mass) of the bar be $\vec{a}_C = a_x \hat{i} + a_y \hat{j}$ and let its counterclockwise angular acceleration be α. Newton's second law of motion gives

$$\sum F_x = A_x = ma_x, \tag{1}$$

$$\sum F_y = A_y - mg = ma_y. \tag{2}$$

Using the principle of angular momentum, the equation of angular motion about the fixed point A is

$$\frac{l}{2} mg = I_A \alpha. \tag{3}$$

We know that the M.I. of the bar about its centre of mass is

$$I = \frac{1}{12} ml^2.$$

Using parallel-axis theorem, the M.I. of the bar about A is

$$I_A = I + \left(\frac{l}{2}\right)^2 m = \frac{1}{12} ml^2 + \frac{l^2}{4} m = \frac{1}{3} ml^2.$$

Substituting this value of I_A into Eq. (3), we get

$$\alpha = \frac{mg \, l/2}{ml^2/3} = \frac{3}{2} \frac{g}{l}. \tag{4}$$

To compute the reactions at A, we shall express the acceleration of the centre of mass C, in terms of the acceleration of A. Thus using Eq. [2.36(a)], we have

$$\vec{a}_C = \vec{a}_A + (\alpha \hat{k} \times \vec{r}_{C/A}) - \omega^2 \vec{r}_{C/A}$$

Noting that $\omega = 0$ and $\vec{a}_A = 0$ at the instant when the bar is released, the above equation becomes

$$\vec{a}_C = a_x \hat{i} + a_y \hat{j} = \alpha \hat{k} \times \left(-\frac{l}{2} \hat{i}\right) = -\frac{l}{2} \alpha \hat{j}.$$

Equating $\hat{i}$ and $\hat{j}$ components, we obtain

$$a_x = 0, \qquad a_y = -\frac{l}{2}\alpha = -\frac{3}{4} g \qquad \text{[using Eq. (4)]}$$

Now, substituting these acceleration components into Eqs. (1) and (2), the reactions at A at the instant the bar is released are found to be

$$A_x = 0, \qquad A_y = mg + m \frac{-3g}{4} = \frac{mg}{4}.$$

3.4 EULERIAN ANGLES

The orientation of a rotating body or vehicle can be completely specified in terms of three angles known as Eulerian angles. A point on a body can be specified in terms of body fixed axes *oxyz*. In order to determine the orientation of the body itself, we introduce Eulerian angles ψ, θ, ϕ, which are three independent quantities capable of defining the body axes *oxyz* relative to the inertial or fixed axes OXYZ. These are three successive angular displacements which can adequately carry out the transformation from one Cartesian system of axes to another. To demonstrate, how Eulerian angles are capable of describing the orientation of the body, consider a fixed frame of reference OXYZ and a body-fixed coordinate frame *oxyz*, which assumes different orientations in space, the two sets being initially coincident. Now, allow the X, Y, Z set to rotate about the Z-axis through an angle ψ to a new position which may be denoted by α', β', γ' as shown in Fig. 3.24.

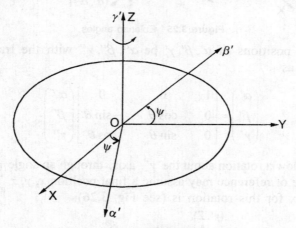

Figure 3.24 Eulerian angles.

The relationship between the two coordinate systems X, Y, Z and α', β', γ' can be seen as

$$X = \alpha' \cos\psi - \beta' \sin\psi$$
$$Y = \alpha' \sin\psi + \beta' \cos\psi$$
$$Z = \gamma'.$$

In the language of aerospace, ψ is called the *yaw angle*. Using matrix notation, the above equations can be recast as

$$\begin{pmatrix} X \\ Y \\ Z \end{pmatrix} = \begin{bmatrix} \cos\psi & -\sin\psi & 0 \\ \sin\psi & \cos\psi & 0 \\ 0 & 0 & 1 \end{bmatrix} \begin{pmatrix} \alpha' \\ \beta' \\ \gamma' \end{pmatrix}. \qquad (3.52)$$

We next allow the rotation θ about α'-axis as shown in Fig. 3.25. Here θ is called the *pitch angle* or *altitude angle*, which usually lies in the range $-\pi/2 \leq \theta \leq \pi/2$.

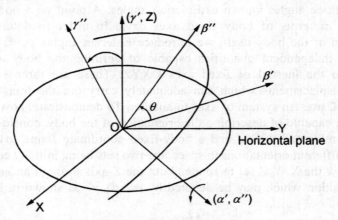

Figure 3.25 Eulerian angles.

Let the new positions of α', β', γ' be $\alpha'', \beta'', \gamma''$ with the transformation matrix given as

$$\begin{pmatrix} \alpha' \\ \beta' \\ \gamma' \end{pmatrix} = \begin{bmatrix} 1 & 0 & 0 \\ 0 & \cos\theta & -\sin\theta \\ 0 & \sin\theta & \cos\theta \end{bmatrix} \begin{pmatrix} \alpha'' \\ \beta'' \\ \gamma'' \end{pmatrix}. \qquad (3.53)$$

Finally, we allow a rotation about the γ''-axis, through an angle ϕ, so that the moving frame of reference may assume a final position x, y, z. The transformation matrix for this rotation is (see Fig. 3.26)

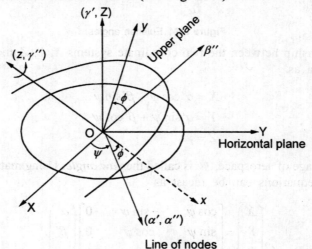

Figure 3.26 Eulerian angles.

$$\begin{pmatrix} \alpha'' \\ \beta'' \\ \gamma'' \end{pmatrix} = \begin{bmatrix} \cos\phi & -\sin\phi & 0 \\ \sin\phi & \cos\phi & 0 \\ 0 & 0 & 1 \end{bmatrix} \begin{pmatrix} x \\ y \\ z \end{pmatrix}. \qquad (3.54)$$

Here, ϕ is called the *roll angle*, which is limited to the range $0 \le \phi \le 2\pi$. Thus, we see that the body-fixed axes is related to the fixed frame of reference (inertial frame) by the matrix equation obtained with the help of Eqs. (3.52) to (3.54) as

$$\begin{pmatrix} X \\ Y \\ Z \end{pmatrix} = \begin{bmatrix} \cos\psi & -\sin\psi & 0 \\ \sin\psi & \cos\psi & 0 \\ 0 & 0 & 1 \end{bmatrix} \begin{bmatrix} 1 & 0 & 0 \\ 0 & \cos\theta & -\sin\theta \\ 0 & \sin\theta & \cos\theta \end{bmatrix} \begin{bmatrix} \cos\phi & -\sin\phi & 0 \\ \sin\phi & \cos\phi & 0 \\ 0 & 0 & 1 \end{bmatrix} \begin{pmatrix} x \\ y \\ z \end{pmatrix}. \qquad (3.55)$$

Here, ψ, θ and ϕ are respectively called yaw, pitch and roll angles. On simplification, we obtain

$$\begin{pmatrix} X \\ Y \\ Z \end{pmatrix} = \begin{bmatrix} \begin{pmatrix} \cos\phi\cos\psi - \\ \sin\phi\cos\theta\sin\psi \end{pmatrix} & \begin{pmatrix} -\sin\phi\cos\psi - \\ \sin\psi\cos\theta\cos\phi \end{pmatrix} & \sin\theta\sin\psi \\ \begin{pmatrix} \cos\phi\sin\psi + \\ \sin\phi\cos\theta\cos\psi \end{pmatrix} & \begin{pmatrix} -\sin\phi\sin\psi + \\ \cos\phi\cos\theta\cos\psi \end{pmatrix} & -\sin\theta\cos\psi \\ \sin\theta\sin\phi & \sin\theta\cos\phi & \cos\theta \end{bmatrix} \begin{pmatrix} x \\ y \\ z \end{pmatrix}$$

$$(3.56)$$

We can notice that the transformation matrix is an orthogonal matrix. Therefore, the inverse transformation from (X, Y, Z) to (x, y, z) can be straightaway written as

$$\begin{pmatrix} x \\ y \\ z \end{pmatrix} = \begin{bmatrix} \begin{pmatrix} \cos\psi\cos\phi - \\ \sin\psi\cos\theta\sin\phi \end{pmatrix} & \begin{pmatrix} \sin\psi\cos\phi + \\ \cos\psi\cos\theta\sin\phi \end{pmatrix} & \sin\theta\sin\phi \\ \begin{pmatrix} -\cos\psi\sin\phi - \\ \sin\psi\cos\theta\cos\phi \end{pmatrix} & \begin{pmatrix} -\sin\psi\sin\phi + \\ \cos\psi\cos\theta\cos\phi \end{pmatrix} & \sin\theta\cos\phi \\ \sin\psi\sin\theta & -\cos\psi\sin\theta & \cos\theta \end{bmatrix} \begin{pmatrix} X \\ Y \\ Z \end{pmatrix}$$

$$(3.57)$$

The axis (α', α'') is called the *line of nodes*. Assuming that the Eulerian angles are limited to ranges $0 \le \psi \le 2\pi$, $-\pi/2 \le \theta \le \pi/2$, $0 \le \phi \le 2\pi$, it can be verified that any possible orientation of the body can be attained by performing proper rotations in the given order. Since the Cartesian coordinates

x, y, z and the Eulerian angles ψ, θ, ϕ describe the configuration of a rigid body completely; a rigid body freely moving in space has six degrees of freedom.

It must be clear that, in arriving at the final position of the body axis, we have encountered four sets of orthogonal axis (X, Y, Z), $(\alpha', \beta', \gamma')$, $(\alpha'', \beta'', \gamma'')$, (x, y, z). Some of these axes coincide such as (Z, γ'), (γ'', z) and (α', α'') the line of nodes. ϕ is measured from the line of nodes on the upper plane. The conventional sequence is ψ, θ, ϕ. If the sequence changes, the final result also will change. This important result is demonstrated in the following example:

Example 3.20 In introducing the concept of Eulerian angles, we stated that the final result depends on the sequence in which the rotations are carried out. Illustrate by means of a diagram if $\psi = \theta = \phi = 90°$, the result of conventional sequence ψ, θ, ϕ and a sequence ψ, ϕ, θ.

Solution The conventional sequence ψ, θ, ϕ gives the situation as depicted in Fig. 3.27.

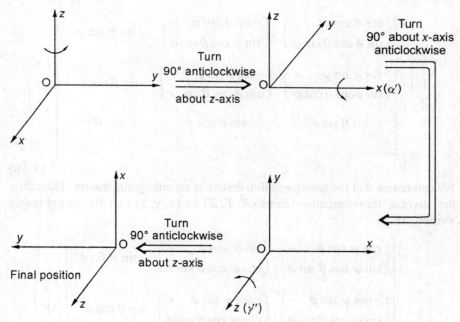

Figure 3.27 Illustration for Example 3.20.

The sequence of rotations ψ, ϕ, θ is illustrated in Fig. 3.28. Obviously, Figs. 3.27 and 3.28 demonstrate that the final result depends upon the order of sequence of rotations caried out. Hence proved.

3.4.1 Angular Velocities Expressed in Terms of Eulerian Angles

Very often, we need to express the angular velocity components $\omega_x, \omega_y, \omega_z$ of a rotating rigid body about x, y, z axes, in terms of Eulerian angles. It is known that the motion of a rigid body can be determined when its

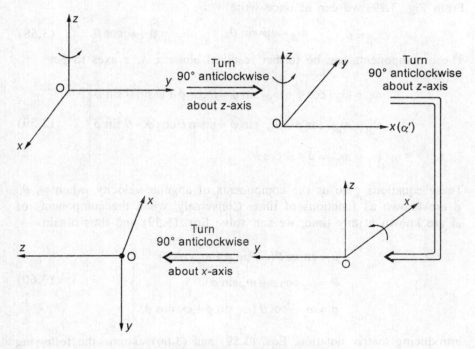

Figure 3.28 Illustration of Example 3.20.

angular velocity $\vec{\omega}(t)$ is known. Similarly the motion of a rigid body is described when Eulerian angles ψ, θ, ϕ are known as functions of time. Therefore, we now seek expressions for $\omega_x, \omega_y, \omega_z$ in terms of ψ, θ, ϕ and their rates of change.

The angular velocity $\dot{\psi}$ acts along the z-axis which represents 'precessional' velocity. $\dot{\theta}$ acts along the α'-axis or line of nodes which corresponds to 'nutation', and $\dot{\phi}$ acts about the γ''-axis which represents 'spin' angular velocity. These are shown in Fig. 3.29.

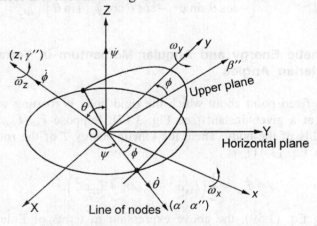

Figure 3.29 Motion of a rigid body in general.

From Fig. 3.29, we can at once write

$$\omega_{\alpha''} = \dot{\theta}, \qquad \omega_{\beta''} = \dot{\psi}\sin\theta, \qquad \omega_{\gamma''} = \dot{\phi} + \dot{\psi}\cos\theta. \qquad (3.58)$$

These components can be further resolved along x, y, z axes to get

$$\left.\begin{aligned}\omega_x &= \omega_{\alpha''}\cos\phi + \omega_{\beta''}\sin\phi = \dot{\theta}\cos\phi + \dot{\psi}\sin\theta\sin\phi \\ \omega_y &= \omega_{\beta''}\cos\phi - \omega_{\alpha''}\sin\phi = \dot{\psi}\sin\theta\cos\phi - \dot{\theta}\sin\phi \\ \omega_z &= \omega_{\gamma''} = \dot{\phi} + \dot{\psi}\cos\theta.\end{aligned}\right\} \qquad (3.59)$$

These equations give us the components of angular velocity, when ψ, θ, ϕ are known as functions of time. Conversely, when the components of $\vec{\omega}$ are known at any time, we can solve Eqs. (3.59) and thus obtain

$$\left.\begin{aligned}\dot{\psi} &= \operatorname{cosec}\theta(\omega_x\sin\phi + \omega_y\cos\phi) \\ \dot{\theta} &= \omega_x\cos\phi - \omega_y\sin\phi \\ \dot{\phi} &= \omega_z - \cot\theta\,(\omega_x\sin\phi + \omega_y\cos\phi).\end{aligned}\right\} \qquad (3.60)$$

Introducing matrix notation, Eqs. (3.59) and (3.60) assume the following forms:

$$\begin{pmatrix}\omega_x \\ \omega_y \\ \omega_z\end{pmatrix} = \begin{bmatrix}\sin\theta\sin\phi & \cos\phi & 0 \\ \sin\theta\cos\phi & -\sin\phi & 0 \\ \cos\theta & 0 & 1\end{bmatrix}\begin{pmatrix}\dot{\psi} \\ \dot{\theta} \\ \dot{\phi}\end{pmatrix} \qquad (3.61)$$

$$\begin{pmatrix}\dot{\psi} \\ \dot{\theta} \\ \dot{\phi}\end{pmatrix} = \operatorname{cosec}\theta\begin{bmatrix}\sin\phi & \cos\phi & 0 \\ \sin\theta\cos\phi & -\sin\theta\sin\phi & 0 \\ -\cos\theta\sin\phi & -\cos\theta\cos\phi & \sin\theta\end{bmatrix}\begin{pmatrix}\omega_x \\ \omega_y \\ \omega_z\end{pmatrix}. \qquad (3.62)$$

3.4.2 Kinetic Energy and Angular Momentum in Terms of Eulerian Angles

Let O be a fixed point about which the rigid body is rotating with angular velocity $\vec{\omega}$ at a given instant (see Fig. 3.17). Suppose I_{xx}, I_{yy}, I_{zz} are the principal M.Is of the body. Then, the kinetic energy T of the rotating body, as given in Eq. (3.41), is

$$T_{\text{rot}} = \frac{1}{2}\left(I_{xx}\omega_x^2 + I_{yy}\omega_y^2 + I_{zz}\omega_z^2\right).$$

Now, using Eq. (3.59), the above expression in terms of Eulerian angles becomes

$$T_{\rm rot} = \frac{1}{2}\Big[I_{xx}(\dot\theta\cos\phi + \dot\psi\sin\theta\sin\phi)^2 +$$

$$I_{yy}(\dot\psi\sin\theta\cos\phi - \dot\theta\sin\phi)^2 + I_{zz}(\dot\phi + \dot\psi\cos\theta)^2 \Big]. \quad (3.63)$$

For a symmetric top, where $I_{xx} = I_{yy}$, the above expression simplifies further to

$$T_{\rm rot} = \frac{1}{2}\Big[I_{xx}(\dot\theta^2 + \dot\psi^2\sin^2\theta) + I_{zz}(\dot\phi + \dot\psi\cos\theta)^2 \Big]. \quad (3.64)$$

The angular momentum vector is

$$\vec{H} = H_x\hat{i} + H_y\hat{j} + H_z\hat{k},$$

where $\hat{i}, \hat{j}, \hat{k}$ are the unit vectors directed along the principal axes at O. Then, using Eq. (3.50), we have

$$H^2 = |\vec{H}|^2 = I_{xx}^2\,\omega_x^2 + I_{yy}^2\,\omega_y^2 + I_{zz}^2\,\omega_z^2.$$

In terms of Eulerian angles, using Eq. (3.59), the above equation can be recast as

$$H^2 = I_{xx}^2\Big[(\dot\psi\sin\theta\sin\phi)^2 + (\dot\theta\cos\phi)^2\Big] + I_{yy}^2\Big[(\dot\psi\sin\theta\cos\phi)^2 + (\dot\theta\sin\phi)^2\Big]$$

$$+ I_{zz}^2\Big[(\dot\psi\cos\theta)^2 + \dot\phi^2\Big]. \quad (3.65)$$

3.5 THE COMPOUND PENDULUM

The *compound pendulum* consists of a rigid body which is free to oscillate under the influence of gravity about a fixed horizontal axis. Suppose we assume that the vertical plane of oscillation of the pendulum is chosen as XY-plane, so that the Z-axis through the origin O is the horizontal axis of suspension, as shown in Fig. 3.30.

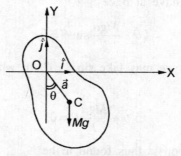

Figure 3.30 Compound pendulum.

Let C be the centre of mass whose position vector relative to O is $\vec{a}$. Since the body is rigid, $|\vec{a}| = a$ is constant. The only force acting on the body is its weight, acting vertically downwards, that is,

$$M\mathbf{g} = -Mg\hat{j},$$

where M is the mass of the pendulum. Now, the potential energy of the system is given by

$$-\nabla V = -\frac{\partial V}{\partial x}\hat{i} - \frac{\partial V}{\partial y}\hat{j} - \frac{\partial V}{\partial z}\hat{k} = -Mg\hat{j},$$

which gives us

$$\frac{\partial V}{\partial x} = 0, \qquad \frac{\partial V}{\partial y} = Mg, \qquad \frac{\partial V}{\partial z} = 0.$$

On integration, we have

$$V = Mgy + c = -Mga\cos\theta + c. \tag{3.66}$$

In Section 3.2, we noted that the kinetic energy T of a rigid body which is constrained to rotate about a fixed axis through O is given by Eq. (3.42) as

$$T = \frac{1}{2}I\omega^2 = \frac{1}{2}I\dot{\theta}^2, \tag{3.67}$$

where I is the M.I. of the rigid body about the instantaneous axis of rotation and ω the instantaneous angular velocity of the rigid body.

Following the principle of conservation of energy, we have

$$T + V = \frac{1}{2}I\dot{\theta}^2 - Mga\cos\theta = \text{constant}. \tag{3.68}$$

Now, differentiating Eq. (3.68) with respect to t, we have

$$I\dot{\theta}\ddot{\theta} + Mga\sin\theta\,\dot{\theta} = 0.$$

As $\dot{\theta}$ is not zero, we have at once

$$\ddot{\theta} + \frac{Mga}{I}\sin\theta = 0. \tag{3.69}$$

For small oscillations, we may take $\sin\theta = \theta$, in which case, the equation of motion becomes

$$\ddot{\theta} + \frac{Mga}{I}\theta = 0. \tag{3.70}$$

The period of oscillation is thus found to be

$$\tau = 2\pi\sqrt{\frac{I}{Mga}} \tag{3.71}$$

Comparing this period with the period of a simple pendulum of length l as given in Eq. (1.46), we find that the length of the equivalent simple pendulum is given by $l = I/Ma$.

EXERCISES

3.1 Find the moment of inertia of a solid homogeneous cube with edge-length $2a$, about the concurrent axes and also their products of inertia.

3.2 Show that the M.Is of a solid cone of mass M and semi-vertical angle θ at its vertex are

$$I_{xx} = (3/10)\, Mh^2 \tan^2\theta,$$

$$I_{yy} = I_{zz} = (3/20)\, Mh^2\, (4 + \tan^2\theta)$$

3.3 Obtain the M.Is of a uniform rectangular block about a central axes parallel to the two faces with dimensions of the block as $2a \times 2b \times 2c$

3.4 Show that the M.I. of a uniform solid cone of radius a, height h, mass M about a line which lies on its surface is

$$\frac{3}{20} Ma^2 \frac{a^2 + 6h^2}{a^2 + h^2}.$$

3.5 The moments and products of inertia of a rigid body about x, y, z axes are $I_{xx} = 3, I_{yy} = 10/3, I_{zz} = 8/3, I_{xy} = 4/3, I_{yz} = 0, I_{zx} = -4/3$. Find the principal moments of inertia and the principal directions.

3.6 Show that the principal moments of inertia of a uniform cylinder of radius a and height h are given by

$$I^*_{xx} = I^*_{yy} = \frac{M}{12}(3a^2 + h^2), \quad I^*_{zz} = \frac{Ma^2}{2}.$$

3.7 Two circular metal discs have the same mass M and the same thickness t. Disc 1 has a uniform density ρ_1 which is less than ρ_2, the uniform density of disc 2. Which disc has the larger moment of inertia?

3.8 A string is wrapped around the periphery of a thin circular disc of radius 0.5 metre and of mass 10 kg, as shown in Fig. 3.31. At a particular instant, the string is pulled up with a force of 200 N. Determine the acceleration of the centre of the disc and the angular acceleration of the disc.

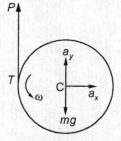

Fig. 3.31 Description of Exercise 3.8.

3.9 A uniform slender rod of length $L = 750$ mm and mass $m = 2$ kg hangs freely from a hinge at A. If a force P of magnitude 12 newton is applied at B horizontally to the left so that $AB = L$, determine (i) the angular acceleration of the rod and (ii) the components of reaction at A.

4

Dynamics of a Rigid Body Motion in Space

Things not understood are admired.

—THOMAS FULLER

4.1 EULER'S DYNAMICAL EQUATIONS FOR A RIGID BODY ROTATING ABOUT A FIXED POINT

Having learnt about the various basic kinematic concepts of a rigid body motion in the previous chapter, we shall now focus on understanding the various applications. To begin with, we shall derive the Euler's dynamical equations involving the motion of a rigid body.

Imagine a rigid body which is constrained to rotate about a fixed point O (may be its centre of gravity), with angular velocity vector $\vec{\omega}$. Suppose the origin of the body-fixed coordinate system is O. Let $\hat{i}, \hat{j}, \hat{k}$ be the unit vectors along OX, OY, OZ respectively (see Fig. 4.1).

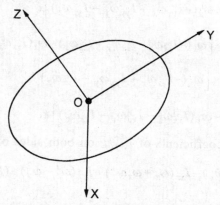

Figure 4.1 Motion of a rigid body in space.

If the components of $\vec{\omega}$ are ω_x, ω_y and ω_z, then, we write

$$\vec{\omega} = \omega_x \hat{i} + \omega_y \hat{j} + \omega_z \hat{k}. \tag{4.1}$$

From the principle of angular momentum, which states that the rate of change of angular momentum of a body about a point, fixed or moving with the centre of mass, is equal to the total moment of the external forces about that point. Symbolically, we can express it as

$$\frac{d\vec{H}}{dt} = \vec{G} \text{ (Torque)}. \qquad (4.2)$$

Following Eq. (3.48), we have

$$\vec{H} = \begin{bmatrix} I_{xx} & -I_{xy} & -I_{xz} \\ -I_{yx} & I_{yy} & -I_{yz} \\ -I_{zx} & -I_{zy} & I_{zz} \end{bmatrix} \begin{pmatrix} \omega_x \\ \omega_y \\ \omega_z \end{pmatrix}. \qquad (4.3)$$

However, with reference to the body fixed frame, the operator relation allows us to write

$$\vec{G} = \frac{d\vec{H}}{dt} = \frac{\delta \vec{H}}{\delta t} + \vec{\omega} \times \vec{H}$$

$$= \dot{H}_x \hat{i} + \dot{H}_y \hat{j} + \dot{H}_z \hat{k} + \begin{vmatrix} \hat{i} & \hat{j} & \hat{k} \\ \omega_x & \omega_y & \omega_z \\ H_x & H_y & H_z \end{vmatrix}.$$

That is,

$$G_x \hat{i} + G_y \hat{j} + G_z \hat{k} = (I_{xx}\dot{\omega}_x - I_{xy}\dot{\omega}_y - I_{xz}\dot{\omega}_z)\hat{i} + (-I_{yx}\dot{\omega}_x + I_{yy}\dot{\omega}_y - I_{yz}\dot{\omega}_z)\hat{j}$$

$$+ (-I_{zx}\dot{\omega}_x - I_{zy}\dot{\omega}_y + I_{zz}\dot{\omega}_z)\hat{k} + \Big[\omega_y(-I_{zx}\omega_x - I_{zy}\omega_y + I_{zz}\omega_z)$$

$$- \omega_z(-I_{yx}\omega_x + I_{yy}\omega_y - I_{yz}\omega_z)\Big]\hat{i}$$

$$- \Big[\omega_x(-I_{zx}\omega_x - I_{zy}\omega_y + I_{zz}\omega_z) - \omega_z(I_{xx}\omega_x - I_{xy}\omega_y - I_{xz}\omega_z)\Big]\hat{j}$$

$$+ \Big[\omega_x(-I_{yx}\omega_x + I_{yy}\omega_y - I_{yz}\omega_z)$$

$$- \omega_y(I_{xx}\omega_x - I_{xy}\omega_y - I_{xz}\omega_z)\Big]\hat{k}. \qquad (4.4)$$

Now, equating the coefficients of $\hat{i}, \hat{j}, \hat{k}$ on both sides of Eq. (4.4), we get

$$I_{xx}\dot{\omega}_x + I_{xy}(\omega_z\omega_x - \dot{\omega}_y) - I_{xz}(\dot{\omega}_z + \omega_x\omega_y) + I_{yz}(\omega_z^2 - \omega_y^2) + (I_{zz} - I_{yy})\omega_y\omega_z = G_x$$

$$(4.5a)$$

$$I_{yy}\dot{\omega}_y + I_{yz}(\omega_x\omega_y - \dot{\omega}_z) - I_{yx}(\dot{\omega}_x + \omega_y\omega_z) + I_{xz}(\omega_x^2 - \omega_z^2) + (I_{xx} - I_{zz})\omega_x\omega_z = G_y$$

$$(4.5b)$$

$$I_{zz}\dot{\omega}_z + I_{xz}(\omega_y\omega_z - \dot{\omega}_x) - I_{yz}(\dot{\omega}_y + \omega_z\omega_x) + I_{xy}(\omega_y^2 - \omega_x^2) + (I_{yy} - I_{xx})\omega_y\omega_x = G_z$$

$$(4.5c)$$

This set of equations of motion has very important applications, specially in space-related problems. Some of them are presented in the following sections. These equations are known as *Euler's dynamical equations of motion for a rigid body rotating about its centre of mass*. Of course, these equations are very complex and extremely difficult to manipulate. However, considerable simplification of these equations can be achieved by allowing the body-fixed coordinate axes to coincide with its principal axes. In such a case, all the P.Is become zero and Eqs. (4.5) simplify to

$$\left. \begin{array}{l} I_{xx}\dot{\omega}_x - (I_{yy} - I_{zz})\omega_y\omega_z = G_x \\ I_{yy}\dot{\omega}_y - (I_{zz} - I_{xx})\omega_z\omega_x = G_y \\ I_{zz}\dot{\omega}_z - (I_{xx} - I_{yy})\omega_x\omega_y = G_z. \end{array} \right\} \quad (4.6)$$

These are *Euler's equations of motion for a rigid body with a fixed point*, in a body-fixed frame of reference. These equations are widely used in solving the rotational motion of a rigid body. These are also nonlinear ordinary differential equations and found to be difficult to have analytical solution and, therefore, require the use of numerical techniques, such as Runge–Kutta method of fourth order* and a fast electronic computer for their solution. However, for a given set of initial conditions and assuming that $\vec{G}$ is a known function of time, position and velocity of the body, Eqs. (4.6) can be solved, in principle, to determine the components of $\vec{\omega}$ as a function of time.

4.1.1 Body-fixed Translational Equations

It is convenient to solve both the translational as well as rotational motions of a rigid body, in a body-fixed coordinate system, especially, when the applied forces can be specified in a body-fixed coordinate system.

In the Newtonian frame of reference, the translational equations of motion of a rigid body can be written as

$$\vec{F} = m\dot{\vec{V}}, \quad (4.7)$$

where $\vec{F}$ is the total external force acting on the rigid body and $\vec{V}$ the velocity of the centre of mass of the body. In a body-fixed frame, $\vec{V}$ and $\vec{F}$ may be expressed as

$$\vec{V} = V_x\hat{i} + V_y\hat{j} + V_z\hat{k}$$

$$\vec{F} = F_x\hat{i} + F_y\hat{j} + F_z\hat{k}.$$

Now, following Newton's second law of motion, we have

---

*See Sankara Rao, 2004

Dynamics of a Rigid Body Motion in Space (Ch. 4)

$$\vec{F} = m\left(\frac{\delta \vec{V}}{\delta t} + \vec{\omega} \times \vec{V}\right) = m\left(\dot{V}_x\hat{i} + \dot{V}_y\hat{j} + \dot{V}_z\hat{k} + \begin{vmatrix} \hat{i} & \hat{j} & \hat{k} \\ \omega_x & \omega_y & \omega_z \\ V_x & V_y & V_z \end{vmatrix}\right). \quad (4.8)$$

In component form, we may also write as

$$\left. \begin{aligned} F_x &= m\left(\frac{dV_x}{dt} + \omega_y V_z - \omega_z V_y\right) \\ F_y &= m\left(\frac{dV_y}{dt} + \omega_z V_x - \omega_x V_z\right) \\ F_z &= m\left(\frac{dV_z}{dt} + \omega_x V_y - \omega_y V_x\right). \end{aligned} \right\} \quad (4.9)$$

Here, the components of $\vec{\omega}$ are either given directly or may be obtained from the rotational Eqs. (4.6). Thus, Eqs. (4.9) can be solved to get V_x, V_y and V_z as a function of time. To sum up, the set of Eqs. (4.6) and (4.9) together constitute the dynamical equations of motion of a rigid body in a body-fixed coordinate frame. For illustration, we present below a few examples.

Example 4.1 Show directly from the Euler's dynamical equations of motion (Eqs. 4.6) that if $\vec{G} = 0$ and $I_{xx} = I_{yy}$, then ω is constant.

Solution We are given the data $I_{xx} = I_{yy}$ and therefore the Euler's dynamical equations reduce to

$$I_{xx}\dot{\omega}_x - (I_{yy} - I_{zz})\omega_y\omega_z = 0 \quad (1)$$

$$I_{yy}\dot{\omega}_y - (I_{zz} - I_{xx})\omega_z\omega_x = 0 \quad (2)$$

$$I_{zz}\dot{\omega}_z = 0. \quad (3)$$

Since $I_{zz} \neq 0$, Eq. (3) becomes $\dot{\omega}_z = 0$. Therefore,

$$\omega_z = \text{constant}. \quad (4)$$

Now, multiplying Eq. (1) by ω_x and Eq. (2) by ω_y and adding, we get

$$I_{xx}\dot{\omega}_x\omega_x + I_{yy}\dot{\omega}_y\omega_y - (I_{yy} - I_{zz} + I_{zz} - I_{xx})\omega_x\omega_y\omega_z = 0.$$

Using the fact that $I_{xx} = I_{yy}$, the above equation reduces to

$$I_{xx}(\dot{\omega}_x\omega_x + \dot{\omega}_y\omega_y) = 0.$$

Since $I_{xx} \neq 0$, we have $\dot{\omega}_x\omega_x + \dot{\omega}_y\omega_y = 0$, which can be rewritten as

$$\frac{1}{2}\frac{d}{dt}(\omega_x^2 + \omega_y^2) = 0.$$

Let $I_{xx}^*, I_{yy}^*, I_{zz}^*$ be the roots of the characteristic equation $|H - \lambda I| = 0$, where I is a unit matrix, which on expansion can be seen as

$$\begin{vmatrix} I_{xx} - \lambda & -I_{xy} & -I_{xz} \\ -I_{yx} & I_{yy} - \lambda & -I_{yz} \\ -I_{zx} & -I_{zy} & I_{zz} - \lambda \end{vmatrix} = 0. \qquad (3.35)$$

This is a cubic equation in λ, the solution of which yields three roots, namely $I_{xx}^*, I_{yy}^*, I_{zz}^*$ which are called *principal moments of inertia* at a point in question and the corresponding directions are the principal directions, such as OX*, OY*, OZ*. The important observation here is that all P.Is about the principal axes at a given point vanish. In other words,

$$I_{yz} = I_{zx} = I_{xy} = 0.$$

Corresponding to the principal M.I., that is, I_{xx}^*, the direction cosines l, m, n of the principal direction can be obtained by solving the system of equations

$$\left. \begin{array}{r} (I_{xx} - I_{xx}^*)l - I_{xy}m - I_{xz}n = 0 \\ -I_{yx}l + (I_{yy} - I_{yy}^*)m - I_{yz}n = 0 \\ -I_{zx}l - I_{zy}m + (I_{zz} - I_{zz}^*)n = 0 \end{array} \right\} \qquad (3.36)$$

subject to the condition

$$l^2 + m^2 + n^2 = 1.$$

Similarly, the directions corresponding to I_{yy}^* and I_{zz}^* can be obtained by replacing I_{xx}^* by I_{yy}^* and I_{zz}^* respectively in Eqs. (3.36). To conclude, it can be stated that for any point on the body, there exists three mutually perpendicular principal directions about which the M.Is are I_{xx}^*, I_{yy}^*, I_{zz}^*, while the corresponding P.Is are zero. To illustrate this important concept, we shall consider the following example.

Example 3.11 Find the principal M.Is and the principal directions at a corner of a uniform cube of length $2a$.

Solution In Example 3.6, we obtained the M.Is of the cube about OX, OY, OZ as

$$I_{xx} = I_{yy} = I_{zz} = \frac{8}{3} Ma^2,$$

and the corresponding P.Is as

$$I_{yz} = I_{zx} = I_{xy} = Ma^2,$$

where M is the mass of the cube. Let the M.Is at the corner O be $I_{xx}^*, I_{yy}^*, I_{zz}^*$, which can be obtained as the roots of the characteristic equation

$$\begin{vmatrix} \frac{8}{3}Ma^2 - \lambda & -Ma^2 & -Ma^2 \\ -Ma^2 & \frac{8}{3}Ma^2 - \lambda & -Ma^2 \\ -Ma^2 & -Ma^2 & \frac{8}{3}Ma^2 - \lambda \end{vmatrix} = 0.$$

Its expansion gives

$$\left(\frac{8}{3}Ma^2 - \lambda\right)^3 - 3(Ma^2)^2\left(\frac{8}{3}Ma^2 - \lambda\right) - 2(Ma^2)^3 = 0,$$

which, when factorized, yields

$$\left(\frac{11}{3}Ma^2 - \lambda\right)^2 \left(\frac{2}{3}Ma^2 - \lambda\right) = 0.$$

Hence the required principal M.Is are

$$I_{xx}^* = \frac{11}{3}Ma^2, \quad I_{yy}^* = \frac{11}{3}Ma^2, \quad I_{zz}^* = \frac{2}{3}Ma^2. \tag{1}$$

The principal direction corresponding to, say, I_{zz}^* is obtained from Eqs. (3.36), that is, from equations

$$\left.\begin{aligned} \left(\frac{8}{3}Ma^2 - \frac{2}{3}Ma^2\right)l - Ma^2 m - Ma^2 n &= 0, \\ -Ma^2 l + \left(\frac{8}{3}Ma^2 - \frac{2}{3}Ma^2\right)m - Ma^2 n &= 0, \\ -Ma^2 l - Ma^2 m + \left(\frac{8}{3}Ma^2 - \frac{2}{3}Ma^2\right)n &= 0 \end{aligned}\right\} \tag{2}$$

which simplify to

$$2l - m - n = 0, \quad -l + 2m - n = 0, \quad -l - m + 2n = 0. \tag{3}$$

and whose solution is found to be

$$l = m = n.$$

Since the direction cosines have to satisfy the constraint

$$l^2 + m^2 + n^2 = 1$$

we get,

$$l = m = n = \frac{1}{\sqrt{3}}$$

which indicates that the principal direction is that of the diagonal of a cube. Similarly, the other principal directions associated with I_{xx}^*, I_{yy}^* can be found.

3.2 KINETIC ENERGY OF A RIGID BODY ROTATING ABOUT A FIXED POINT

Let OXYZ be any rectangular system of coordinate axes. Imagine a rigid body rotating about a fixed point O with angular velocity $\vec{\omega}$ at a given instant, as shown in Fig. 3.17.

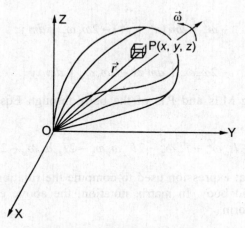

Fig. 3.17 A rotating rigid body.

Now, consider a typical particle P(x, y, z) of mass dm. We note that the speed v of this particle P is equal to the magnitude of the velocity given by the expression

$$\vec{V} = \vec{\omega} \times \vec{r}. \tag{3.37}$$

Then, the kinetic energy of the particle P is

$$T = \frac{1}{2} dm\, v^2.$$

Hence, the total rotational kinetic energy T_{rot} of the body at a given instant is

$$T_{\text{rot}} = \frac{1}{2} \int dm\, v^2 = \frac{1}{2} \int dm\, |\vec{\omega} \times \vec{r}|^2. \tag{3.38}$$

Now, let us seek an alternate expression for rotational kinetic energy, involving angular velocity $\vec{\omega}$ and the M.Is of the rotating body.

Let $\hat{i}, \hat{j}, \hat{k}$ be the unit vectors along OX, OY, OZ so that $\overline{OP} = \vec{r}$. Thus,

$$\vec{r} = \hat{i}x + \hat{j}y + \hat{k}z \quad \text{and} \quad \vec{\omega} = \hat{i}\omega_x + \hat{j}\omega_y + \hat{k}\omega_z.$$

Then,

$$\vec{\omega} \times \vec{r} = \begin{vmatrix} \hat{i} & \hat{j} & \hat{k} \\ \omega_x & \omega_y & \omega_z \\ x & y & z \end{vmatrix} = (\omega_y z - \omega_z y)\hat{i} + (\omega_z x - \omega_x z)\hat{j} + (\omega_x y - \omega_y x)\hat{k}.$$

Substituting the preceding expression into Eq. (3.38), we get

$$T_{\text{rot}} = \frac{1}{2} \int dm \left[(\omega_y z - \omega_z y) + (\omega_z x - \omega_x z)^2 + (\omega_x y - \omega_y x)^2 \right]$$

$$= \frac{1}{2} \left[\omega_x^2 \int dm (y^2 + z^2) + \omega_y^2 \int dm (z^2 + x^2) \right.$$

$$+ \omega_z^2 \int dm (x^2 + y^2) - 2\omega_y \omega_z \int dm \, yz$$

$$\left. - 2\omega_z \omega_x \int dm \, zx - 2\omega_x \omega_y \int dm \, xy \right].$$

Now, introducing M.Is and P.Is of the body through Eqs. (3.4) and (3.5), we get

$$T_{\text{rot}} = \frac{1}{2} \left(I_{xx} \omega_x^2 + I_{yy} \omega_y^2 + I_{zz} \omega_z^2 - 2I_{yz} \omega_y \omega_z - 2I_{zx} \omega_z \omega_x - 2I_{xy} \omega_x \omega_y \right) \quad (3.39)$$

This is the general expression used to compute the rotational kinetic energy of a rotating rigid body. In matrix notation, the above expression can be recast into the form

$$T_{\text{rot}} = \frac{1}{2} \begin{pmatrix} \omega_x \\ \omega_y \\ \omega_z \end{pmatrix}^T \begin{bmatrix} I_{xx} & -I_{xy} & -I_{xz} \\ -I_{yx} & I_{yy} & -I_{yz} \\ -I_{zx} & -I_{zy} & I_{zz} \end{bmatrix} \begin{pmatrix} \omega_x \\ \omega_y \\ \omega_z \end{pmatrix}. \quad (3.40)$$

In case the coordinate axes coincide with the principal axes, all P.Is vanish and the corresponding expression for the rotational kinetic energy reduces to

$$T_{\text{rot}} = \frac{1}{2} (I_{xx} \omega_x^2 + I_{yy} \omega_y^2 + I_{zz} \omega_z^2). \quad (3.41)$$

Suppose, the rigid body is constrained to rotate about a fixed axis through O, then Eq. (3.41) further simplifies to

$$T_{\text{rot}} = \frac{1}{2} I \omega^2, \quad (3.42)$$

where I is the M.I. of the rigid body about the instantaneous axis of rotation and ω is the instantaneous angular velocity of the rigid body. In practice, such a situation can be visualized in the case of motion of a rocket spinning about its axis, during its initial lift off.

It may be noted that the variables ω_x, ω_y and ω_z in Eq. (3.40) do not correspond to the time derivatives of any set of coordinates that specify the orientation of the rigid body. Thus, we are led to search for a new set of three coordinates by which the orientation of the rigid body can be specified as required in many practical applications. To meet this requirement, we have introduced a set of generalized coordinates known as Eulerian angles, as described in Section 3.4.

Kinetic energy of a rigid body in general

To find the kinetic energy of a rigid body moving generally in space which has both translational and rotational motion we use the theorem of Konig which is stated below:

Theorem 3.3 (*Theorem of Konig*) The kinetic energy of a moving body quite generally in space is equal to the sum of

(a) the kinetic energy of a fictitious particle whose mass is equal to the total mass of the body moving with its centre of mass and
(b) the kinetic energy of the body relative to its centre of mass.

Proof Let OXYZ be any rectangular frame. Suppose $\vec{R}$ be the position vector of the centre of mass G of the rigid body. Let dm be the mass of the typical element located at P, whose position vector with respect to G is $\vec{s}$. Also, let $\vec{r}$ be the position vector of P with respect to O. Then, from Fig. 3.18, we have

$$\vec{r} = \vec{R} + \vec{s}.$$

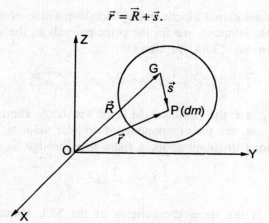

Fig. 3.18 General motion of a rigid body.

Hence, the kinetic energy of the rigid body is

$$T = \frac{1}{2} \int dm \, v^2$$

$$= \frac{1}{2} \int dm \, (\dot{\vec{r}} \cdot \dot{\vec{r}})$$

$$= \frac{1}{2} \int (\dot{\vec{R}} + \dot{\vec{s}}) \cdot (\dot{\vec{R}} + \dot{\vec{s}}) \, dm$$

$$= \frac{1}{2} \int (\dot{\vec{R}} \cdot \dot{\vec{R}}) \, dm + \frac{1}{2} \int (\dot{\vec{s}} \cdot \dot{\vec{s}}) \, dm + \dot{\vec{R}} \cdot \int \dot{\vec{s}} \, dm.$$

That is,

$$T = \frac{1}{2} M v_c^2 + \frac{1}{2} \int \dot{s}^2 \, dm + \dot{\vec{R}} \cdot \int \dot{\vec{s}} \, dm,$$

where v_c is the speed of the centre of mass, $\dot{\vec{s}}$ is the relative velocity of dm as viewed by a non-rotating observer moving with the centre of mass. Now, as a consequence of the definition of centre of mass

$$\int \dot{\vec{s}} \, dm = 0.$$

Hence,

$$T = \frac{1}{2} M v_c^2 + \frac{1}{2} \int \dot{s}^2 \, dm. \tag{3.43}$$

Here, the first term on the right-hand side corresponds to the translational kinetic energy, while the second term corresponds to the rotational kinetic energy of the rigid body. Thus, the kinetic energy of the rigid body moving generally in space is given by

$$T = \frac{1}{2} M v_c^2 + T_{\text{rot}}, \tag{3.44}$$

where T_{rot} is the rotational kinetic energy due to rotation of the body about its centre of mass. Suppose, we fix the principal axis at the centre of mass of the body, then Eq. (3.44) becomes

$$T = \frac{1}{2} M v_c^2 + \frac{1}{2} (I_{xx}^* \omega_x^2 + I_{yy}^* \omega_y^2 + I_{zz}^* \omega_z^2), \tag{3.45}$$

where $I_{xx}^*, I_{yy}^*, I_{zz}^*$ are the principal M.Is of the body about its centre of mass and $\omega_x, \omega_y, \omega_z$ are the components of angular velocity $\vec{\omega}$. If the rigid body is constrained to rotate about a fixed line through O, then

$$T = \frac{1}{2} M v_c^2 + \frac{1}{2} I \omega^2. \tag{3.46}$$

Definition 3.1 If the three components of the M.I. about the principal axis are equal, that is, if

$$I_{xx}^* = I_{yy}^* = I_{zz}^*,$$

then the body is called a *spherical top*. If two of them are equal, that is, if

$$I_{xx}^* = I_{yy}^* \neq I_{zz}^*,$$

then the body is called a *symmetrical top*. If $I_{xx}^*, I_{yy}^*, I_{zz}^*$ are unequal, the body is called an *asymmetrical top*.

In order to illustrate the ideas developed thus far, we shall consider the following examples:

Example 3.12 Find the kinetic energy of a homogeneous circular cylinder of mass m and radius a, rolling on a plane with linear velocity v.

Solution We know that the kinetic energy of a rigid body moving quite generally in space can be obtained from the Eq. (3.44), that is,

$$T = \frac{1}{2}mv_c^2 + T_{\text{rot}},$$

where T_{rot} is the kinetic energy of the rotating body about its centre of mass. Since the cylinder is constrained to roll about its axis, we have

$$T = \frac{1}{2}mv_c^2 + \frac{1}{2}I\omega^2,$$

where I is the M.I. of the cylinder of mass m and radius a about its axis, which is known to be

$$I = \frac{1}{2}ma^2.$$

Hence, the required kinetic energy is

$$T = \frac{1}{2}mv^2 + \frac{1}{2}\frac{ma^2}{2}\frac{v^2}{a^2} = \frac{3}{4}mv^2.$$

3.3 ANGULAR MOMENTUM OF A RIGID BODY

It is known that the general motion of a rigid body possesses both translational and rotational motion. In rotational motion, angular momentum and torque will come into play. Before we study them in detail, let us recall our understanding of angular momentum of a particle. Consider the motion of a particle of mass m, moving with velocity $\vec{V}$, relative to some fixed frame of reference OXYZ. Then, the linear momentum of the particle is defined as $m\vec{V}$. Now, we can define the angular momentum $\vec{h}$ of the particle about any point, say O, as the moment of $m\vec{V}$ about O. Thus,

$$\vec{h} = \vec{r} \times m\vec{V},$$

where $\vec{r}$ is the position vector of the particle with respect to O. To extend this concept to a rigid body—a rigid body in rotation about a fixed point possesses momentum. Let us consider a typical mass dm situated at P, whose position vector is $\vec{r}$ relative to the origin O (see Fig. 3.17). Then, the angular momentum of the typical mass dm at P about O is

$$d\vec{h} = \vec{r} \times dm\vec{V} = \vec{r} \times (\vec{\omega} \times \vec{r}) \, dm.$$

Hence, the total angular momentum of the rigid body about O can be found by integrating the momenta of each differential element. That is,

$$\vec{H} = \int \vec{r} \times (\vec{\omega} \times \vec{r}) \, dm = \int (\vec{r} \cdot \vec{r}) \vec{\omega} \, dm - \int (\vec{r} \cdot \vec{\omega}) \vec{r} \, dm.$$

In other words,

$$\vec{H} = \int \vec{\omega} r^2 \, dm - \int \vec{r} (\vec{\omega} \cdot \vec{r}) \, dm. \tag{3.47}$$

Here, the integral is over the volume of the body.

Let $\hat{i}, \hat{j}, \hat{k}$ be the unit vectors along the coordinate axes, then,

$$\vec{r} = \hat{i}x + \hat{j}y + \hat{k}z, \quad \text{(therefore, } r^2 = x^2 + y^2 + z^2\text{)}$$

$$\vec{\omega} = \hat{i}\omega_x + \hat{j}\omega_y + \hat{k}\omega_z,$$

$$\vec{H} = \hat{i}H_x + \hat{j}H_y + \hat{k}H_z.$$

Now, Eq. (3.47) gives the component in the direction of $\hat{i}$ as

$$H_x = \int (x^2 + y^2 + z^2)\omega_x \, dm - \int (x\omega_x + y\omega_y + z\omega_z)x \, dm$$

$$= \omega_x \int (y^2 + z^2) \, dm - \omega_y \int xy \, dm - \omega_z \int xz \, dm$$

$$= I_{xx}\omega_x - I_{xy}\omega_y - I_{xz}\omega_z.$$

Similarly, the components in the $\hat{j}$ and $\hat{k}$ directions yield

$$H_y = -I_{yx}\omega_x + I_{yy}\omega_y - I_{yz}\omega_z,$$

$$H_z = -I_{zx}\omega_x - I_{zy}\omega_y + I_{zz}\omega_z.$$

Thus, the complete set of components of angular momentum vector of a rotating rigid body can be written in matrix notation as

$$\begin{pmatrix} H_x \\ H_y \\ H_z \end{pmatrix} = \begin{bmatrix} I_{xx} & -I_{xy} & -I_{xz} \\ -I_{yx} & I_{yy} & -I_{yz} \\ -I_{zx} & -I_{zy} & I_{zz} \end{bmatrix} \begin{pmatrix} \omega_x \\ \omega_y \\ \omega_z \end{pmatrix} \quad (3.48)$$

or simply

$$\vec{H} = [I]\vec{\omega}. \quad (3.49)$$

Here, the matrix $[I]$ is called *inertia tensor* of a rigid body and is a dyadic operator which changes angular velocity into angular momentum. Thus, we observe that the angular momentum vector of a rotating rigid body does not coincide with its angular velocity vector. It may be noted that the elements of the dyadic or inertia tensor depend on the shape and mass distribution of the body and the coordinate system to which they are referred. In fact, when a body fixed coordinate system is employed in the study of angular motions of a rigid body such as a missile or a rocket, the elements of inertia matrix would be a function of time.

In case $\hat{i}, \hat{j}, \hat{k}$ coincides with the principal axes, then,

$$H_x = I_{xx}^* \omega_x, \qquad H_y = I_{yy}^* \omega_y, \qquad H_z = I_{zz}^* \omega_z. \quad (3.50)$$

where $I_{xx}^*, I_{yy}^*, I_{zz}^*$ are the principal M.Is of the body. Suppose, the rigid body is constrained to rotate about a fixed axis, say Z-axis, with angular velocity ω, then Eq. (3.48) simply reduces to

$$H_x = -I_{xz}\,\omega, \qquad H_y = -I_{yz}\omega, \qquad H_z = I_{zz}\,\omega. \qquad (3.51)$$

and then the angular momentum vector does not lie along the axis of rotation, unless the latter is a principal axis of inertia. However, the component H_z along the axis of rotation is equal to the product of the M.I. about that axis and the angular velocity. This idea is of fundamental importance to understand the phenomenon of gyroscopic motion.

Example 3.13 If a rigid body rotates about a fixed point with an angular velocity $\vec{\omega}$ and has an angular momentum $\vec{H}$, show that the kinetic energy T is given by

$$T = \frac{1}{2}\vec{\omega} \cdot \vec{H}.$$

Solution We know that the kinetic energy in the given situation is given by

$$T = \frac{1}{2}\int dm\, v^2,$$

where dm is the mass of the typical element of the body which can be rewritten as

$$T = \frac{1}{2}\int dm\,(\vec{V}\cdot\vec{V}) = \frac{1}{2}\int dm\,(\vec{\omega}\times\vec{r})\cdot(\vec{\omega}\times\vec{r}),$$

where $\vec{r}$ is the position vector of the typical particle. Since '·' and '×' can be interchanged, we have

$$T = \frac{1}{2}\int dm\,\{\vec{\omega}\cdot[\vec{r}\times(\vec{\omega}\times\vec{r})]\}.$$

Hence,

$$T = \frac{1}{2}\vec{\omega}\cdot\vec{H}.$$

Example 3.14 A body turns about a fixed point. Show that the angle between its angular velocity vector and its angular momentum vector about the fixed point is always acute.

Solution Let $\vec{\omega} = \omega_x\hat{i} + \omega_y\hat{j} + \omega_z\hat{k}$ be the angular velocity vector with which the given body is turning about a fixed point, then the angular momentum vector about the fixed point is given by Eq. (3.49) as

$$\vec{H} = [I]\,\vec{\omega}, \qquad (1)$$

where $[I]$ is an inertia tensor. Thus,

$$\vec{\omega}\cdot\vec{H} = (\omega_x\hat{i}+\omega_y\hat{j}+\omega_z\hat{k})\cdot[I](\omega_x\hat{i}+\omega_y\hat{j}+\omega_z\hat{k})$$
$$= [I](\omega_x^2+\omega_y^2+\omega_z^2).$$

We have seen in Example 3.13, that

$$\vec{\omega}\cdot\vec{H} = 2T, \qquad (2)$$

But,

$$\vec{\omega}\cdot\vec{H} = |\vec{\omega}||\vec{H}|\cos\theta, \qquad (3)$$

where θ is the angle between $\vec{\omega}$ and $\vec{H}$. Therefore, from Eqs. (2) and (3), we find that

$$\cos\theta = \frac{2T}{|\vec{\omega}||\vec{H}|} \quad \text{(always positive)}$$

implying $\theta \leq 90°$, which means θ is always acute.

Example 3.15 A uniform sphere of mass M and radius a is released from rest on a plane inclined at an angle α to the horizontal. If the sphere rolls down without slipping, show that the acceleration of the centre of the sphere is constant and equal to $(5/7)g\sin\alpha$.

Solution Let O be the point of contact of the sphere with the plane, initially. Since rolling is without slipping, imagine a sphere of radius a, rolling down a plane inclined at an angle α to the horizontal (Fig. 3.19).

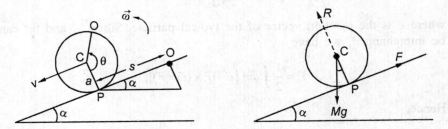

Figure 3.19 Illustration for Example 3.15.

When the body has descended a distance s, let P be the point of contact of the sphere with the plane. Let $\angle PCO = \theta$, then $\dot{s} = a\omega$, where $\dot{s}$ is the instantaneous velocity of C, the centre of the sphere, and ω is the instantaneous angular velocity about C.

At any time, the forces acting on the sphere are

(i) its weight Mg acting vertically downwards through its centre of mass C,

(ii) reaction $\vec{R}$ of the plane perpendicular to the plane itself, and

(iii) the frictional force $\vec{F}$ acting upwards on the plane.

Since $\vec{R}$ and $\vec{F}$ both act through P which is instantaneously at rest and they do no work during the motion and hence the energy equation holds true, the kinetic energy of the system T is given by Eq. (3.46) as

$$T = \frac{1}{2} M v_c^2 + \frac{1}{2} I \omega^2,$$

that is,

$$T = \frac{1}{2} M \dot{s}^2 + \frac{1}{2} I \left(\frac{\dot{s}}{a}\right)^2. \quad (1)$$

When the sphere is at P, there is a loss of potential energy given by

$$V = -Mgs \sin \alpha. \quad (2)$$

For energy conservation, we must have from Eqs. (1) and (2)

$$\frac{1}{2} M \dot{s}^2 + \frac{1}{2} I \frac{\dot{s}^2}{a^2} - Mgs \sin \alpha = E \text{ (constant)}.$$

Using the initial conditions, $s = \dot{s} = 0$ when $t = 0$ implying $E = 0$ and thus, we have

$$\frac{1}{2} M \dot{s}^2 + \frac{1}{2} I \frac{\dot{s}^2}{a^2} = Mgs \sin \alpha$$

or

$$\left(M + \frac{I}{a^2}\right) \dot{s}^2 = 2 Mgs \sin \alpha. \quad (3)$$

Now, differentiating Eq. (3) with respect to t, we get after cancelling $2\dot{s}$ on both sides

$$\ddot{s} = \frac{Mg \sin \alpha}{M + I/a^2}.$$

But, for sphere, the moment of inertia $I = (2/5)Ma^2$. Hence,

$$\ddot{s} = \frac{Mg \sin \alpha}{M + (2/5)M} = \frac{5}{7} g \sin \alpha \quad (4)$$

from which, we can infer that the acceleration of the centre of the sphere is constant.

Example 3.16 A uniform circular disc of radius R and mass M is rigidly mounted on one end of a thin light shaft CD of length L. The shaft is normal to the disc at the centre C. The disc rolls on a rough horizontal plane, D, being fixed in this plane by a smooth universal joint. If the centre of the disc rotates without slipping about the vertical through D with constant angular velocity Ω, find the angular velocity, the kinetic energy and the angular momentum of the disc about D.

Solution The description of the problem can be seen as in Fig. 3.20.

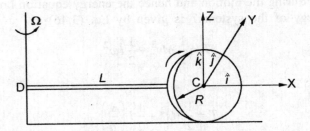

Figure 3.20 Illustration for Example 3.16.

The linear velocity of the centre of the wheel is

$$\vec{V} = L\Omega \hat{j}. \tag{1}$$

Therefore,

$$\text{Linear momentum} = M\vec{V} = ML\Omega \hat{j}, \tag{2}$$

where M is the mass of the wheel. Since the wheel is rolling without slipping, its angular velocity $\vec{\omega}$ in component form is

$$\vec{\omega} = \left(-\frac{L\Omega}{R} \hat{i},\ 0,\ \Omega \hat{k} \right). \tag{3}$$

If the wheel has principal moments of inertia I_{xx}, $I_{yy} = I_{zz}$, I_{zz}, then the angular momentum about the point C is found to be

$$H_x = I_{xx}\omega_x = -\frac{I_{xx}L\Omega}{R}\hat{i}$$

$$H_y = I_{yy}\omega_y = 0$$

$$H_z = I_{zz}\omega_z = I_{zz}\omega_z.$$

Therefore,

$$\vec{H}_C = -\frac{I_{xx}L\Omega}{R}\hat{i} + I_{zz}\Omega \hat{k}. \tag{4}$$

The angular momentum about D can then be obtained as

$$\vec{H}_D = \vec{H}_C + \vec{r}_{DC} \times M\vec{V} = \vec{H}_C + \vec{r}_{DC} ML\Omega \hat{j}$$

Now, using Eq. (4), we obtain

$$\vec{H}_D = -I_{xx}\frac{L}{R}\Omega \hat{i} + I_{zz}\Omega \hat{k} + L\hat{i} \times ML\Omega \hat{j}$$

$$= -\frac{I_{xx}L\Omega}{R}\hat{i} + (I_{zz} + ML^2)\Omega \hat{k}. \tag{5}$$

But, we know that the M.I. of the wheel about CX and that about CZ are given by

$$I_{xx} = \frac{MR^2}{4}, \qquad I_{zz} = \frac{MR^2}{2}.$$

Hence Eq. (5) becomes

$$\vec{H}_D = -\frac{MR^2 L\Omega}{4R} i + \left(\frac{MR^2}{2} + ML^2\right)\Omega \hat{k}. \qquad (6)$$

Finally, the kinetic energy of the wheel using Konig's theorem is

$$T = \frac{1}{2} M v_c^2 + \frac{1}{2} I \omega^2$$

$$= \frac{M}{2} L^2 \Omega^2 + \frac{1}{2} \frac{MR^2}{2} \left(\frac{L^2}{R^2} \Omega^2 + \Omega^2\right)$$

$$= \frac{M}{2} L^2 \Omega^2 + \frac{M\Omega^2}{4} (L^2 + R^2). \qquad (7)$$

Example 3.17 A uniform circular disc of mass M and radius R is so mounted that it can turn freely about its centre, which is fixed. It is spinning with angular velocity $\vec{\omega}$ about the perpendicular to its plane at the centre, the plane being horizontal. A particle of mass m, falling vertically, hits the disc near the edge and adheres to it. Prove that immediately afterwards the particle is moving in a direction inclined to the horizontal at an angle α given by

$$\tan \alpha = 4 \frac{m(M+2m)}{M(M+4m)} \frac{v}{R\omega},$$

where v is the speed of the particle just before impact.

Solution Let $\hat{i}, \hat{j}, \hat{k}$ be the unit vectors along the fixed coordinate axes OX, OY, OZ fixed at O, the centre of the circular disc, as shown in Fig. 3.21. Since the impulsive force acting on the particle and the disc are

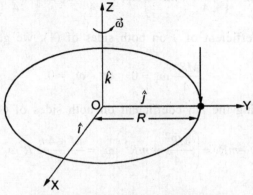

Figure 3.21 Illustration for Example 3.17.

equal and opposite, the resultant moment turns out to be zero. Similarly, the moment of the impulsive reaction at O is also zero. Therefore, the angular momentum of the system about O will be equal to that after the impact about O.

Before the disc is struck, its angular momentum about O is $I\omega \hat{k}$, where I is the moment of inertia of the disc about its axis of rotation. Hence the angular momentum of the disc, before the impact, about O is found to be $(MR^2/2)\omega\hat{k}$. Immediately after the particle strikes the disc, let $\omega_x, \omega_y, \omega_z$ be the components of its angular velocity, then the angular momentum of the disc about O, after impact, is given by

$$\vec{H}_{\text{disc}} = I_{xx}\omega_x \hat{i} + I_{yy}\omega_y \hat{j} + I_{zz}\omega_z \hat{k},$$

that is,

$$\vec{H}_{\text{disc}} = \frac{MR^2}{4}\omega_x \hat{i} + \frac{MR^2}{4}\omega_y \hat{j} + \frac{MR^2}{2}\omega_z \hat{k}.$$

Now, the angular momentum of the particle, before impact, about O is

$$R\hat{j} \times m(-v)\hat{k} = -mRv\hat{i}$$

Similarly, the angular momentum of the particle about O, after impact, is

$$\vec{H}_{\text{particle}} = R\hat{j} \times [m(\omega_x \hat{i} + \omega_y \hat{j} + \omega_z \hat{k}) \times R\hat{j}]$$
$$= R^2 m(\omega_x \hat{i} + \omega_y \hat{j} + \omega_z \hat{k}) - mR\omega_y R\hat{j},$$

that is,

$$\vec{H}_{\text{particle}} = R^2 m(\omega_x \hat{i} + \omega_z \hat{k}).$$

Using the fact that the angular momentum of the system before impact is equal to that after impact, we obtain

$$-mRv\hat{i} + \frac{MR^2}{2}\omega\hat{k} = \left(\frac{MR^2}{4} + mR^2\right)\omega_x \hat{i} + \frac{MR^2}{4}\omega_y \hat{j} + \left(\frac{MR^2}{4} + mR^2\right)\omega_z \hat{k}. \quad (1)$$

Equating the coefficient of $\hat{j}$ on both sides of (1), we get

$$\frac{MR^2}{4}\omega_y = 0 \quad \text{or} \quad \omega_y = 0. \tag{2}$$

Similarly, equating the $\hat{i}$th coefficient on both sides of (1), we have

$$-mRv = \left(\frac{MR^2}{4} + mR^2\right)\omega_x = \frac{M+4m}{4}R^2\omega_x$$

which gives

$$\omega_x = -\frac{4mv}{(M+4m)R}. \quad (3)$$

Equating the $\hat{k}$ th coefficient on both sides of (1), we get

$$\frac{MR^2}{2}\omega = \left(\frac{MR^2}{2} + mR^2\right)\omega_z$$

that is,

$$\omega_z = \frac{M\omega}{M+2m}. \quad (4)$$

Thus, using Eqs. (2)–(4), the velocity of the particle immediately after impact is found to be

$$\vec{v} = \vec{R} \times \vec{\omega} = R\hat{j} \times (\omega_x \hat{i} + \omega_y \hat{j} + \omega_z \hat{k})$$

$$= -R\omega_x \hat{k} + R\omega_z \hat{i}$$

$$= \frac{4mv}{M+4m}\hat{k} + \frac{M\omega R}{M+2m}\hat{i}. \quad (5)$$

Hence, the required angle is obtained as

$$\tan\alpha = \frac{4mv}{M+4m}\frac{M+2m}{M\omega R},$$

that is,

$$\tan\alpha = \frac{4m}{M}\frac{M+2m}{M+4m}\frac{v}{R\omega}.$$

Example 3.18 A slender bar of length l and mass m slides on the smooth floor and wall and has counterclockwise angular velocity $\vec{\omega}$ at the instant shown in Fig. 3.22(a). What is the bar's acceleration.

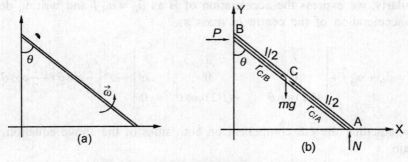

Figure 3.22 Description of Example 3.18.

Solution Let P and N be the normal forces acting on the rod exerted by the wall and the floor respectively, as shown in Fig. 3.22(b). Let C be the point of centre of mass of the rod. Writing the acceleration of the point C (centre of mass) as

$$\vec{a}_C = a_x \hat{i} + a_y \hat{j}$$

Newton's second law gives

$$\sum F_x = P = ma_x \tag{1}$$

$$\sum F_y = N - mg = ma_y \tag{2}$$

Let $\vec{\alpha}$ denote the bar's counterclockwise angular acceleration, then the principle of angular momentum, that is

$$\vec{G} = \dot{\vec{H}} = I\vec{\alpha}$$

gives

$$G_C = N\frac{l}{2}\sin\theta - P\frac{l}{2}\cos\theta = I\alpha, \tag{3}$$

where I is the M.I. of the bar about its centre of mass. Now, we have only three equations in five unknowns P, N, a_x, a_y and α. To have a solution, we must relate the acceleration of the centre of mass C of the bar to its angular acceleration. Expressing acceleration of A as $\vec{a}_A = a_A \hat{i}$ and using Eq. (2.36), the acceleration of the centre of mass C is given by

$$\vec{a}_C = \vec{a}_A + \vec{\alpha} r_{C/A} - \omega^2 r_{C/A}$$

Therefore,

$$a_x\hat{i} + a_y\hat{j} = a_A\hat{i} + \begin{vmatrix} \hat{i} & \hat{j} & \hat{k} \\ 0 & 0 & \alpha \\ (-l/2)\sin\theta & (l/2)\cos\theta & 0 \end{vmatrix} - \omega^2\left(-\frac{l}{2}\sin\theta\,\hat{i} + \frac{l}{2}\cos\theta\,\hat{j}\right).$$

Equating the $\hat{j}$ th component on both sides of the above equation, we get

$$a_y = -\frac{l}{2}(\alpha\sin\theta + \omega^2\cos\theta). \tag{4}$$

Similarly, we express the acceleration of B as $\vec{a}_B = a_B \hat{j}$ and writing down the acceleration of the centre of mass as

$$\vec{a}_C = \vec{a}_B + a_B\hat{j} + \begin{vmatrix} \hat{i} & \hat{j} & \hat{k} \\ 0 & 0 & \alpha \\ (l/2)\sin\theta & -(l/2)\cos\theta & 0 \end{vmatrix} - \omega^2\left(\frac{l}{2}\sin\theta\,\hat{i} - \frac{l}{2}\cos\theta\,\hat{j}\right).$$

Now equating the $\hat{i}$ th component on both sides of the above equation, we obtain

$$a_x = \frac{l}{2}\left(\alpha \cos\theta - \omega^2 \sin\theta\right). \qquad (5)$$

Further, we know that the M.I. of the rod about its C.M. is given as

$$I = \frac{1}{12} ml^2.$$

With the two kinematical Eqs. (4) and (5) we have five equations in five unknowns. Equations (1), (2), (4) and (5) give us

$$P = ma_x = \frac{mL}{2}\left(\alpha \cos\theta - \omega^2 \sin\theta\right) \qquad (6)$$

$$N = mg - \frac{ml}{2}\left(\alpha \sin\theta + \omega^2 \cos\theta\right). \qquad (7)$$

Now Eq. (3) gives

$$I\alpha = \frac{mgl}{2}\sin\theta - \frac{ml^2\alpha}{4}\sin^2\theta - \frac{ml^2}{4}\omega^2 \sin\theta \cos\theta$$

$$-\frac{ml^2}{4}\alpha \cos^2\theta + \frac{ml^2}{4}\sin\theta \cos\theta,$$

which simplifies to

$$\alpha\left(\frac{ml^2}{12} + \frac{ml^2}{4}\right) = \frac{mgl}{2}\sin\theta \qquad \left[\text{since } I = \frac{ml^2}{12}\right]$$

or

$$\alpha = \frac{3}{2}\frac{g}{l}\sin\theta, \qquad (8)$$

which is the required angular acceleration of the bar.

Example 3.19 A slender bar of mass m is released from rest in the horizontal position, as shown in Fig. 3.23(a), at an instant. Determine the bar's angular acceleration and the force exerted on the bar by the support A.

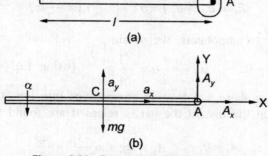

Figure 3.23 Description of Example 3.19.

Solution Suppose A_x and A_y are the reactions at A when the bar is released. Let the acceleration of the point C (centre of mass) of the bar be $\vec{a}_C = a_x \hat{i} + a_y \hat{j}$ and let its counterclockwise angular accleration be α. Newton's second law of motion gives

$$\sum F_x = A_x = ma_x, \qquad (1)$$

$$\sum F_y = A_y - mg = ma_y. \qquad (2)$$

Using the principle of angular momentum, the equation of angular motion about the fixed point A is

$$\frac{l}{2} mg = I_A \alpha. \qquad (3)$$

We know that the M.I. of the bar about its centre of mass is

$$I = \frac{1}{12} ml^2.$$

Using parallel-axis theorem, the M.I. of the bar about A is

$$I_A = I + \left(\frac{l}{2}\right)^2 m = \frac{1}{12} ml^2 + \frac{l^2}{4} m = \frac{1}{3} ml^2.$$

Substituting this value of I_A into Eq. (3), we get

$$\alpha = \frac{mg\, l/2}{ml^2/3} = \frac{3}{2} \frac{g}{l}. \qquad (4)$$

To compute the reactions at A, we shall express the acceleration of the centre of mass C, in terms of the acceleration of A. Thus using Eq. [2.36(a)], we have

$$\vec{a}_C = \vec{a}_A + (\alpha \hat{k} \times \vec{r}_{C/A}) - \omega^2 \vec{r}_{C/A}$$

Noting that $\omega = 0$ and $\vec{a}_A = 0$ at the instant when the bar is released, the above equation becomes

$$\vec{a}_C = a_x \hat{i} + a_y \hat{j} = \alpha \hat{k} \times \left(-\frac{l}{2}\hat{i}\right) = -\frac{l}{2}\alpha \hat{j}.$$

Equating $\hat{i}$ and $\hat{j}$ components, we obtain

$$a_x = 0, \qquad a_y = -\frac{l}{2}\alpha = -\frac{3}{4} g \qquad \text{[using Eq. (4)]}$$

Now, substituting these acceleration components into Eqs. (1) and (2), the reactions at A at the instant the bar is released are found to be

$$A_x = 0, \qquad A_y = mg + m\frac{-3g}{4} = \frac{mg}{4}.$$

3.4 EULERIAN ANGLES

The orientation of a rotating body or vehicle can be completely specified in terms of three angles known as Eulerian angles. A point on a body can be specified in terms of body fixed axes *oxyz*. In order to determine the orientation of the body itself, we introduce Eulerian angles ψ, θ, ϕ, which are three independent quantities capable of defining the body axes *oxyz* relative to the inertial or fixed axes OXYZ. These are three successive angular displacements which can adequately carry out the transformation from one Cartesian system of axes to another. To demonstrate, how Eulerian angles are capable of describing the orientation of the body, consider a fixed frame of reference OXYZ and a body-fixed coordinate frame *oxyz*, which assumes different orientations in space, the two sets being initially coincident. Now, allow the X, Y, Z set to rotate about the Z-axis through an angle ψ to a new position which may be denoted by α', β', γ' as shown in Fig. 3.24.

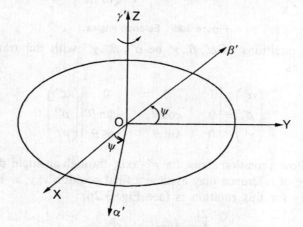

Figure 3.24 Eulerian angles.

The relationship between the two coordinate systems X, Y, Z and α', β', γ' can be seen as

$$X = \alpha' \cos \psi - \beta' \sin \psi$$
$$Y = \alpha' \sin \psi + \beta' \cos \psi$$
$$Z = \gamma'.$$

In the language of aerospace, ψ is called the *yaw angle*. Using matrix notation, the above equations can be recast as

$$\begin{pmatrix} X \\ Y \\ Z \end{pmatrix} = \begin{bmatrix} \cos \psi & -\sin \psi & 0 \\ \sin \psi & \cos \psi & 0 \\ 0 & 0 & 1 \end{bmatrix} \begin{pmatrix} \alpha' \\ \beta' \\ \gamma' \end{pmatrix}. \qquad (3.52)$$

104 Kinematics of a Rigid Body Motion (Ch. 3)

We next allow the rotation θ about α'-axis as shown in Fig. 3.25. Here θ is called the *pitch angle* or *altitude angle*, which usually lies in the range $-\pi/2 \leq \theta \leq \pi/2$.

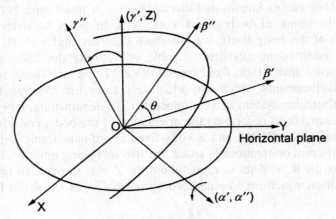

Figure 3.25 Eulerian angles.

Let the new positions of α', β', γ' be $\alpha'', \beta'', \gamma''$ with the transformation matrix given as

$$\begin{pmatrix} \alpha' \\ \beta' \\ \gamma' \end{pmatrix} = \begin{bmatrix} 1 & 0 & 0 \\ 0 & \cos\theta & -\sin\theta \\ 0 & \sin\theta & \cos\theta \end{bmatrix} \begin{pmatrix} \alpha'' \\ \beta'' \\ \gamma'' \end{pmatrix}. \quad (3.53)$$

Finally, we allow a rotation about the γ''-axis, through an angle ϕ, so that the moving frame of reference may assume a final position x, y, z. The transformation matrix for this rotation is (see Fig. 3.26)

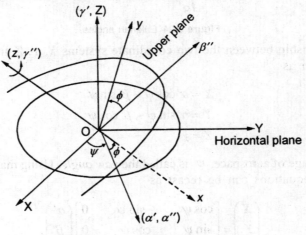

Figure 3.26 Eulerian angles.

$$\begin{pmatrix} \alpha'' \\ \beta'' \\ \gamma'' \end{pmatrix} = \begin{bmatrix} \cos\phi & -\sin\phi & 0 \\ \sin\phi & \cos\phi & 0 \\ 0 & 0 & 1 \end{bmatrix} \begin{pmatrix} x \\ y \\ z \end{pmatrix}. \qquad (3.54)$$

Here, ϕ is called the *roll angle*, which is limited to the range $0 \le \phi \le 2\pi$. Thus, we see that the body-fixed axes is related to the fixed frame of reference (inertial frame) by the matrix equation obtained with the help of Eqs. (3.52) to (3.54) as

$$\begin{pmatrix} X \\ Y \\ Z \end{pmatrix} = \begin{bmatrix} \cos\psi & -\sin\psi & 0 \\ \sin\psi & \cos\psi & 0 \\ 0 & 0 & 1 \end{bmatrix} \begin{bmatrix} 1 & 0 & 0 \\ 0 & \cos\theta & -\sin\theta \\ 0 & \sin\theta & \cos\theta \end{bmatrix} \begin{bmatrix} \cos\phi & -\sin\phi & 0 \\ \sin\phi & \cos\phi & 0 \\ 0 & 0 & 1 \end{bmatrix} \begin{pmatrix} x \\ y \\ z \end{pmatrix}.$$
$$(3.55)$$

Here, ψ, θ and ϕ are respectively called yaw, pitch and roll angles. On simplification, we obtain

$$\begin{pmatrix} X \\ Y \\ Z \end{pmatrix} = \begin{bmatrix} \begin{pmatrix} \cos\phi\cos\psi & - \\ \sin\phi\cos\theta\sin\psi \end{pmatrix} & \begin{pmatrix} -\sin\phi\cos\psi & - \\ \sin\psi\cos\theta\cos\phi \end{pmatrix} & \sin\theta\sin\psi \\ \begin{pmatrix} \cos\phi\sin\psi & + \\ \sin\phi\cos\theta\cos\psi \end{pmatrix} & \begin{pmatrix} -\sin\phi\sin\psi & + \\ \cos\phi\cos\theta\cos\psi \end{pmatrix} & -\sin\theta\cos\psi \\ \sin\theta\sin\phi & \sin\theta\cos\phi & \cos\theta \end{bmatrix} \begin{pmatrix} x \\ y \\ z \end{pmatrix}$$
$$(3.56)$$

We can notice that the transformation matrix is an orthogonal matrix. Therefore, the inverse transformation from (X, Y, Z) to (x, y, z) can be straightaway written as

$$\begin{pmatrix} x \\ y \\ z \end{pmatrix} = \begin{bmatrix} \begin{pmatrix} \cos\psi\cos\phi & - \\ \sin\psi\cos\theta\sin\phi \end{pmatrix} & \begin{pmatrix} \sin\psi\cos\phi & + \\ \cos\psi\cos\theta\sin\phi \end{pmatrix} & \sin\theta\sin\phi \\ \begin{pmatrix} -\cos\psi\sin\phi & - \\ \sin\psi\cos\theta\cos\phi \end{pmatrix} & \begin{pmatrix} -\sin\psi\sin\phi & + \\ \cos\psi\cos\theta\cos\phi \end{pmatrix} & \sin\theta\cos\phi \\ \sin\psi\sin\theta & -\cos\psi\sin\theta & \cos\theta \end{bmatrix} \begin{pmatrix} X \\ Y \\ Z \end{pmatrix}$$
$$(3.57)$$

The axis (α', α'') is called the *line of nodes*. Assuming that the Eulerian angles are limited to ranges $0 \le \psi \le 2\pi$, $-\pi/2 \le \theta \le \pi/2$, $0 \le \phi \le 2\pi$, it can be verified that any possible orientation of the body can be attained by performing proper rotations in the given order. Since the Cartesian coordinates

x, y, z and the Eulerian angles ψ, θ, ϕ describe the configuration of a rigid body completely; a rigid body freely moving in space has six degrees of freedom.

It must be clear that, in arriving at the final position of the body axis, we have encountered four sets of orthogonal axis (X, Y, Z), $(\alpha', \beta', \gamma')$, $(\alpha'', \beta'', \gamma'')$, (x, y, z). Some of these axes coincide such as (Z, γ'), (γ'', z) and (α', α'') the line of nodes. ϕ is measured from the line of nodes on the upper plane. The conventional sequence is ψ, θ, ϕ. If the sequence changes, the final result also will change. This important result is demonstrated in the following example:

Example 3.20 In introducing the concept of Eulerian angles, we stated that the final result depends on the sequence in which the rotations are carried out. Illustrate by means of a diagram if $\psi = \theta = \phi = 90°$, the result of conventional sequence ψ, θ, ϕ and a sequence ψ, ϕ, θ.

Solution The conventional sequence ψ, θ, ϕ gives the situation as depicted in Fig. 3.27.

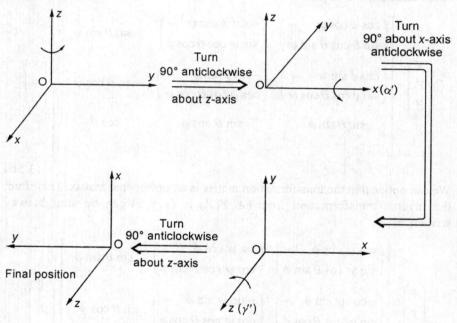

Figure 3.27 Illustration for Example 3.20.

The sequence of rotations ψ, ϕ, θ is illustrated in Fig. 3.28. Obviously, Figs. 3.27 and 3.28 demonstrate that the final result depends upon the order of sequence of rotations caried out. Hence proved.

3.4.1 Angular Velocities Expressed in Terms of Eulerian Angles

Very often, we need to express the angular velocity components $\omega_x, \omega_y, \omega_z$ of a rotating rigid body about x, y, z axes, in terms of Eulerian angles. It is known that the motion of a rigid body can be determined when its

Eulerian Angles (Sec. 3.4) **107**

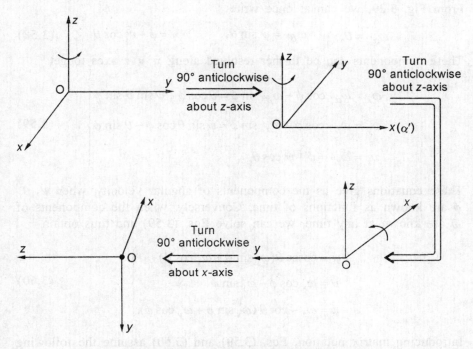

Figure 3.28 Illustration of Example 3.20.

angular velocity $\vec{\omega}(t)$ is known. Similarly the motion of a rigid body is described when Eulerian angles ψ, θ, ϕ are known as functions of time. Therefore, we now seek expressions for ω_x, ω_y, ω_z in terms of ψ, θ, ϕ and their rates of change.

The angular velocity $\dot{\psi}$ acts along the z-axis which represents 'precessional' velocity. $\dot{\theta}$ acts along the α'-axis or line of nodes which corresponds to 'nutation', and $\dot{\phi}$ acts about the γ''-axis which represents 'spin' angular velocity. These are shown in Fig. 3.29.

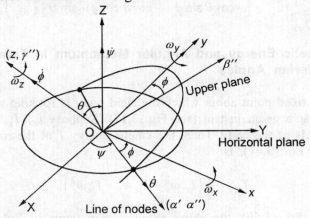

Figure 3.29 Motion of a rigid body in general.

108 | Kinematics of a Rigid Body Motion (Ch. 3)

From Fig. 3.29, we can at once write

$$\omega_{\alpha''} = \dot{\theta}, \qquad \omega_{\beta''} = \dot{\psi}\sin\theta, \qquad \omega_{\gamma''} = \dot{\phi} + \dot{\psi}\cos\theta. \qquad (3.58)$$

These components can be further resolved along x, y, z axes to get

$$\left.\begin{array}{l}\omega_x = \omega_{\alpha''}\cos\phi + \omega_{\beta''}\sin\phi = \dot{\theta}\cos\phi + \dot{\psi}\sin\theta\sin\phi \\ \omega_y = \omega_{\beta''}\cos\phi - \omega_{\alpha''}\sin\phi = \dot{\psi}\sin\theta\cos\phi - \dot{\theta}\sin\phi \\ \omega_z = \omega_{\gamma''} = \dot{\phi} + \dot{\psi}\cos\theta.\end{array}\right\} \qquad (3.59)$$

These equations give us the components of angular velocity, when ψ, θ, ϕ are known as functions of time. Conversely, when the components of $\vec{\omega}$ are known at any time, we can solve Eqs. (3.59) and thus obtain

$$\left.\begin{array}{l}\dot{\psi} = \operatorname{cosec}\theta(\omega_x\sin\phi + \omega_y\cos\phi) \\ \dot{\theta} = \omega_x\cos\phi - \omega_y\sin\phi \\ \dot{\phi} = \omega_z - \cot\theta\,(\omega_x\sin\phi + \omega_y\cos\phi).\end{array}\right\} \qquad (3.60)$$

Introducing matrix notation, Eqs. (3.59) and (3.60) assume the following forms:

$$\begin{pmatrix}\omega_x \\ \omega_y \\ \omega_z\end{pmatrix} = \begin{bmatrix}\sin\theta\sin\phi & \cos\phi & 0 \\ \sin\theta\cos\phi & -\sin\phi & 0 \\ \cos\theta & 0 & 1\end{bmatrix}\begin{pmatrix}\dot{\psi} \\ \dot{\theta} \\ \dot{\phi}\end{pmatrix} \qquad (3.61)$$

$$\begin{pmatrix}\dot{\psi} \\ \dot{\theta} \\ \dot{\phi}\end{pmatrix} = \operatorname{cosec}\theta\begin{bmatrix}\sin\phi & \cos\phi & 0 \\ \sin\theta\cos\phi & -\sin\theta\sin\phi & 0 \\ -\cos\theta\sin\phi & -\cos\theta\cos\phi & \sin\theta\end{bmatrix}\begin{pmatrix}\omega_x \\ \omega_y \\ \omega_z\end{pmatrix}. \qquad (3.62)$$

3.4.2 Kinetic Energy and Angular Momentum in Terms of Eulerian Angles

Let O be a fixed point about which the rigid body is rotating with angular velocity $\vec{\omega}$ at a given instant (see Fig. 3.17). Suppose I_{xx}, I_{yy}, I_{zz} are the principal M.Is of the body. Then, the kinetic energy T of the rotating body, as given in Eq. (3.41), is

$$T_{\text{rot}} = \frac{1}{2}\left(I_{xx}\omega_x^2 + I_{yy}\omega_y^2 + I_{zz}\omega_z^2\right).$$

Now, using Eq. (3.59), the above expression in terms of Eulerian angles becomes

$$T_{\text{rot}} = \frac{1}{2}\Big[I_{xx} (\dot\theta \cos\phi + \dot\psi \sin\theta \sin\phi)^2 +$$

$$I_{yy}(\dot\psi \sin\theta \cos\phi - \dot\theta \sin\phi)^2 + I_{zz}(\dot\phi + \dot\psi \cos\theta)^2 \Big]. \quad (3.63)$$

For a symmetric top, where $I_{xx} = I_{yy}$, the above expression simplifies further to

$$T_{\text{rot}} = \frac{1}{2}\Big[I_{xx}(\dot\theta^2 + \dot\psi^2 \sin^2\theta) + I_{zz}(\dot\phi + \dot\psi \cos\theta)^2 \Big]. \quad (3.64)$$

The angular momentum vector is

$$\vec{H} = H_x \hat{i} + H_y \hat{j} + H_z \hat{k},$$

where $\hat{i}, \hat{j}, \hat{k}$ are the unit vectors directed along the principal axes at O. Then, using Eq. (3.50), we have

$$H^2 = |\vec{H}|^2 = I_{xx}^2 \omega_x^2 + I_{yy}^2 \omega_y^2 + I_{zz}^2 \omega_z^2.$$

In terms of Eulerian angles, using Eq. (3.59), the above equation can be recast as

$$H^2 = I_{xx}^2 \Big[(\dot\psi \sin\theta \sin\phi)^2 + (\dot\theta \cos\phi)^2 \Big] + I_{yy}^2 \Big[(\dot\psi \sin\theta \cos\phi)^2 + (\dot\theta \sin\phi)^2 \Big]$$

$$+ I_{zz}^2 \Big[(\dot\psi \cos\theta)^2 + \dot\phi^2 \Big]. \quad (3.65)$$

3.5 THE COMPOUND PENDULUM

The *compound pendulum* consists of a rigid body which is free to oscillate under the influence of gravity about a fixed horizontal axis. Suppose we assume that the vertical plane of oscillation of the pendulum is chosen as XY-plane, so that the Z-axis through the origin O is the horizontal axis of suspension, as shown in Fig. 3.30.

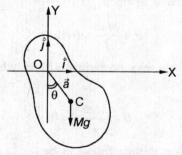

Figure 3.30 Compound pendulum.

Let C be the centre of mass whose position vector relative to O is $\vec{a}$. Since the body is rigid, $|\vec{a}|=a$ is constant. The only force acting on the body is its weight, acting vertically downwards, that is,

$$M\vec{g} = -Mg\hat{j},$$

where M is the mass of the pendulum. Now, the potential energy of the system is given by

$$-\nabla V = -\frac{\partial V}{\partial x}\hat{i} - \frac{\partial V}{\partial y}\hat{j} - \frac{\partial V}{\partial z}\hat{k} = -Mg\hat{j},$$

which gives us

$$\frac{\partial V}{\partial x} = 0, \qquad \frac{\partial V}{\partial y} = Mg, \qquad \frac{\partial V}{\partial z} = 0.$$

On integration, we have

$$V = Mgy + c = -Mga\cos\theta + c. \tag{3.66}$$

In Section 3.2, we noted that the kinetic energy T of a rigid body which is constrained to rotate about a fixed axis through O is given by Eq. (3.42) as

$$T = \frac{1}{2}I\omega^2 = \frac{1}{2}I\dot\theta^2, \tag{3.67}$$

where I is the M.I. of the rigid body about the instantaneous axis of rotation and ω the instantaneous angular velocity of the rigid body.

Following the principle of conservation of energy, we have

$$T + V = \frac{1}{2}I\dot\theta^2 - Mga\cos\theta = \text{constant}. \tag{3.68}$$

Now, differentiating Eq. (3.68) with respect to t, we have

$$I\dot\theta\ddot\theta + Mga\sin\theta\,\dot\theta = 0.$$

As $\dot\theta$ is not zero, we have at once

$$\ddot\theta + \frac{Mga}{I}\sin\theta = 0. \tag{3.69}$$

For small oscillations, we may take $\sin\theta = \theta$, in which case, the equation of motion becomes

$$\ddot\theta + \frac{Mga}{I}\theta = 0. \tag{3.70}$$

The period of oscillation is thus found to be

$$\tau = 2\pi\sqrt{\frac{I}{Mga}} \tag{3.71}$$

Comparing this period with the period of a simple pendulum of length l as given in Eq. (1.46), we find that the length of the equivalent simple pendulum is given by $l = I/Ma$.

EXERCISES

3.1 Find the moment of inertia of a solid homogeneous cube with edge-length $2a$, about the concurrent axes and also their products of inertia.

3.2 Show that the M.Is of a solid cone of mass M and semi-vertical angle θ at its vertex are

$$I_{xx} = (3/10)\, Mh^2 \tan^2\theta,$$

$$I_{yy} = I_{zz} = (3/20)\, Mh^2\, (4 + \tan^2\theta)$$

3.3 Obtain the M.Is of a uniform rectangular block about a central axes parallel to the two faces with dimensions of the block as $2a \times 2b \times 2c$

3.4 Show that the M.I. of a uniform solid cone of radius a, height h, mass M about a line which lies on its surface is

$$\frac{3}{20} Ma^2\, \frac{a^2 + 6h^2}{a^2 + h^2}.$$

3.5 The moments and products of inertia of a rigid body about x, y, z axes are $I_{xx} = 3, I_{yy} = 10/3, I_{zz} = 8/3, I_{xy} = 4/3, I_{yz} = 0, I_{zx} = -4/3$. Find the principal moments of inertia and the principal directions.

3.6 Show that the principal moments of inertia of a uniform cylinder of radius a and height h are given by

$$I_{xx}^* = I_{yy}^* = \frac{M}{12}(3a^2 + h^2), \quad I_{zz}^* = \frac{Ma^2}{2}.$$

3.7 Two circular metal discs have the same mass M and the same thickness t. Disc 1 has a uniform density ρ_1 which is less than ρ_2, the uniform density of disc 2. Which disc has the larger moment of inertia?

3.8 A string is wrapped around the periphery of a thin circular disc of radius 0.5 metre and of mass 10 kg, as shown in Fig. 3.31. At a particular instant, the string is pulled up with a force of 200 N. Determine the acceleration of the centre of the disc and the angular acceleration of the disc.

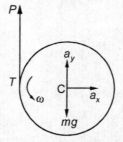

Fig. 3.31 Description of Exercise 3.8.

3.9 A uniform slender rod of length $L = 750$ mm and mass $m = 2$ kg hangs freely from a hinge at A. If a force P of magnitude 12 newton is applied at B horizontally to the left so that $AB = L$, determine (i) the angular acceleration of the rod and (ii) the components of reaction at A.

4

Dynamics of a Rigid Body Motion in Space

Things not understood are admired.
—Thomas Fuller

4.1 EULER'S DYNAMICAL EQUATIONS FOR A RIGID BODY ROTATING ABOUT A FIXED POINT

Having learnt about the various basic kinematic concepts of a rigid body motion in the previous chapter, we shall now focus on understanding the various applications. To begin with, we shall derive the Euler's dynamical equations involving the motion of a rigid body.

Imagine a rigid body which is constrained to rotate about a fixed point O (may be its centre of gravity), with angular velocity vector $\vec{\omega}$. Suppose the origin of the body-fixed coordinate system is O. Let $\hat{i}, \hat{j}, \hat{k}$ be the unit vectors along OX, OY, OZ respectively (see Fig. 4.1).

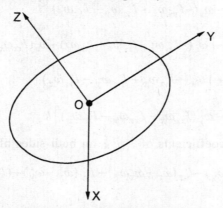

Figure 4.1 Motion of a rigid body in space.

If the components of $\vec{\omega}$ are ω_x, ω_y and ω_z, then, we write

$$\vec{\omega} = \omega_x \hat{i} + \omega_y \hat{j} + \omega_z \hat{k}. \qquad (4.1)$$

From the principle of angular momentum, which states that the rate of change of angular momentum of a body about a point, fixed or moving with the centre of mass, is equal to the total moment of the external forces about that point. Symbolically, we can express it as

$$\frac{d\vec{H}}{dt} = \vec{G} \text{ (Torque)}. \tag{4.2}$$

Following Eq. (3.48), we have

$$\vec{H} = \begin{bmatrix} I_{xx} & -I_{xy} & -I_{xz} \\ -I_{yx} & I_{yy} & -I_{yz} \\ -I_{zx} & -I_{zy} & I_{zz} \end{bmatrix} \begin{pmatrix} \omega_x \\ \omega_y \\ \omega_z \end{pmatrix}. \tag{4.3}$$

However, with reference to the body fixed frame, the operator relation allows us to write

$$\vec{G} = \frac{d\vec{H}}{dt} = \frac{\delta\vec{H}}{\delta t} + \vec{\omega} \times \vec{H}$$

$$= \dot{H}_x \hat{i} + \dot{H}_y \hat{j} + \dot{H}_z \hat{k} + \begin{vmatrix} \hat{i} & \hat{j} & \hat{k} \\ \omega_x & \omega_y & \omega_z \\ H_x & H_y & H_z \end{vmatrix}.$$

That is,

$$G_x \hat{i} + G_y \hat{j} + G_z \hat{k} = (I_{xx}\dot{\omega}_x - I_{xy}\dot{\omega}_y - I_{xz}\dot{\omega}_z)\hat{i} + (-I_{yx}\dot{\omega}_x + I_{yy}\dot{\omega}_y - I_{yz}\dot{\omega}_z)\hat{j}$$

$$+ (-I_{zx}\dot{\omega}_x - I_{zy}\dot{\omega}_y + I_{zz}\dot{\omega}_z)\hat{k} + \left[\omega_y(-I_{zx}\omega_x - I_{zy}\omega_y + I_{zz}\omega_z)\right.$$

$$\left. - \omega_z(-I_{yx}\omega_x + I_{yy}\omega_y - I_{yz}\omega_z)\right]\hat{i}$$

$$- \left[\omega_x(-I_{zx}\omega_x - I_{zy}\omega_y + I_{zz}\omega_z) - \omega_z(I_{xx}\omega_x - I_{xy}\omega_y - I_{xz}\omega_z)\right]\hat{j}$$

$$+ \left[\omega_x(-I_{yx}\omega_x + I_{yy}\omega_y - I_{yz}\omega_z)\right.$$

$$\left. - \omega_y(I_{xx}\omega_x - I_{xy}\omega_y - I_{xz}\omega_z)\right]\hat{k}. \tag{4.4}$$

Now, equating the coefficients of $\hat{i}, \hat{j}, \hat{k}$ on both sides of Eq. (4.4), we get

$$I_{xx}\dot{\omega}_x + I_{xy}(\omega_z\omega_x - \dot{\omega}_y) - I_{xz}(\dot{\omega}_z + \omega_x\omega_y) + I_{yz}(\omega_z^2 - \omega_y^2) + (I_{zz} - I_{yy})\omega_y\omega_z = G_x \tag{4.5a}$$

$$I_{yy}\dot{\omega}_y + I_{yz}(\omega_x\omega_y - \dot{\omega}_z) - I_{yx}(\dot{\omega}_x + \omega_y\omega_z) + I_{xz}(\omega_x^2 - \omega_z^2) + (I_{xx} - I_{zz})\omega_x\omega_z = G_y \tag{4.5b}$$

$$I_{zz}\dot{\omega}_z + I_{xz}(\omega_y\omega_z - \dot{\omega}_x) - I_{yz}(\dot{\omega}_y + \omega_z\omega_x) + I_{xy}(\omega_y^2 - \omega_x^2) + (I_{yy} - I_{xx})\omega_y\omega_x = G_z \tag{4.5c}$$

This set of equations of motion has very important applications, specially in space-related problems. Some of them are presented in the following sections. These equations are known as *Euler's dynamical equations of motion for a rigid body rotating about its centre of mass*. Of course, these equations are very complex and extremely difficult to manipulate. However, considerable simplification of these equations can be achieved by allowing the body-fixed coordinate axes to coincide with its principal axes. In such a case, all the P.Is become zero and Eqs. (4.5) simplify to

$$\left. \begin{array}{l} I_{xx}\dot{\omega}_x - (I_{yy} - I_{zz})\omega_y\omega_z = G_x \\ I_{yy}\dot{\omega}_y - (I_{zz} - I_{xx})\omega_z\omega_x = G_y \\ I_{zz}\dot{\omega}_z - (I_{xx} - I_{yy})\omega_x\omega_y = G_z. \end{array} \right\} \quad (4.6)$$

These are *Euler's equations of motion for a rigid body with a fixed point*, in a body-fixed frame of reference. These equations are widely used in solving the rotational motion of a rigid body. These are also nonlinear ordinary differential equations and found to be difficult to have analytical solution and, therefore, require the use of numerical techniques, such as Runge–Kutta method of fourth order* and a fast electronic computer for their solution. However, for a given set of initial conditions and assuming that $\vec{G}$ is a known function of time, position and velocity of the body, Eqs. (4.6) can be solved, in principle, to determine the components of $\vec{\omega}$ as a function of time.

4.1.1 Body-fixed Translational Equations

It is convenient to solve both the translational as well as rotational motions of a rigid body, in a body-fixed coordinate system, especially, when the applied forces can be specified in a body-fixed coordinate system.

In the Newtonian frame of reference, the translational equations of motion of a rigid body can be written as

$$\vec{F} = m\dot{\vec{V}}, \quad (4.7)$$

where $\vec{F}$ is the total external force acting on the rigid body and $\vec{V}$ the velocity of the centre of mass of the body. In a body-fixed frame, $\vec{V}$ and $\vec{F}$ may be expressed as

$$\vec{V} = V_x\hat{i} + V_y\hat{j} + V_z\hat{k}$$

$$\vec{F} = F_x\hat{i} + F_y\hat{j} + F_z\hat{k}.$$

Now, following Newton's second law of motion, we have

---

*See Sankara Rao, 2004

$$\vec{F} = m\left(\frac{\delta \vec{V}}{\delta t} + \vec{\omega} \times \vec{V}\right) = m\left(\dot{V}_x \hat{i} + \dot{V}_y \hat{j} + \dot{V}_z \hat{k} + \begin{vmatrix} \hat{i} & \hat{j} & \hat{k} \\ \omega_x & \omega_y & \omega_z \\ V_x & V_y & V_z \end{vmatrix}\right). \quad (4.8)$$

In component form, we may also write as

$$\left. \begin{array}{l} F_x = m\left(\dfrac{dV_x}{dt} + \omega_y V_z - \omega_z V_y\right) \\[4pt] F_y = m\left(\dfrac{dV_y}{dt} + \omega_z V_x - \omega_x V_z\right) \\[4pt] F_z = m\left(\dfrac{dV_z}{dt} + \omega_x V_y - \omega_y V_x\right). \end{array} \right\} \quad (4.9)$$

Here, the components of $\vec{\omega}$ are either given directly or may be obtained from the rotational Eqs. (4.6). Thus, Eqs. (4.9) can be solved to get V_x, V_y and V_z as a function of time. To sum up, the set of Eqs. (4.6) and (4.9) together constitute the dynamical equations of motion of a rigid body in a body-fixed coordinate frame. For illustration, we present below a few examples.

Example 4.1 Show directly from the Euler's dynamical equations of motion (Eqs. 4.6) that if $\vec{G} = 0$ and $I_{xx} = I_{yy}$, then ω is constant.

Solution We are given the data $I_{xx} = I_{yy}$ and therefore the Euler's dynamical equations reduce to

$$I_{xx}\dot{\omega}_x - (I_{yy} - I_{zz})\omega_y \omega_z = 0 \quad (1)$$

$$I_{yy}\dot{\omega}_y - (I_{zz} - I_{xx})\omega_z \omega_x = 0 \quad (2)$$

$$I_{zz}\dot{\omega}_z = 0. \quad (3)$$

Since $I_{zz} \neq 0$, Eq. (3) becomes $\dot{\omega}_z = 0$. Therefore,

$$\omega_z = \text{constant}. \quad (4)$$

Now, multiplying Eq. (1) by ω_x and Eq. (2) by ω_y and adding, we get

$$I_{xx}\dot{\omega}_x \omega_x + I_{yy}\dot{\omega}_y \omega_y - (I_{yy} - I_{zz} + I_{zz} - I_{xx})\omega_x \omega_y \omega_z = 0.$$

Using the fact that $I_{xx} = I_{yy}$, the above equation reduces to

$$I_{xx}(\dot{\omega}_x \omega_x + \dot{\omega}_y \omega_y) = 0.$$

Since $I_{xx} \neq 0$, we have $\dot{\omega}_x \omega_x + \dot{\omega}_y \omega_y = 0$, which can be rewritten as

$$\frac{1}{2}\frac{d}{dt}(\omega_x^2 + \omega_y^2) = 0.$$

That is,

$$\omega_x^2 + \omega_y^2 = \text{constant}. \tag{5}$$

From Eqs. (4) and (5), we can at once write

$$\omega_x^2 + \omega_y^2 + \omega_z^2 = \text{constant},$$

which means $\omega = |\vec{\omega}| = \omega_x^2 + \omega_y^2 + \omega_z^2 = \text{constant}$. Hence the result follows.

Example 4.2 Imagine that a rigid body is rotating about a fixed point with angular velocity $\vec{\omega}$. Assuming that the coordinate axes coincide with the principal axes, if T stands for kinetic energy and $\vec{G}$ for external torque acting on the body, show that

$$\frac{dT}{dt} = \vec{G} \cdot \vec{\omega}.$$

Solution Assuming that the body-fixed coordinate axes coincide with the principal axes, let

$$\vec{\omega} = \omega_x \hat{i} + \omega_y \hat{j} + \omega_z \hat{k}$$

and

$$\vec{G} = G_x \hat{i} + G_y \hat{j} + G_z \hat{k}.$$

Then, the expression for the kinetic energy of a rigid body rotating about a fixed point is given as

$$T = \frac{1}{2}\left(I_{xx}\omega_x^2 + I_{yy}\omega_y^2 + I_{zz}\omega_z^2\right). \tag{1}$$

Therefore,

$$\frac{dT}{dt} = I_{xx}\dot{\omega}_x \omega_x + I_{yy}\dot{\omega}_y \omega_y + I_{zz}\dot{\omega}_z \omega_z. \tag{2}$$

Recalling Euler's dynamical equations of motion such as

$$I_{xx}\dot{\omega}_x - (I_{yy} - I_{zz})\omega_y \omega_z = G_x, \text{ etc...}$$

Equation (2) becomes

$$\frac{dT}{dt} = \omega_x \left[G_x + (I_{yy} - I_{zz})\omega_y \omega_z\right] + \omega_y \left[G_y + (I_{zz} - I_{xx})\omega_z \omega_x\right]$$

$$+ \omega_z \left[G_z + (I_{xx} - I_{yy})\omega_x \omega_y\right].$$

On simplification, the above equation becomes

$$\frac{dT}{dt} = G_x \omega_x + G_y \omega_y + G_z \omega_z = \vec{\omega} \cdot \vec{G}.$$

Hence, the result follows.

Example 4.3 Show that the kinetic energy and angular momentum of the torque-free motion of a rigid body is constant.

Solution If the torque $\vec{G}$ is set to zero, the Euler's equations of motion for a rigid body can be written as

$$\left.\begin{array}{l} I_{xx}\dot{\omega}_x - (I_{yy} - I_{zz})\omega_y\omega_z = 0 \\ I_{yy}\dot{\omega}_y - (I_{zz} - I_{xx})\omega_z\omega_x = 0 \\ I_{zz}\dot{\omega}_z - (I_{xx} - I_{yy})\omega_x\omega_y = 0. \end{array}\right\} \quad (1)$$

Multiplying these equations respectively by ω_x, ω_y and ω_z and adding, we get

$$I_{xx}\dot{\omega}_x\omega_x + I_{yy}\dot{\omega}_y\omega_y + I_{zz}\dot{\omega}_z\omega_z = 0.$$

On integration with respect to t, we have at once

$$\frac{1}{2}(I_{xx}\omega_x^2 + I_{yy}\omega_y^2 + I_{zz}\omega_z^2) = \text{constant}. \quad (2)$$

Recalling Eq. (3.41), we observe that

$$I_{xx}\omega_x^2 + I_{yy}\omega_y^2 + I_{zz}\omega_z^2 = 2T = \text{constant}. \quad (3)$$

Hence, the kinetic energy, T, of a torque-free motion of a rigid body is constant. This is also called the *energy integral of the system*.

Multiplying Eqs. (1) respectively by $I_{xx}\omega_x$, $I_{yy}\omega_y$ and $I_{zz}\omega_z$ and adding, we obtain

$$I_{xx}^2\dot{\omega}_x\omega_x + I_{yy}^2\dot{\omega}_y\omega_y + I_{zz}^2\dot{\omega}_z\omega_z = 0.$$

Now, integrating this equation with respect to time, we get at once

$$(I_{xx}\omega_x)^2 + (I_{yy}\omega_y)^2 + (I_{zz}\omega_z)^2 = \text{constant}.$$

However, if we recall Eq. (3.50), we notice that

$$(I_{xx}\omega_x)^2 + (I_{yy}\omega_y)^2 + (I_{zz}\omega_z)^2 = |\vec{H}|^2 = \text{constant}. \quad (4)$$

This equation means that the square of the modulus of the angular momentum of a torque-free motion of a rigid body is also a constant. This is also called the *angular momentum integral* of the system considered.

4.1.2 Gyroscopic Motion

Gyroscopic motion occurs whenever the axis about which a body is spinning is itself rotating about another axis. The motion of a top and the motion of the Earth are well-known examples to mention a few.

4.2 MOTION OF A SYMMETRICAL SPINNING TOP

A 'top' is a symmetric rigid body terminating at a sharp point on the axis of symmetry called *vertex*. The top may be assumed to be spinning on a floor or plane rough enough to prevent slipping about the fixed vertex coinciding with the point of contact on the plane. In this section, we shall study the motion of the top, subjected to a uniform gravitational field. It may be noted that the top is a rigid body spinning about its vertex which is different from the centre of mass of the body. This situation includes a variety of physical systems, a child's spinning top, rotating earth, motion of a space capsule, navigational instruments like gyroscope, and so on.

We shall present below the mathematical modelling and explain the phenomena of its gyroscopic motion using Eulerian angles ψ, θ, ϕ, as shown in Fig. 4.2.

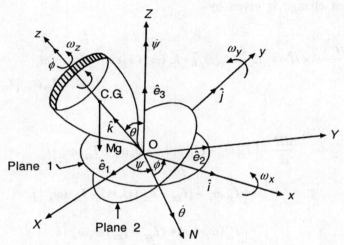

Figure 4.2 A symmetrical top with fixed point at O.

Let OXYZ be a fixed frame S with the origin at the vertex O of the top. Let Oxyz be the principal coordinate system S' fixed with the top having the same origin O. Here, θ measures the inclination of the axis of the top from the vertical Z-axis. The intersection of planes 1 and 2 gives the line of nodes ON, which is located through an angle ψ from the X-axis. The y-axis lies on plane 2 and the z-axis coincides with the axis of the top. The angles θ and ψ completely specify the position of the axis of the top. The angle ϕ is measured on plane 2 from x-axis to ON-axis. We choose the orientation of the xy-plane in such a way that OZ and Oz are coplanar. As the gravitational field is uniform, the gravitational force Mg passes through the centre of gravity which is, say, at a distance l from O along the axis of the top and acts parallel to Z-axis but in opposite direction so as to form a positive torque about O. Now, let $\bar{\omega}$ be the angular velocity vector with which the S' frame is rotating with reference to the fixed frame or inertial frame of reference S.

Dynamics of a Rigid Body Motion in Space (Ch. 4)

Also, let $\hat{i}, \hat{j}, \hat{k}$ be the unit vectors along the coordinate axes of the rotating frame S' and $\hat{e}_1, \hat{e}_2, \hat{e}_3$ be the corresponding unit vectors along the coordinate axes in S frame, then

$$\vec{\omega} = \omega_x \hat{i} + \omega_y \hat{j} + \omega_z \hat{k}. \qquad (4.10)$$

The top spins about its axis of symmetry with uniform angular velocity and at the same time, the axis of symmetry revolves about the vertical line uniformly generating a right circular cone about Z-axis which is referred to as precession and hence, the angular momentum vector should also include an additional component $s\hat{k}$, where s is the spin of the top. Thus, recalling Eq. (3.50), the angular momentum vector is given as

$$\vec{H} = I_{xx}\omega_x \hat{i} + I_{yy}\omega_y \hat{j} + I_{zz}(\omega_z + s)\hat{k}. \qquad (4.11)$$

Its rate of change is given by

$$\frac{d\vec{H}}{dt} = \frac{\delta \vec{H}}{\delta t} + \vec{\omega} \times \vec{H} = I_{xx}\dot{\omega}_x \hat{i} + I_{yy}\dot{\omega}_y \hat{j} + I_{zz}(\dot{\omega}_z + \dot{s})\hat{k} + \begin{vmatrix} \hat{i} & \hat{j} & \hat{k} \\ \omega_x & \omega_y & \omega_z \\ I_{xx}\omega_x & I_{yy}\omega_y & I_{zz}(\omega_z + s) \end{vmatrix}.$$

That is,

$$\frac{d\vec{H}}{dt} = \left[I_{xx}\dot{\omega}_x + (I_{zz} - I_{yy})\omega_y\omega_z + I_{zz}s\omega_y \right]\hat{i}$$

$$+ \left[I_{yy}\dot{\omega}_y + (I_{xx} - I_{zz})\omega_x\omega_z - I_{zz}s\omega_x \right]\hat{j}$$

$$+ \left[I_{zz}(\dot{\omega}_z + \dot{s}) + (I_{yy} - I_{xx})\omega_x\omega_y \right]\hat{k}. \qquad (4.12)$$

Now, the total torque about O or the moment of the weight of the top about O, designated by $\vec{G}$ is

$$\vec{G} = (l\hat{k}) \times (-Mg\hat{e}_3).$$

But

$$\hat{e}_3 = \sin\theta \hat{j} + \cos\theta \hat{k}.$$

Therefore, the torque

$$\vec{G} = (l\hat{k}) \times \left[-Mg\left(\sin\theta \hat{j} + \cos\theta \hat{k}\right) \right] = Mgl\sin\theta \hat{i}. \qquad (4.13)$$

Following the principle of angular momentum, we have

$$\frac{d\vec{H}}{dt} = \vec{G}. \qquad (4.14)$$

Using Eqs. (4.12) and (4.13) and also the fact that the top is symmetrical about the z-axis and, therefore, $I_{xx} = I_{yy}$, we get at once

$$\left.\begin{array}{r}I_{xx}\dot{\omega}_x - (I_{xx} - I_{zz})\omega_y\omega_z + I_{zz}\omega_y s = Mgl\sin\theta = G_x \\ I_{xx}\dot{\omega}_y - (I_{zz} - I_{xx})\omega_z\omega_x - I_{zz}\omega_x s = 0 = G_y \\ I_{zz}(\dot{\omega}_z + \dot{s}) = 0 = G_z.\end{array}\right\} \quad (4.15)$$

Since $I_{zz} \neq 0$, the third of Eq. (4.15), on integration with respect to time yields

$$\omega_z + \dot{\phi} = \omega_z + s = \text{constant} = A \text{ (say)}. \quad (4.16)$$

Multiplying the first two equations of (4.15) by ω_x, ω_y and the third by $(\omega_z + s)$ and adding, the result is

$$I_{xx}(\omega_x\dot{\omega}_x + \omega_y\dot{\omega}_y) + I_{zz}(\omega_z + s)(\dot{\omega}_z + \dot{s}) = Mgl\sin\theta\,\omega_x,$$

which can be rewritten as

$$\frac{d}{dt}\left[\frac{I_{xx}}{2}(\omega_x^2 + \omega_y^2) + \frac{I_{zz}}{2}(\omega_z + s)^2\right] = Mgl\omega_x\sin\theta. \quad (4.17)$$

Using the result given by Eq. (4.16), Eq. (4.17) can be recast into

$$\frac{d}{dt}\left[\frac{I_{xx}}{2}(\omega_x^2 + \omega_y^2) + \frac{I_{zz}}{2}A^2\right] = Mgl\omega_x\sin\theta. \quad (4.18)$$

Now, recalling relations (3.59), which connect the components of angular velocity in terms of Eulerian angles ψ, θ, ϕ, that is,

$$\left.\begin{array}{r}\omega_x = \dot{\theta}\cos\phi + \dot{\psi}\sin\theta\sin\phi \\ \omega_y = \dot{\psi}\sin\theta\cos\phi - \dot{\theta}\sin\phi \\ \omega_z = \dot{\phi} + \dot{\psi}\cos\theta.\end{array}\right\} \quad (4.19)$$

we can specify the orientation of the xyz-system using Eulerian angles ψ and θ (see Fig. 4.2). In this case, it is clear that the spin s must correspond to $\dot{\phi}$, and, therefore,

$$s = \dot{\phi}.$$

The rotational motion of the xyz-system does not involve ϕ, so by setting $\phi = \dot{\phi} = 0$, Eqs. (4.19) simplify to

$$\omega_x = \dot{\theta}, \qquad \omega_y = \dot{\psi}\sin\theta, \qquad \omega_z = \dot{\psi}\cos\theta. \quad (4.20)$$

Utilizing this result, Eq. (4.18) can be written as

$$\frac{d}{dt}\left[\frac{I_{xx}}{2}(\dot{\theta}^2 + \dot{\psi}^2\sin^2\theta) + \frac{I_{zz}}{2}A^2\right] = Mgl\sin\theta\,\dot{\theta} = -\frac{d}{dt}(Mgl\cos\theta).$$

On integration with respect to t, we get at once

$$\frac{I_{xx}}{2}(\dot{\theta}^2 + \dot{\psi}^2 \sin^2\theta) + \frac{I_{zz}}{2}A^2 + Mgl\cos\theta = E \text{ (constant)}. \quad (4.21)$$

In fact, we can recognize that this result is equivalent to the principle of conservation of energy, that is, the sum of kinetic energy (T) and potential energy V. Thus,

$$T = \frac{I_{xx}}{2}(\dot{\theta}^2 + \dot{\psi}^2 \sin^2\theta) + \frac{I_{zz}}{2}A^2 \quad (4.22)$$

and

$$V = Mgl\cos\theta. \quad (4.23)$$

Here, $\dot{\psi}$ is referred to *precessional velocity* and $\dot{\theta}$ corresponds to *nutation*. Now, using Eqs. (4.20) in Eqs. (4.15), they can be written as

$$\left. \begin{array}{c} I_{xx}\ddot{\theta} - (I_{xx} - I_{zz})\dot{\psi}^2 \sin\theta\cos\theta + I_{zz}\dot{\psi}s\sin\theta = Mgl\sin\theta \\ I_{xx}(\ddot{\psi}\sin\theta + \dot{\psi}\cos\theta\dot{\theta}) - (I_{zz} - I_{xx})\dot{\theta}\dot{\psi}\cos\theta - I_{zz}\dot{\theta}s = 0 \\ I_{zz}(\ddot{\psi}\cos\theta - \dot{\psi}\sin\theta\dot{\theta} + \dot{s}) = 0. \end{array} \right\} \quad (4.24)$$

These are general equations of motion for a spinning top about its centre of mass.

Let us now investigate the nutational motion $\dot{\theta}$ of the symmetrical top. For, the second of Eq. (4.24) gives

$$I_{xx}(\ddot{\psi}\sin\theta + 2\dot{\psi}\dot{\theta}\cos\theta) - I_{zz}\dot{\theta}(s + \dot{\psi}\cos\theta) = 0.$$

But, noting that $s + \dot{\psi}\cos\theta = s + \omega_z = A$, the above equation becomes

$$I_{xx}(\ddot{\psi}\sin\theta + 2\dot{\psi}\dot{\theta}\cos\theta) - I_{zz}A\dot{\theta} = 0. \quad (4.25)$$

Multiplying Eq. (4.25) by $\sin\theta$, we can rewrite it as

$$I_{xx}(\ddot{\psi}\sin^2\theta + 2\dot{\psi}\dot{\theta}\sin\theta\cos\theta) - I_{zz}A\dot{\theta}\sin\theta = 0$$

that is,

$$\frac{d}{dt}(I_{xx}\dot{\psi}\sin^2\theta + I_{zz}A\cos\theta) = 0$$

which on integration yields

$$I_{xx}\dot{\psi}\sin^2\theta + I_{zz}A\cos\theta = \text{constant} = K\text{(say)}. \quad (4.26)$$

This equation gives us

$$\dot{\psi} = \frac{K - I_{zz}A\cos\theta}{I_{xx}\sin^2\theta}. \quad (4.27)$$

Substituting $\dot{\psi}$ into Eq. (4.21), we get

$$\frac{1}{2}I_{xx}\dot{\theta}^2 + \frac{(K - I_{zz}A\cos\theta)^2}{2I_{xx}\sin^2\theta} + \frac{1}{2}I_{zz}A^2 + Mgl\cos\theta = E. \tag{4.28}$$

We observe that this is a differential equation involving only θ, and therefore, its solution helps us understand the nutational motion of the top as explained below.

Let $\cos\theta = u$, so that

$$-\sin\theta\,\dot{\theta} = \dot{u} \quad \text{and} \quad \sin^2\theta = 1 - u^2$$

Thus, Eq. (4.28) reduces to

$$\frac{1}{2}I_{xx}\frac{\dot{u}^2}{1-u^2} + \frac{(K - I_{zz}Au)^2}{2I_{xx}(1-u^2)} + Mglu = E - \frac{1}{2}I_{zz}A^2,$$

which can be rewritten as

$$\dot{u}^2 + \left(\frac{K - I_{zz}Au}{I_{xx}}\right)^2 + \frac{2Mglu(1-u^2)}{I_{xx}} = \frac{2(1-u^2)}{I_{xx}}\left(E - \frac{I_{zz}A^2}{2}\right). \tag{4.29}$$

Let us now define

$$\left.\begin{aligned}\alpha &= (2E - I_{zz}A^2)/I_{xx} \\ \beta &= 2Mgl/I_{xx} \\ \gamma &= K/I_{xx} \\ \delta &= I_{zz}A/I_{xx}.\end{aligned}\right\} \tag{4.30}$$

Introducing these quantities, Eqs. (4.29) and (4.27), assume the following simple forms:

$$\dot{u}^2 = (\alpha - \beta u)(1 - u^2) - (\gamma - \delta u)^2 = f(u) \text{ (say)} \tag{4.31}$$

and

$$\dot{\psi} = \frac{\gamma - \delta u}{1 - u^2}. \tag{4.32}$$

Had θ been known as a function of time, Eq. (4.32) could be integrated to find the dependence of precession ψ on time.

From Eq. (4.31), we notice that $f(u)$ is a polynomial of degree 3 and it can be written as

$$\frac{du}{dt} = \sqrt{f(u)} \quad \text{or} \quad dt = \frac{du}{\sqrt{f(u)}}.$$

On integration, we have

$$t = \int \frac{du}{\sqrt{f(u)}} + \text{constant}, \qquad (4.33)$$

which is having the representation of an elliptic integral.

Going back to the cubic polynomial given by Eq. (4.31), we note that this polynomial in principle can take any value in the range $(-\infty, \infty)$. But the acceptable range of u is given by the range $-1 \leq u \leq 1$. The roots of $f(u)$ correspond to points of zero nutation, that is, $\dot{\theta} = 0$. If we examine carefully $f(u)$, we notice that for large u, the polynomial (4.31) behaves like βu^3. Therefore, $f(u)$ is positive for large positive value of u and negative for large negative value of u. Also for $u = \pm 1$, the first term in Eq. (4.31) vanishes and $f(u) < 0$, because the second term in Eq. (4.31) is always negative. Since $\beta > 0$, we have

$$f(\infty) = \infty \qquad \text{(positive)}$$

$$f(-\infty) = -\infty \qquad \text{(negative)}$$

$$f(1) = -(\gamma - \delta)^2 \quad \text{(negative)}$$

$$f(-1) = -(\gamma + \delta)^2 \quad \text{(negative)}$$

If the motion of the top has to take place, we must have $\dot{u} \geq 0$, implying $f(u) \geq 0$. Also, since $0 \leq \theta \leq \pi/2$, we have $0 \leq u \leq 1$. Therefore, it gives us a clue that there must be two roots u_1 and u_2 in the range $0 \leq u \leq 1$. We also observe that as u goes from 1 to ∞, there is a change of sign and hence by Descartes rule of signs, there must be another root, say u_3. With this brief information, the typical curve of $f(u)$ is depicted in Fig. 4.3.

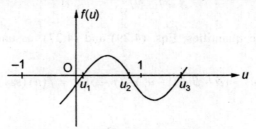

Figure 4.3 Graph of the function $f(u)$.

Hence, the physical interpretation is that the symmetric axis of the top must nutate periodically in the annular region bounded by

$$\cos\theta_1 = u_1 \quad \text{and} \quad \cos\theta_2 = u_2.$$

This bobbling motion, which is up and down of the axis between the limits θ_1 and θ_2, is described as *nutation*. The nutational motion takes place simultaneously as the precessional motion of the axis of the top about the vertical and the spinning of the top about its axis. Of course, the whole motion is periodic.

We can visualize the actual motion of the top through the spherical coordinates ψ and θ of the axis of symmetry which themselves are functions of time. To have a clear idea of the way the top behaves, we shall visualize the essential features of the motion, by fixing our attention on the axis of the type that traces out a curve on a unit sphere with the origin at O. This type of curve traced out by the axis of symmetry depends both on nutation $\dot\theta$ and precession $\dot\psi$. From Eq. (4.32), the precession is given by

$$\dot\psi = \frac{\gamma - \delta u}{1 - u^2}.$$

The denominator is in general positive and the type of precession has three possibilities:

Case (i): If the root u' of the equation $\gamma = \delta u$ lies outside of (u_1, u_2), then $\gamma - \delta u > 0$, then the angle ψ varies monotonically and the axis traces a curve like a sinusoid on the unit sphere, as shown in Fig. 4.4(a).

Case (ii): If the root u' of equation $\gamma = \delta u$ lies inside (u_1, u_2), then the rate of change of ψ is in opposite directions on the parallels θ_1 and θ_2 and the axis traces a looping curve in the sphere, as shown in Fig. 4.4(b).

Case (iii): If the root u' of the equation $\gamma = \delta u$ lies on the boundary, for example $u' = u_2$ (say), then the axis traces a curve with cusps, as shown in Fig. 4.4(c).

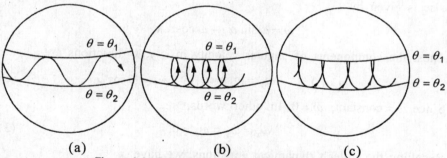

Figure 4.4 Path of the top's axis on the unit sphere.

Thus, the complete motion of the top consists of rotation around its own axis, nutation and precession. Each of the three motions has its own frequency.

Example 4.4 A thin uniform disc of radius a and mass m is rotating freely on a frictionless bearing with uniform angular velocity ω about a fixed vertical axis passing through its centre and is inclined at an angle α to the axis of symmetry of the disc. What is the magnitude and direction of the torque of the net force acting between the disc and the axis?

Solution Suppose, we take the coordinate axes OXYZ attached to the disc with the origin at its centre O in such a way that OZ is along the normal to the disc, while the OX and OY axes are along the plane of the disc. Let the disc be rotating about the vertical OZ′, as shown in Fig. 4.5.

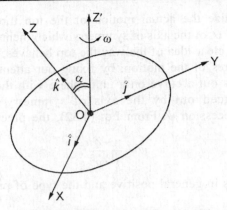

Figure 4.5 Description of Example 4.4.

Now, taking OXYZ axes to be the principal axes of the disc, the principal moments of inertia are given by

$$I_{xx} = I_{yy} = \frac{1}{4}ma^2, \quad I_{zz} = \frac{1}{2}ma^2, \tag{1}$$

where m is the mass of the disc and a is its radius. Let $\hat{i}, \hat{j}, \hat{k}$, be the unit vectors along the principal axes OX, OY, OZ, respectively. If α is the angle between $\hat{k}$ and the axis of rotation OZ′, then the angular velocity of the disc is given by

$$\vec{\omega} = \omega \sin \alpha \, \hat{j} + \omega \cos \alpha \, \hat{k}.$$

Thus, the components of angular velocity in $\hat{i}, \hat{j}, \hat{k}$ directions are

$$\omega_x = 0, \quad \omega_y = \omega \sin \alpha, \quad \omega_z = \omega \cos \alpha. \tag{2}$$

Since $\vec{\omega}$ = constant, $\dot{\vec{\omega}} = 0$. In other words,

$$\dot{\omega}_x = \dot{\omega}_y = \dot{\omega}_z = 0. \tag{3}$$

Recalling the Euler's dynamical equations, we have

$$\left. \begin{array}{l} I_{xx}\dot{\omega}_x - (I_{yy} - I_{zz})\omega_y \omega_z = G_x \\ I_{yy}\dot{\omega}_y - (I_{zz} - I_{xx})\omega_z \omega_x = G_y \\ I_{zz}\dot{\omega}_z - (I_{xx} - I_{yy})\omega_x \omega_y = G_z. \end{array} \right\} \tag{4}$$

Now, using Eqs. (1)–(3) into Eqs. (4), we obtain

$$\frac{1}{4}ma^2\omega^2 \sin \alpha \cos \alpha = G_x$$

$$-\frac{1}{4}ma^2\omega \cos \alpha (0) = G_y$$

$$0 = G_z.$$

Hence, the required torque is

$$\vec{G} = \frac{1}{4} ma^2 \omega^2 \sin\alpha \cos\alpha \, \hat{i}. \tag{5}$$

Therefore, the torque has only ith direction and is in the plane of the disc and is perpendicular to the plane formed by the normal to the disc and the axis of rotation. Its magnitude is given by G_x.

Example 4.5 A rectangular plate spins with constant angular velocity $\vec{\omega}$ about a diagonal. Find the couple which must act on the plate in order to maintain this motion.

Solution Let m be the mass of the plate, and $2a$, $2b$ be the length and breadth of the plate respectively. Let O be the centre of mass of the plate. Suppose $\hat{i}, \hat{j}, \hat{k}$ are the unit vectors along the principal axes of inertia at O, such that $\hat{k}$ is normal to the plate, while $\hat{i}$ and $\hat{j}$ lie in the plane of the plate. The principal moments of inertia are found to be

$$I_{xx} = \frac{1}{3} mb^2, \qquad I_{yy} = \frac{1}{3} ma^2, \qquad I_{zz} = \frac{1}{3} m(a^2 + b^2). \tag{1}$$

If α is the angle between $\hat{i}$ and the axis of rotation (see Fig. 4.6), then the angular velocity of the plate is

$$\vec{\omega} = \omega \cos\alpha \, \hat{i} + \omega \sin\alpha \, \hat{j}$$

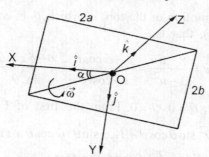

Figure 4.6 Description of Example 4.5.

and the components of angular velocity are therefore, given by

$$\omega_x = \omega \cos\alpha, \qquad \omega_y = \omega \sin\alpha, \qquad \omega_z = 0. \tag{2}$$

Since $\vec{\omega} = $ constant, $\dot{\vec{\omega}} = 0$, which means

$$\dot{\omega}_x = \dot{\omega}_y = \dot{\omega}_z = 0. \tag{3}$$

Now, recalling the Euler's dynamical equations, that is

$$\left. \begin{array}{l} I_{xx}\dot{\omega}_x - (I_{yy} - I_{zz})\omega_y \omega_z = G_x \\ I_{yy}\dot{\omega}_y - (I_{zz} - I_{xx})\omega_z \omega_x = G_y \\ I_{zz}\dot{\omega}_z - (I_{xx} - I_{yy})\omega_x \omega_y = G_z \end{array} \right\} \tag{4}$$

Substituting Eqs. (1), (2) and (3) into Eqs. (4), we get at once

$$\left.\begin{aligned} G_x &= 0 \\ G_y &= 0 \\ G_z &= -\frac{1}{3} m (b^2 - a^2) \omega^2 \sin \alpha \cos \alpha. \end{aligned}\right\} \quad (5)$$

But, we observe from Fig. 4.6 that

$$\cos \alpha = \frac{a}{\sqrt{a^2 + b^2}}, \quad \sin \alpha = \frac{b}{\sqrt{a^2 + b^2}}.$$

Therefore,

$$G_z = -\frac{1}{3} \frac{mab\omega^2 (b^2 - a^2)}{a^2 + b^2}. \quad (6)$$

Hence, the couple has only kth direction and acts in the plane of the plate and the axis of the couple is normal to the plate and rotates with it. The magnitude of the couple is given by Eq. (6). Hence the result.

Steady precession of the top

Let us consider the motion of the top, in which θ is constant (see Fig. 4.2). Recalling Eq. (4.32), that is,

$$\dot{\psi} = \frac{\gamma - \delta u}{1 - u^2} = \frac{\gamma - \delta \cos\theta}{1 - \cos^2\theta} = \frac{\gamma - \delta \cos\theta}{\sin^2\theta}.$$

Since θ is constant, $\dot{\theta} = 0$, $\ddot{\theta} = 0$. From the first of Eq. (4.24), we have

$$I_{xx}\ddot{\theta} - I_{xx}\dot{\psi}^2 \sin\theta \cos\theta + I_{zz}\dot{\psi} \sin\theta (\dot{\psi}\cos\theta + s) = Mgl \sin\theta.$$

Now, using the condition $\ddot{\theta} = 0$ and the fact

$$\dot{\psi} \cos\theta + s = \omega_z + s = A,$$

the above equation can be rewritten as

$$(I_{xx}\dot{\psi}^2 \cos\theta - I_{zz}A\dot{\psi} + Mgl) \sin\theta = 0,$$

which gives

$$I_{xx}\dot{\psi}^2 \cos\theta - I_{zz}A\dot{\psi} + Mgl = 0 \quad (4.34)$$

being a quadratic in $\dot{\psi}$, the precession rate is found to be

$$\dot{\psi} = \frac{I_{zz}A \pm \sqrt{I_{zz}^2 A^2 - 4MglI_{xx} \cos\theta}}{2I_{xx} \cos\theta}. \quad (4.35)$$

Thus, two distinct precessional frequencies are possible, provided the square root term in Eq. (4.35) in real. That is, if

$$I_{zz}^2 A^2 - 4Mgl\, I_{xx} \cos\theta > 0.$$

In other words,

$$I_{zz}^2 A^2 > 4Mgl\, I_{xx} \cos\theta. \tag{4.36}$$

This is the required condition for the steady precession of the top.

Example 4.6 (Motion of a rolling disc) Here, we shall consider the motion of a thin circular disc of radius r and mass m as it rolls on a horizontal plane and derive the governing differential equation of motion to analyze its rolling motion in general.

The general position of the disc as it rolls on a horizontal plane is shown in Fig. 4.7. Let P be the point of contact with the plane which is supposed to be rough, in order to prevent slipping. Let OQ be the vertical and θ be the inclination of the plane of the disc with the vertical. Let the angle between the fixed horizontal direction and the tangent at P be denoted by ψ. Let $\hat{i}, \hat{j}, \hat{k}$ be an orthogonal triad of unit vectors directed along X, Y, Z axes respectively, such that $\hat{k}$ is perpendicular to the disc at its centre O, $\hat{i}$ is along OP and $\hat{j}$ obviously is horizontal and lies on the plane of the disc.

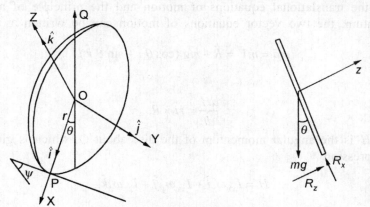

Figure 4.7 A circular disc rolling on a plane.

Let $\vec{V}$ be the velocity of the centre and $\vec{\omega}$ the angular velocity of the disc. Since the particle at P is instantaneously at rest and further, in view of our assumption that the disc rolls without slipping, we have at once

$$\vec{V} + \vec{\omega} \times r\hat{i} = 0, \tag{1}$$

where

$$\vec{V} = V_x \hat{i} + V_y \hat{j} + V_z \hat{k} \quad \text{and} \quad \vec{\omega} = \omega_x \hat{i} + \omega_y \hat{j} + \omega_z \hat{k}.$$

Now, Eq. (1) can be rewritten as

$$V_x \hat{i} + V_y \hat{j} + V_z \hat{k} + (\omega_x \hat{i} + \omega_y \hat{j} + \omega_z \hat{k}) \times r\hat{i} = 0$$

or
$$V_x\hat{i} + V_y\hat{j} + V_z\hat{k} + r(-\omega_y\hat{k} + \omega_z\hat{j}) = 0$$
which gives
$$V_x = 0, \quad V_y + r\omega_z = 0, \quad V_z - r\omega_y = 0. \qquad (2)$$

Thus, the velocity of the centre of the disc can be computed if its angular velocity $\vec{\omega}$ is given.

Let $\vec{\Omega}$ be the angular velocity of the triad, which arises solely by the changes in θ and ψ. Thus, we find
$$\vec{\Omega} = -\cos\theta\,\dot{\psi}\hat{i} - \dot{\theta}\hat{j} + \sin\theta\,\dot{\psi}\hat{k}. \qquad (3)$$

Since the angular velocity of the disc, ω, and that of the triad, Ω, differ only in $\hat{k}$ component, we can at once write
$$\omega_x = -\cos\theta\,\dot{\psi}, \quad \omega_y = -\dot{\theta}. \qquad (4)$$

Now, noting that the force acting on the disc at the contact point P is the reaction of the plane $\vec{R}$, which can be written in the form
$$\vec{R} = R_x\hat{i} + R_y\hat{j} + R_z\hat{k}. \qquad (5)$$

Using the translational equations of motion and the principle of angular momentum, the two vector equations of motion can be written as
$$m\vec{a} = m\dot{\vec{V}} = \vec{R} + mg(\cos\theta\,\hat{i} - \sin\theta\,\hat{k}) \qquad (6a)$$
and
$$\frac{d\vec{H}}{dt} = r\hat{i} \times \vec{R}. \qquad (6b)$$

Here, $\vec{H}$ is the angular momentum of the disc about O, which is given by the expression
$$\vec{H} = I_{xx}\omega_x\hat{i} + I_{xx}\omega_y\hat{j} + I_{zz}\omega_z\hat{k}. \qquad (7)$$

Since
$$\frac{d\vec{V}}{dt} = \frac{\delta\vec{V}}{\delta t} + \vec{\Omega} \times \vec{V} \qquad (8)$$
or
$$\dot{\vec{V}} = \dot{V}_x\hat{i} + \dot{V}_y\hat{j} + \dot{V}_z\hat{k} + \begin{vmatrix} \hat{i} & \hat{j} & \hat{k} \\ -\cos\theta\,\dot{\psi} & -\dot{\theta} & \sin\theta\,\dot{\psi} \\ V_x & V_y & V_z \end{vmatrix}$$

$$= (\dot{V}_x - \dot{\theta}V_z - \sin\theta \, \dot{\psi} V_y)\hat{i} + (\dot{V}_y + \cos\theta \, \dot{\psi}V_z + \sin\theta \, \dot{\psi}V_x)\hat{j}$$
$$+ (\dot{V}_z - \cos\theta \, \dot{\psi} V_y + \dot{\theta}V_x)\hat{k}.$$

Substituting this expression, into Eq. (6a) we obtain the equations of motion in component form as

$$\left. \begin{array}{l} m(\dot{V}_x - \dot{\theta}V_z - \sin\theta \, \dot{\psi}V_y) = R_x + mg\cos\theta \\ m(\dot{V}_y + \sin\theta \, \dot{\psi}V_x + \cos\theta \, \dot{\psi}V_z) = R_y \\ m(\dot{V}_z - \cos\theta \, \dot{\psi}V_y + \dot{\theta}V_x) = R_z - mg\sin\theta. \end{array} \right\} \quad (9)$$

Now, using Eqs. (2) and (4), to eliminate V_x, V_y and V_z, Eqs. (9) can be recast into the form

$$\left. \begin{array}{l} mr\,(\dot{\theta}^2 + \sin\theta \, \dot{\psi} \, \omega_z) = R_x + mg\cos\theta \\ -mr\,(\dot{\omega}_z + \cos\theta \, \dot{\theta}\dot{\psi}) = R_y \\ -mr\,(\ddot{\theta} - \cos\theta \, \dot{\psi} \, \omega_z) = R_z - mg\sin\theta. \end{array} \right\} \quad (10)$$

Similarly, Eq. (6b) can be written in component form as

$$\dot{\vec{H}} = I_{xx}\dot{\omega}_x\hat{i} + I_{xx}\dot{\omega}_y\hat{j} + I_{zz}\dot{\omega}_z\hat{k} + \vec{\Omega} \times \vec{H}$$
$$= r\hat{i} \times (R_x\hat{i} + R_y\hat{j} + R_z\hat{k})$$
$$= r(R_y\hat{k} - R_z\hat{j})$$

that is,

$$I_{xx}\dot{\omega}_x\hat{i} + I_{xx}\dot{\omega}_y\hat{j} + I_{zz}\dot{\omega}_z\hat{k} + \begin{vmatrix} \hat{i} & \hat{j} & \hat{k} \\ -\cos\theta \, \dot{\psi} & -\dot{\theta} & \sin\theta \, \dot{\psi} \\ I_{xx}\omega_x & I_{xx}\omega_y & I_{zz}\omega_z \end{vmatrix} = r(R_y\hat{k} - R_z\hat{j}).$$

Its expansion gives Eq. (6b) in component form as

$$\left. \begin{array}{l} I_{xx}\dot{\omega}_x - I_{zz}\dot{\theta}\omega_z - I_{xx}\sin\theta \, \dot{\psi} \, \omega_y = 0 \\ I_{xx}\dot{\omega}_y + I_{xx}\sin\theta \, \dot{\psi} \, \omega_x + I_{zz}\cos\theta \, \dot{\psi}\omega_z = -rR_z \\ I_{zz}\dot{\omega}_z - I_{xx}\cos\theta\dot{\psi} \, \omega_y + I_{xx}\dot{\theta}\omega_x = rR_y. \end{array} \right\} \quad (11)$$

Now, using Eq. (4) we can eliminate ω_x and ω_y from the above set of equations to get

$$\left. \begin{array}{l} I_{xx}\dfrac{d}{dt}(\cos\theta\,\dot\psi) + I_{zz}\dot\theta\omega_z - I_{xx}\sin\theta\,\dot\theta\dot\psi = 0 \\[4pt] I_{xx}\ddot\theta + I_{xx}\sin\theta\cos\theta\,\dot\psi^2 - I_{zz}\cos\theta\,\dot\psi\omega_z = rR_z \\[4pt] I_{zz}\dot\omega_z = rR_y. \end{array} \right\} \qquad (12)$$

Thus we have six equations (10) and (12) together in six unknowns $\theta, \psi, \omega_z, R_x, R_y, R_z$ whose solution in principle gives us the time history of Euler angles and the reaction components. It is in fact difficult to get the analytical solution of Eqs. (10) and (12).

However, as a special case, if we consider that the disc is rolling steadily along a circular path, then the motion is described by

$$\theta = \text{constant}, \qquad \dot\psi = \text{constant}, \qquad \omega_z = \text{constant}. \qquad (13)$$

Substituting these values along with the corresponding values $\dot\theta = \ddot\theta = 0$, $\ddot\psi = 0$, $\dot\omega_z = 0$, Eqs. (10) give us the reactions as

$$\left. \begin{array}{l} R_x = m(r\sin\theta\,\dot\psi\,\omega_z - g\cos\theta) \\[4pt] R_y = 0 \\[4pt] R_z = m(r\cos\theta\,\dot\psi\,\omega_z + g\sin\theta). \end{array} \right\} \qquad (14)$$

The third of Eq. (12) is satisfied automatically. The second of Eq. (12) using Eq. (14) gives

$$I_{xx}\sin\theta\cos\theta\,\dot\psi^2 - I_{zz}\cos\theta\,\dot\psi\,\omega_z = mr(r\cos\theta\,\dot\psi\,\omega_z + g\sin\theta)$$

or

$$(I_{zz} + mr^2)\cos\theta\,\dot\psi\,\omega_z + mgr\sin\theta = I_{xx}\sin\theta\cos\theta\,\dot\psi^2. \qquad (15)$$

This is the required condition that must be satisfied by the constant values of θ, $\dot\psi$ and ω_z to describe a circular path by a circular disc of radius r.

4.3 TORQUE-FREE MOTION OF A SYMMETRICAL RIGID BODY—ROTATIONAL MOTION OF THE EARTH

One of the most important problems of practical value, where the Euler's dynamical equations can be directly applied is the rotational motion of the Earth. To a first approximation, we may assume that the Earth is a symmetric rigid body with symmetry about Z-axis (of course assumed for convenience) so that $I_{xx} = I_{yy}$. In addition, its motion is torque-free. Under these conditions and choosing the axes of symmetry coincident with one of the principal axes, the governing Euler's equations of motion in a body-fixed coordinate system can be written as

$$\left. \begin{array}{l} I_{xx}\dot{\omega}_x - (I_{yy} - I_{zz})\omega_y\omega_z = 0 \\ I_{yy}\dot{\omega}_y - (I_{zz} - I_{xx})\omega_z\omega_x = 0 \\ I_{zz}\dot{\omega}_z = 0. \end{array} \right\} \quad (4.37)$$

Since $I_{zz} \neq 0$, we get from the third of Eq. (4.37) that

$$\omega_z = \text{constant} = A \text{ (say)}. \quad (4.38)$$

That is, the z-component of the angular velocity is constant. Now, let us define

$$\mu = \frac{I_{xx} - I_{zz}}{I_{xx}}\omega_z. \quad (4.39)$$

Then, the first-two equations of Eq. (4.37) can be rewritten as (since $I_{xx} = I_{yy}$)

$$\dot{\omega}_x = \mu\omega_y, \qquad \dot{\omega}_y = -\mu\omega_x. \quad (4.40)$$

Differentiating the first of Eq. (4.40) with respect to time t and using the second, we get

$$\ddot{\omega}_x = -\mu^2 \omega_x, \quad (4.41)$$

which is a differential equation representing simple harmonic motion, whose solution is known to be

$$\omega_x = K \sin(\mu t + \theta_0), \quad (4.42)$$

where μ and θ_0 are constants. Substituting this value of ω_x into the second of Eq. (4.40) and integrating with respect to t, we get

$$\omega_y = K \cos(\mu t + \theta_0). \quad (4.43)$$

Thus, we notice that the components ω_x, ω_y of the angular velocity vector, $\vec{\omega}$, change in such a way that their resultant

$$\omega_{xy} = \omega_x \hat{i} + \omega_y \hat{j} \quad (4.44)$$

is in the xy-plane, whose magnitude is

$$|\omega_{xy}| = (\omega_x^2 + \omega_y^2)^{1/2} = K \text{ (say)} = \text{a constant}.$$

Now, to find the angular velocity ω_{xy} with respect to the body, let us represent it by a complex vector

$$\omega_{xy} = \omega_x + i\omega_y.$$

Differentiating this vector with respect to time and using Eqs. (4.40), we get

$$\dot{\omega}_{xy} = \dot{\omega}_x + i\dot{\omega}_y = \mu\omega_y - i\mu\omega_x = -i\mu(\omega_x + i\omega_y)$$

that is,
$$\dot{\omega}_{xy} + i\mu\omega_{xy} = 0.$$

Its solution is
$$\omega_{xy}(t) = \omega_{xy}(0)e^{-i\mu t},$$

where $\omega_{xy}(0)$ is the initial value of the vector ω_{xy}.

This solution clearly indicates that the complex vector ω_{xy} rotates with angular velocity μ relative to the body. Also, the total angular velocity vector $\vec{\omega}$ has magnitude

$$|\vec{\omega}| = (\omega_x^2 + \omega_y^2 + \omega_z^2)^{1/2} = (K^2 + A^2)^{1/2} = \text{constant}.$$

Thus, the angular velocity vector $\vec{\omega}$ rotates about Z-axis describing a cone with the vertex at the origin O. This motion is called precession and the body is said to precess about Z-axis with precessional velocity μ, as shown in Fig. 4.8. The cone described by the angular velocity vector $\vec{\omega}$ is known as *body cone*, whose semi-vertical denoted by β is given by

$$\tan\beta = \frac{|\omega_x + i\omega_y|}{|\omega_z|} = \frac{K}{A} = \text{constant}. \tag{4.45}$$

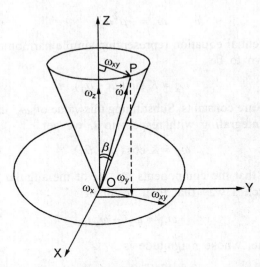

Figure 4.8 Precession of the rotating rigid body or Earth.

Also, for a torque-free or force-free motion of a rigid body, the principle of angular momentum states

$$\frac{d\vec{H}}{dt} = \vec{G} = 0,$$

which means that the angular momentum vector $\vec{H}$ is constant and fixed in space. The angular velocity vector $\vec{\omega}$ precesses about the body Z-axis. To understand the connection between $\vec{\omega}$ and $\vec{H}$, let us recall the relation which is already established in Chapter 3, such as

$$\begin{pmatrix} H_x \\ H_y \\ H_z \end{pmatrix} = \begin{bmatrix} I_{xx} & 0 & 0 \\ 0 & I_{yy} & 0 \\ 0 & 0 & I_{zz} \end{bmatrix} \begin{pmatrix} \omega_x \\ \omega_y \\ \omega_z \end{pmatrix}. \qquad (4.46)$$

Thus, the projection of the angular momentum vector in the xy-plane becomes

$$H_{xy} = I_{xx}(\omega_x \hat{i} + \omega_y \hat{j}) = I_{xx}\omega_{xy}. \qquad (4.47)$$

From Eqs. (4.44) and (4.47), it is clear that the projections of $\vec{H}$ and $\vec{\omega}$ in the XY-plane lie along the same line, while $\vec{H}$ has a component $I_{zz}\omega_z$ in the Z-direction. In component form $\vec{H}$ and $\vec{\omega}$ can be written as

$$\vec{\omega} = K \sin(\mu t + \theta_0)\hat{i} + K \cos(\mu t + \theta_0)\hat{j} + A\hat{k}$$

$$\vec{H} = I_{xx}K[\sin(\mu t + \theta_0)\hat{i} + \cos(\mu t + \theta_0)\hat{j}] + I_{zz}A\hat{k}$$

If α is the angle between ω_z and $\vec{H}$, then we have

$$\omega_z \hat{k} \cdot \vec{H} = |\omega_z||\vec{H}|\cos\alpha,$$

that is,

$$I_{zz}A^2 = A(I_{xx}^2 K^2 + I_{zz}^2 A^2)^{1/2} \cos\alpha$$

from which, we get

$$\cos\alpha = \frac{I_{zz}A}{(I_{xx}^2 K^2 + I_{zz}^2 A^2)^{1/2}}. \qquad (4.48)$$

Similarly, if β is the angle between ω_z and $\vec{\omega}$, then,

$$\omega_z \hat{k} \cdot \vec{\omega} = |\omega_z||\vec{\omega}|\cos\beta,$$

that is,

$$A^2 = A(K^2 + A^2)^{1/2} \cos\beta$$

which gives us

$$\cos\beta = \frac{A}{(K^2 + A^2)^{1/2}}. \qquad (4.49)$$

Equations (4.48) and (4.49) gives us at once that

$$\tan\alpha = \frac{I_{xx}}{I_{zz}}\frac{K}{A} \quad \text{and} \quad \tan\beta = \frac{K}{A}.$$

Thus, we have

$$\frac{\tan \alpha}{\tan \beta} = \frac{I_{xx}}{I_{zz}}. \qquad (4.50)$$

It may be noted that the direction of $\vec{H}$ is fixed in space, while the direction of $\vec{\omega}$ is not. But, the angle between $\vec{H}$ and $\vec{\omega}$ is constant, while the vector $\vec{\omega}$ must rotate around $\vec{H}$ and in so doing it describes a cone fixed in space called *space cone*.

Now, we shall establish that the vectors $\vec{H}$, $\vec{\omega}$ and Z-axis lie in one plane. For, we have to show that

$$\vec{H} \cdot (\omega_z \times \vec{\omega}) = 0.$$

Consider

$$\omega_z \times \vec{\omega} = \begin{vmatrix} \hat{i} & \hat{j} & \hat{k} \\ 0 & 0 & A \\ K\sin(\mu t + \theta_0) & K\cos(\mu t + \theta_0) & A \end{vmatrix}$$

$$= [-AK\cos(\mu t + \theta_0)]\hat{i} + [AK\sin(\mu t + \theta_0)]\hat{j}$$

Then, it follows that

$$\vec{H} \cdot (\omega_z \times \vec{\omega}) = [I_{xx}K\{\sin(\mu t + \theta_0)\hat{i} + \cos(\mu t + \theta_0)\hat{j}\} + I_{zz}A\hat{k}]$$

$$\cdot \left[-AK\cos(\mu t + \theta_0)\hat{i} + AK\sin(\mu t + \theta_0)\hat{j} \right]$$

$$= 0.$$

In fact, this entire plane containing vectors $\vec{H}, \vec{\omega}$ and Z-axis rotates about the angular momentum vector $\vec{H}$. Since the components ω_x and ω_y change with time, the vector $\vec{\omega}$ also rotates about Z-axis and describes another cone, which being fixed in the body is called a *body cone*. Also, since the plane containing Z-axis, $\vec{H}$ and $\vec{\omega}$ also rotate with body, the body cone and the space cone will have a common generatrix, which coincides with the vector $\vec{\omega}$. Now, two cases arise:

Case I: If the model of the Earth is an oblate spheroid (flattened at the poles), we have $I_{zz} > I_{xx}$ and Eq. (4.50) gives us $\tan \alpha < \tan \beta$, implying $\alpha < \beta$. This situation corresponds to the flat body and can be visualized from Fig. 4.9(a), where the space cone lies inside the body cone and the body cone rolls over the space cone.

Case II: If we assume that the Earth's model is a prolate spheroid, when $I_{zz} < I_{xx}$ (which is true for any slender body). Equation (4.50) then gives

$$\tan \alpha > \tan \beta$$

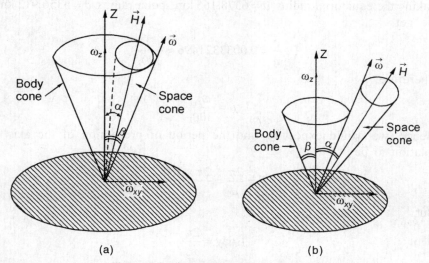

Figure 4.9 Geometrical representation of (a) retrograde precession and (b) direct precession of a freely rotating body.

implying $\alpha > \beta$. In this case, the space cone will be outside the body cone, as shown in Fig. 4.9(b). Of course, in reality, only the first case is applicable as far as the rotational motion of the Earth is concerned.

The results of the present discussion are directly useful to predict the precessional motion of the Earth. We know that the Earth rotates freely about its polar axis and is symmetrical about it and is bulged at the equator. Hence, taking the Earth's Z-axis coincident with its polar axis and also noting $I_{zz} > I_{xx}$, $I_{xx} = I_{yy}$ we give the following interpretation: From observations it is known that for an oblate Earth

$$\frac{I_{zz} - I_{xx}}{I_{xx}} = 0.0033.$$

Hence, the precessional speed

$$|\mu| = \left|\frac{I_{xx} - I_{zz}}{I_{xx}}\right| \omega_z = 0.0033 \omega_z.$$

Taking Earth as an oblate spheroid with dimensions $2a \times 2b \times 2c$, $I_{xx} = M(b^2 + c^2)/5$, $I_{yy} = M(c^2 + a^2)/5$ and $I_{zz} = M(a^2 + b^2)/5$, where M is the mass of the Earth. For oblate spheroid $a = b$. Therefore,

$$\frac{I_{zz} - I_{xx}}{I_{xx}} = \frac{a^2 - c^2}{b^2 + c^2} = \frac{(a-c)(a+c)}{a^2 + c^2}.$$

If c differs only slightly from a, then $a + c = 2a$, $a^2 + c^2 = 2a^2$. Thus,

$$\frac{I_{zz} - I_{xx}}{I_{xx}} = \frac{2a(a-c)}{2a^2} = 1 - \frac{c}{a}.$$

Taking the equatorial radius $a = 6378.165$ km, polar radius $c = 6356.912$ km, we get

$$1 - \frac{c}{a} = 0.0033321496 \approx \frac{1}{300.1}.$$

Therefore,

$$\mu = \frac{\omega_z}{300}.$$

Hence, we would expect to find the period of precession of the axis of rotation to be

$$T = \frac{2\pi}{\mu} = \frac{2\pi \times 300}{\omega_z}.$$

But,

$$1 \text{ day} = \frac{2\pi}{\omega_z}.$$

Therefore,

$$T = 300 \text{ days}.$$

Thus, an observer on the Earth finds that the axis of rotation of the Earth traces out a circle about the North Pole every 300 days. But, it is roughly observed to be 433 days. This discrepancy is due to the fact that the Earth is not perfectly rigid, in fact elastic, and is not force-free, as the presence of the Sun, Moon and the atmosphere around the Earth exert some torque on the Earth, causing precession and nutation of the Earth's polar axis.

To conclude, it is appropriate to add the following comments which may kindle the mind of the readers to think ahead.

Because of the Earth's dynamic climate, winds and atmospheric pressure systems undergo constant change. These fluctuations may affect Earth's rotation about its axis. Earth's rotation may become slightly slow because of stronger winds, increasing the day-length by a fraction of a millisecond. When an additional force acts due to changes in surface winds, the rate of Earth's rotation may change. In view of the law of conservation of angular momentum, small but detectable changes in the Earth's rotation and those in the rotation of the atmosphere are linked.

An example of this principle occurs when a skater pulls his or her arms inwards during a spin, changing the mass distribution to one near the rotation axis, reducing the moment of inertia, thereby increasing the skater's spin to keep the total angular momentum of the system unchanged.

The key to this is that the sum of the angular momentum of the solid Earth together with that of the atmospheric system must remain constant unless an outside force or torque is applied. Thus, if the atmosphere speeds up due to say stronger westerly winds, then the Earth may slow down and thereby the length of day may increase.

The other motions of the atmosphere, such as larger mass in one hemisphere than the other, can also lead to a wobble and the poles of the Earth may thus move following the law of conservation of angular momentum.

4.4 ATTITUDINAL STABILITY OF EARTH SATELLITES

The trajectory of a satellite or a spacecraft flying through space is defined by the motion of its centre of mass and therefore the knowledge of the forces that act on a body in a space environment is necessary to determine its path. For many applications, the Earth-bound satellites are required to maintain a certain prescribed attitude in space during their useful life. For example, a satellite intended for weather observation or communication purposes usually will be of the Earth-pointing type, one of its axes being constantly directed along the local vertical. In such cases, the attitudinal behaviour is consequential to the success of a mission. In the absence of all external force effects, the satellite orientation will remain fixed in inertial space. But, the presence of disturbing forces in a space environment, regardless of how small, changes this ideal situation and may alter the orientation of the spacecraft markedly. In this section, we shall examine the influence of small disturbances upon the attitudinal stability of a satellite spinning about its centre of mass O (see Fig. 4.10). Let OXYZ be the principal coordinate axes fixed with the satellite having the same origin O.

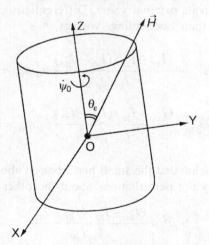

Figure 4.10 Spinning satellite of symmetry displaced slightly from equilibrium.

If no external torques act on the spacecraft, the Euler's equations of motion described by Eqs. (4.6) can be written as

$$\left. \begin{array}{l} I_{xx}\,\dot{\omega}_x = (I_{yy} - I_{zz})\,\omega_y \omega_z \\ I_{yy}\,\dot{\omega}_y = (I_{zz} - I_{xx})\,\omega_z \omega_x \\ I_{zz}\,\dot{\omega}_z = (I_{xx} - I_{yy})\,\omega_x \omega_y \end{array} \right\} \qquad (4.51)$$

If the satellite spins steadily about any one of the three principal axes, say Z-axis, then the angular velocity components relative to the body axes may be written as

$$\omega_z = \dot{\psi}_0 \text{ (constant)}, \quad \omega_x = \omega_y = 0.$$

Under these conditions, all the right-hand terms in Eq. (4.51) vanish. Therefore, we conclude that the attitude of the spin axis will remain invariant in the inertial space. However, let us suppose that the angular rotation of the satellite is perturbed slightly such that

$$\omega_z = \dot{\psi}_0 + \omega_{z\varepsilon}, \quad \omega_x = \omega_{x\varepsilon}, \quad \omega_y = \omega_{y\varepsilon}. \tag{4.52}$$

Neglecting the products of small quantities in the equations of motion (4.51), these equations now become (see Gerlach, 1965)

$$\left.\begin{array}{l} I_{xx}\dot{\omega}_{x\varepsilon} = (I_{yy} - I_{zz})\omega_{y\varepsilon}\dot{\psi}_0 \\ I_{yy}\dot{\omega}_{y\varepsilon} = (I_{zz} - I_{xx})\omega_{x\varepsilon}\dot{\psi}_0 \\ I_{zz}\dot{\omega}_{z\varepsilon} = 0. \end{array}\right\} \tag{4.53}$$

Since $I_{zz} \neq 0$, we see that $\omega_{z\varepsilon}$ = constant. Therefore, the angular velocity about the axis of initial rotation will have no tendency to increase anymore, but will neither return to its original value. Differentiating the first and second of Eqs. (4.53) and then uncoupling, we get

$$\left.\begin{array}{l} \ddot{\omega}_{x\varepsilon} + \dfrac{(I_{xx} - I_{zz})(I_{yy} - I_{zz})}{I_{xx}I_{yy}}\dot{\psi}_0^2\omega_{x\varepsilon} = 0 \\[2mm] \ddot{\omega}_{y\varepsilon} + \dfrac{(I_{xx} - I_{zz})(I_{yy} - I_{zz})}{I_{xx}I_{yy}}\dot{\psi}_0^2\omega_{y\varepsilon} = 0. \end{array}\right\} \tag{4.54}$$

These equations indicate that the small perturbation about Z-axis will remain smaller. But to study the perturbations about the other two axes, we define

$$\mu^2 = \frac{(I_{xx} - I_{zz})(I_{yy} - I_{zz})}{I_{xx}I_{yy}}. \tag{4.55}$$

So that Eqs. (4.54) can be rewritten as

$$\left.\begin{array}{l} \ddot{\omega}_{x\varepsilon} + \mu^2\omega_{x\varepsilon} = 0 \\ \ddot{\omega}_{y\varepsilon} + \mu^2\omega_{y\varepsilon} = 0. \end{array}\right\} \tag{4.56}$$

The motion defined by Eqs. (4.56) is called *simple harmonic* motion. If μ^2 is positive, the differential equations (4.56) possess a periodic solution of type

$$\omega_{x\varepsilon} = A\cos\mu t + B\sin\mu t. \qquad (4.57)$$

To determine the constants of integration A and B, we differentiate Eq. (4.57) with respect to t and get

$$\frac{\dot{\omega}_{x\varepsilon}}{\mu} = -A\sin\mu t + B\cos\mu t.$$

The initial condition $t = 0$ gives

$$A = \omega_{x\varepsilon}(0) \quad \text{and} \quad B = \frac{\dot{\omega}_{x\varepsilon}(0)}{\mu}.$$

Thus, the general periodic solution of Eqs. (4.56) is given by

$$\left.\begin{array}{l} \omega_{x\varepsilon} = \omega_{x\varepsilon}(0)\cos\mu t + \dfrac{[\dot{\omega}_{x\varepsilon}(0)]}{\mu}\sin\omega t \\[2mm] \omega_{y\varepsilon} = \omega_{y\varepsilon}(0)\cos\mu t + \dfrac{[\dot{\omega}_{y\varepsilon}(0)]}{\mu}\sin\omega t \end{array}\right\} \qquad (4.58)$$

which express undamped oscillations. That is, perturbations about X and Y axes will increase with time, but vary harmonically and therefore remain bounded. Thus, we conclude that for $\mu^2 > 0$, the system is stable.

If μ^2 is negative, we see that the resulting motion, from Eq. (4.56), after a small disturbance will have a solution in the form

$$\omega_{x\varepsilon} = Ae^{\mu t} + Be^{-\mu t}, \quad \omega_{y\varepsilon} = Ce^{\mu t} + De^{-\mu t},$$

which consists of two aperiodic parts: a damped converging motion and an undamped diverging motion, which means that the original steady motion is unstable if $\mu^2 < 0$.

Now, the stability condition for $\mu^2 > 0$ is satisfied if:

$$\frac{I_{xx} - I_{zz}}{I_{xx}} \frac{I_{yy} - I_{zz}}{I_{yy}} > 0. \qquad (4.59)$$

Thus, two cases arise:

Case I: If

$$I_{zz} > I_{xx} \quad \text{and} \quad I_{zz} > I_{yy}. \qquad (4.60)$$

In this case, I_{zz} is the greatest moment of inertia.

Case II: If

$$I_{zz} < I_{xx} \quad \text{and} \quad I_{zz} < I_{yy}. \qquad (4.61)$$

In this case, I_{zz} is the smallest moment of inertia. The result of this simple analysis is that for stability the moment of inertia about the axis of initial rotation must be either the smallest or the largest of the three. This conclusion holds good, if the satellite is perfectly rigid.

However, a further investigation reveals that, if the satellite is not perfectly rigid and consequently can dissipate some internal energy, the only stable rotation possible is about the axis of the largest moment of inertia. This fact is illustrated in the example that follows.

Example 4.7 If a satellite is not perfectly rigid and hence can dissipate some kinetic energy internally, show that the only stable rotation possible is about the axis having the largest moment of inertia.

Solution Suppose that the satellite considered is symmetrically spinning about, say, Z-axis and is displaced slightly from its invariant angular momentum vector $\vec{H}$, through a small angle θ_ε (see Fig. 4.10). Recalling the expressions for kinetic energy T and the magnitude of angular momentum $\vec{H}$ of a rotating rigid body, as shown in Example 4.3, we have

$$T = \frac{1}{2}\left(I_{xx}\omega_x^2 + I_{yy}\omega_y^2 + I_{zz}\omega_z^2\right) \tag{1}$$

and

$$|\vec{H}|^2 = \left[(I_{xx}\,\omega_x)^2 + (I_{yy}\,\omega_y)^2 + (I_{zz}\,\omega_z)^2\right] \tag{2}$$

Let us assume that the satellite is symmetrical and rotating about Z-axis, then $I_{xx} = I_{yy}$. Thus the expressions (1) and (2) reduce to

$$T = \frac{1}{2}\left[I_{xx}(\omega_x^2 + \omega_y^2) + I_{zz}\omega_z^2\right] \tag{3}$$

and

$$H^2 = I_{xx}^2(\omega_x^2 + \omega_y^2) + I_{zz}^2\,\omega_z^2. \tag{4}$$

Combining the expressions of T and H, we obtain

$$H^2 - 2I_{xx}T = (I_{zz}^2 - I_{xx}I_{zz})\,\omega_z^2. \tag{5}$$

If the satellite rotates not only about Z-axis but also about X and Y axes, then the Z-axis does not coincide with the angular momentum vector $\vec{H}$. From Fig. 4.10, it follows that

$$I_{zz}\omega_z = H \cos\theta_\varepsilon. \tag{6}$$

Due to the initial disturbance, θ_ε differs from zero. If the original steady motion is stable, θ_ε will return to zero. This means that the condition for stability can be expressed as $\dot\theta_\varepsilon < 0$. Whether or not this condition is satisfied can be verified as follows:

Using Eq. (6) into Eq. (5), it can be written as

$$H^2 - 2I_{xx}T = H^2 \cos^2\theta_\varepsilon\left(1 - \frac{I_{xx}}{I_{zz}}\right) \tag{7}$$

where H is constant, T may decrease due to internal dissipation of energy caused for instance due to deformation of the material of the satellite. Differentiating Eq. (7) with respect to time t, we obtain

$$\frac{dT}{dt} = \dot{T} = H^2 \cos\theta_\varepsilon \sin\theta_\varepsilon \frac{I_{zz} - I_{xx}}{I_{xx}I_{zz}} \dot{\theta}_\varepsilon.$$

As $\dot{T}$ can only be negative, $\dot{\theta}_\varepsilon$ is positive if $I_{zz} < I_{xx}$ and $\dot{\theta}_\varepsilon$ is negative if $I_{zz} > I_{xx}$. For stability, that is, $\dot{\theta}_\varepsilon < 0$ the required condition is

$$I_{zz} > I_{xx}.$$

We, therefore, conclude that the given satellite system is stable and there is a dissipation of kinetic energy if spin occurs about the Z-axis—the axis of symmetry having maximum moment of inertia.

4.5 GYROSCOPE

It is known that the rotation of a body can be induced by the application of an external torque. However, if an external torque is applied to a body that is already spinning and possesses angular momentum, the effect may be to produce a new rotation, instead of increasing its angular velocity. This is the basic principle of a gyroscope. To have a physical insight into gyroscopic action, consider a symmetrical rotor spinning about Z-axis, with angular velocity $\vec{\omega}$ known as *spin velocity*. If we apply two forces $\vec{F}$ to the rotor axle to form a couple $\vec{G}$, whose vector is directed along the x-axis, we observe that the rotor shaft will rotate in the XZ-plane about Y-axis, in the sense indicated with a relatively slow angular velocity $\dot{\psi}$, called *precession velocity*. This situation is portrayed in Fig. 4.11.

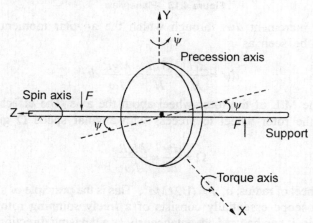

Figure 4.11 Gyroscope.

Thus, we observe that the spin axis, torque axis, and precession axis form a right-handed triad.

We assume that the rotor axle about which it is rotating is sufficiently light, so that the centre of mass is located at R, the centre of the wheel. Gravity acts through the centre of mass of the wheel to produce a torque (see Fig. 4.12)

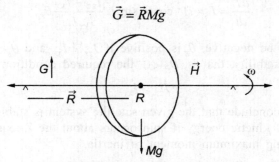

Figure 4.12 Gyroscopic effect.

acting in the direction shown.

In time dt, the torque acts to produce a change in the angular momentum $d\vec{H}$, which is given by

$$\frac{d\vec{H}}{dt} = \vec{G} \quad \text{or} \quad d\vec{H} = \vec{G}\,dt = Mg\,\vec{R}\,dt. \tag{4.62}$$

The direction in which this change occurs is best seen through the plane view described in Fig. 4.13.

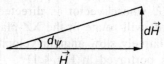

Figure 4.13 Plane view.

The angular increment $d\psi$, through which the angular momentum vector changes can be seen as

$$d\psi = \frac{d\vec{H}}{\vec{H}} = \frac{\vec{G}\,dt}{\vec{H}} = \frac{Mg\vec{R}}{I\omega}\,dt, \tag{4.63}$$

where I is the M.I. of the rotor wheel about the axis. The effect of gravity is to cause the rotor wheel to precess with angular speed Ω given by

$$\Omega = \frac{d\psi}{dt} = \frac{Mg\vec{R}}{I\omega}. \tag{4.64}$$

For a rotor wheel of radius, a, $I = (1/2)Ma^2$. This is the principle of a gyroscope.

The gyroscope essentially consists of a freely spinning rotor about its geometric axis. It can be used advantageously as a devising direction indicating instrument. Gyrocompass is one such instrument. It consists of a high-speed rotor mounted on a platform through two gimbals (see Fig. 4.14). The idea of this arrangement of three mutually orthogonal axes is to isolate the rotor axis from any platform rotation. If the platform is now thought of being fastened to a navigating vehicle such as aircraft or a rocket, the stubborn tendency of the rotor axis to preserve its orientation in inertial space serves as a reliable direction indicator. To be more specific, if the rotor axis of the gyrocompass is set to point south, its orientation remains unchanged in space regardless of any platform rotation. In other words, a rapid rotation imparts stability to a rigid body.

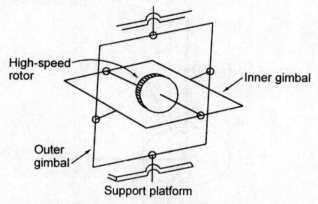

Figure 4.14 Two-gimbal gyro mounted on a platform.

EXERCISES

4.1 Derive the Euler's equations of motion for a rigid body rotating about a fixed point. If the torque $\vec{G} = 0$ and $I_{xx} = I_{yy}$, show that ω is constant.

4.2 Discuss in detail the motion of a symmetrical spinning top and hence obtain general equations of motion of a spinning top in terms of Eulerian angles.

4.3 Describe the precessional and nutational motion of a spinning top.

4.4 Find the condition for a spinning top to move in a steady precession with spin 's' at inclination θ to the vertical.

4.5 An axially symmetric rocket is fired in free space at time $t = 0$. Assume the X, Y, Z axes as principal axes fixed in the body and take Z-axis as the axis of symmetry, as shown in Fig. 4.15. Discuss its torque-free motion.

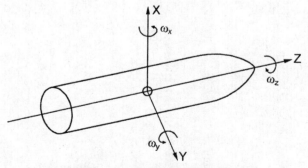

Figure 4.15 Rocket as a body of revolution.

4.6 A rigid wheel has principal M.Is, $I_{xx} = I_{yy} \neq I_{zz}$ about its body principal axes X, Y, Z, as shown in Fig. 4.16. The wheel is attached at its centre of mass to a bearing which allows frictionless rotation about one-space fixed axes. The wheel is dynamically balanced, that is, it can rotate at constant $\omega \neq 0$ and exert no torque on its bearing. What conditions must the components of $\bar{\omega}$ satisfy?

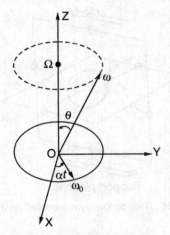

Figure 4.16 Description for Problem 4.6.

4.7 Find the number of degrees of freedom for a rigid body which (a) can move freely in space, (b) has one point fixed, (c) has two points fixed.

4.8 Find the torque needed to rotate a square plate of side $2a$ about its diagonal with constant angular velocity $\bar{\omega}$.

4.9 Find the condition for a sleeping top.

5

Orbital Motion

When you look at the stars and galaxy, you feel that you are not just from any particular piece of land, but from the solar system.

—KALPANA CHAWLA

5.1 KEPLER'S LAWS OF PLANETARY MOTION

Nature has always been a wonder and a source of inspiration for intellectual activity to man. Apart from the physical components of the universe, there are the natural phenomena and a whole variety of rules and mechanisms that are responsible for the entire universe to function, as a simple system. It is a fact that phenomena like natural cycles, planetary movements and a host of others occur with clockwise precision and accuracy. Nothing happens in this universe in an arbitrary manner, but happens according to certain eternal laws, and that is why we are able to formulate principles based on observations that can reflect the natural order. Every aspect of the universe is mathematically describable. For example, the motion of the planets is governed by the following three laws, which are due to Kepler. The first two laws were published in the year 1609, while the third one in the year 1619.

First law (***law of ellipses***) Every planet revolves in an elliptic orbit with the Sun at one of its foci. Mathematically, the equation of the orbit can be written as

$$r = \frac{p}{1 + e \cos \nu}. \qquad (5.1)$$

The second focus called the *empty focus* has significance only in the construction of an ellipse but otherwise has no physical significance. Here, p is the semilatus rectum and e is the eccentricity. It may be noted that, the Indian-born genius Aryabhatta had conceived the idea of elliptical orbits of planets, a thousand years before Kepler.

Second law (***law of areas***) Every planet revolves in its orbit in such a way that the line joining it to the Sun sweeps out equal areas in equal intervals of time. That is,

$$\frac{d\vec{A}}{dt} = \text{constant}. \tag{5.2}$$

To elaborate this point, let us consider that a planet moves in the XY-plane and has polar coordinates, $P(r, \theta)$, at time t as shown in Fig. 5.1.

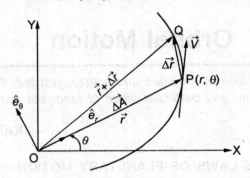

Figure 5.1 Areal velocity.

Let this planet take up the position Q in time $t + \Delta t$. Let $\Delta \vec{A}$ represent vectorially the area of the triangle OPQ swept out by the radius vector in time Δt. Then,

$$\Delta \vec{A} = \frac{1}{2}[\vec{r} \times (\vec{r} + \Delta \vec{r})] = \frac{1}{2}\vec{r} \times \Delta \vec{r}.$$

The rate at which the radius vector joining the fixed origin O to a moving point P sweeps out a surface is called *areal velocity* relative to the fixed origin O. Hence,

$$\frac{\Delta \vec{A}}{\Delta t} = \frac{1}{2}\vec{r} \times \frac{\Delta \vec{r}}{\Delta t}.$$

In the limit as $\Delta t \to 0$, we obtain

$$\frac{d\vec{A}}{dt} = \frac{1}{2}\vec{r} \times \dot{\vec{r}} = \frac{1}{2}\vec{r} \times \vec{V}, \tag{5.3}$$

which is defined as the areal velocity at P. It is a vector perpendicular to the plane containing $\vec{r}$ and $\dot{\vec{r}}$. In Cartesian coordinates

$$\frac{d\vec{A}}{dt} = \frac{1}{2}\vec{r} \times \dot{\vec{r}} = \frac{1}{2}\begin{vmatrix} \hat{i} & \hat{j} & \hat{k} \\ x & y & z \\ \dot{x} & \dot{y} & \dot{z} \end{vmatrix}.$$

Therefore,

$$\dot{A}_x = \frac{y\dot{z} - \dot{y}z}{2}, \quad \dot{A}_y = \frac{z\dot{x} - \dot{z}x}{2}, \quad \dot{A}_z = \frac{x\dot{y} - \dot{x}y}{2}. \tag{5.4}$$

In the radial and transverse coordinate system, the areal velocity can be expressed as

$$\frac{d\vec{A}}{dt} = \frac{1}{2}\vec{r}\times\dot{\vec{r}} = \frac{1}{2}\vec{r}\times\vec{V} = \frac{1}{2}r\hat{e}_r \times (\dot{r}\hat{e}_r + r\dot{\theta}\hat{e}_\theta),$$

that is,

$$\frac{d\vec{A}}{dt} = \frac{1}{2}r^2\dot{\theta}\hat{e}_A = \text{a constant}, \tag{5.5}$$

where $\hat{e}_A$ is a unit vector perpendicular to both $\vec{r}$ and $\vec{V}$, that is, perpendicular to the instantaneous plane defined by these vectors. Therefore, the magnitude of the areal velocity is $(1/2)r^2\dot{\theta}$, and hence

$$\frac{d}{dt}(r^2\dot{\theta}) = 2r\dot{r}\dot{\theta} + r^2\ddot{\theta} = 0. \tag{5.6}$$

But, by Eq. (1.23), $2\dot{r}\dot{\theta} + r\ddot{\theta}$ is the acceleration component perpendicular to the radius vector, which vanishes in view of Eq. (5.6). Hence, there is no acceleration and, therefore, no force perpendicular to r and the line of action of the entire force passes through the origin.

Third law (harmonic law) The squares of the periods of any two planets are in the same proportion as the cubes of their semi-major axes.

5.2 NEWTON'S LAW OF GRAVITATION

If two particles of masses M and m are situated at a distance, r, apart, each particle then attracts the other with a force $\vec{F}$ given by

$$F = G\frac{Mm}{r^2} \text{ in magnitude}, \tag{5.7}$$

where G is a universal gravitational constant, which is known to be $6.668 \times 10^{-11}\,\text{N-m}^2/\text{kg}^2$. Some authors use k^2 for G.

Example 5.1 Find the value of GM and its dimensions in Newton's universal gravitational law, when the attracting body is the Earth.

Solution Let the attracting body be the Earth of mass M and let the attracted body be of mass m which may lie on the surface of the Earth. Then the magnitude of force $F = mg$. Taking the radius of the Earth,

$$r = r_e = 6378 \text{ km},$$

we have from Eq. (5.7),

$$mg = G\frac{Mm}{r_e^2}.$$

Therefore,

$$GM = gr_e^2 = (9.806)(6378 \times 1000)^2 = 3.979 \times 10^{14} \text{ m}^3/\text{s}^2.$$

To get the dimensions of GM, we write the dimensions of other terms in the SI system of Eq. (5.7) as

$$\frac{ML}{T^2} = \frac{(GM)M}{L^2}.$$

Therefore, the dimensions of $GM = L^3/T^2$. Here, GM is called the intensity of the force.

5.3 SOME DEFINITIONS FROM SPHERICAL ASTRONOMY

Celestial sphere If we look at the sky, during nights, the stars, the planets and all other celestial objects appear projected on the interior of a hemispherical dome surrounding us. This imaginary sphere of an arbitrarily large radius on which celestial objects appear projected is called the *celestial sphere*. Normally, we take the centre of the Earth as the centre of the celestial sphere. At times, it may be convenient to have the observer or the Sun as the centre of the celestial sphere. The radius of the celestial sphere can be taken to be so large that the Earth, whose equatorial radius is about 6378 km, will be almost a point at the centre of the celestial sphere.

Solar system Our solar system consists of the Sun and the various planets that revolve around the sun in different elliptic orbits. Mercury, Venus, Earth, Mars, Jupiter, Saturn, Uranus, Neptune and Pluto are the nine known planets respectively in the order of increasing distance from the Sun.

Satellite A satellite is a body revolving around a planet. It is normally of negligible mass when compared with the parent planet. Our artificial satellites such as GSAT, INSAT are of the order of 2000 to 4000 kg. Of course, there is an exception to this rule. The moon, the natural satellite of the Earth, has a mass of about 1/80 of the Earth's mass (mass of the Earth $= 5.98 \times 10^{24}$ kg).

Great circle and small circle Let us imagine a celestial sphere on which the celestial objects like stars, the Sun, the planets and the moon appear to move. The plane section of this sphere passing through its centre is called a *great circle*. All other sections are *small circles*. In Fig. 5.2, X is a great circle, X' is a small circle.

It can be seen that the radius of the great circle R is the radius of the sphere itself. Let AA' be the diameter of the sphere perpendicular to the planes X and X'. The extremities A and A' are called the *poles* of the circles X and X'. Suppose B and C are any two points on the great circle X, then the arc length BC $= R\theta$, where θ is the angle BOC. Since R is constant, the arcs of the great circles on the same sphere are proportional to the angles they subtend at the centre O of the sphere and are measured by these angles.

A great circle passing through the poles of a circle is called a *secondary* to the circle. Thus, great circles ABA', ACA' are secondaries to X, while X is called the *primary*. It may be observed that the planes of these secondaries are all perpendicular to the primary plane X. That is, the angles AOB, AOC are right angles. The arcs of great circles like AB, AC which subtend a right angle at the centre O are called *quadrants*. It can also be observed in Fig. 5.2, that, if two great circles are secondaries to another great circle, the angle between their planes is equal to the arc they intercept on the primary.

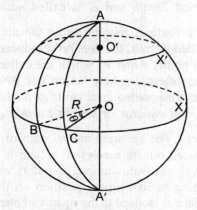

Figure 5.2 Great circle.

Perigee and apogee If a body (satellite) is revolving round another body (say Earth) in an elliptic orbit, the least and greatest separation positions are called *perigee* and *apogee*. For a planet revolving round the Sun, the corresponding positions are called *perihelion* and *aphelion*.

Celestial horizon Let A and B be two places at the ends of the diameter of the Earth. Let us draw tangent planes at A and B on the Earth, as depicted in Fig. 5.3. These planes cut the celestial sphere along the circles NS_0 and

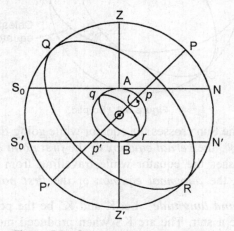

Figure 5.3 Celestial horizon and equator.

$N'S_0'$. In fact, the stars which lie in the upper half of the sphere above NS_0 will be visible at A but not at B. Similarly, those in the lower half are visible at B but not at A. However, the diameter of the Earth AB is very small when compared with that of the celestial sphere and hence the circles NS_0 and $N'S_0'$ can be treated as almost great circles and coincide. This great circle NS_0 which is the section of the celestial sphere by a tangent plane at A to the Earth is called *celestial horizon* of the place A. Suppose ZOZ' be the diameter of the celestial sphere, then Z and Z' are the poles of the celestial horizon. Here Z is called *Zenith* and Z' is called *nadir* of the place.

Celestial equator The Earth is rotating about the diameter pp', as shown in Fig. 5.3, which is called its *axis*, the line pp' produced both ways meets the celestial sphere at two points P and P' which are called *celestial poles*. The great circle QR on the celestial sphere which has PP' as its axis is called *celestial equator*. The corresponding great circle qr on the Earth having pp' as its axis is called *terrestial equator* or simply *Earth's equator*.

Ecliptic and equinoxes The apparent annual path of the Sun with respect to fixed stars when traced out on a celestial sphere is a great circle, called *ecliptic*. The Sun moves eastwards among the stars at a rate of about one degree per day, returning to its original position on the celestial sphere in one year. The ecliptic plane is inclined to the equatorial plane at an angle ε called *obliquity* of the ecliptic (see Fig. 5.4) which at present is $23°27'$. The ecliptic plane cuts the equator at two points, as shown in Fig. 5.4. They are called

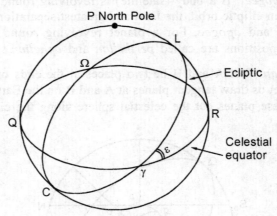

Figure 5.4 Ecliptic.

equinoxes. When the Sun crosses the equator while going from south to north, it passes through 'γ', the *vernal equinox* or the *first point of aries*. Similarly, when the Sun crosses the equator while travelling from north to south, it passes through Ω, the *autumnal equinox* or the *first point of libra*.

Celestial latitude and longitude Let K and K' be the poles of the ecliptic CL. Suppose, S be a star. The arc KS when produced meets CL at M (see Fig. 5.5). The distance SM of the star from the ecliptic is called *celestial latitude* denoted by the symbol β, which is measured positive north of the ecliptic and negative south of the ecliptic. The distance γM measured eastward from the vernal equinox is called *celestial longitude* denoted by the symbol λ.

Some Definitions from Spherical Astronomy (Sec. 5.3)

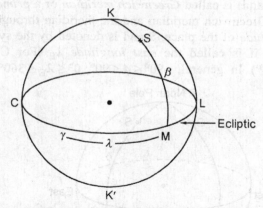

Figure 5.5 Celestial latitude and longitude.

It may be noted that λ and β are measured with respect to vernal equinox and hence do not vary from place to place and also with time.

Right ascension and declination Let γ be the vernal equinox. Suppose S is a star. Let PS produced meet the celestial equator at M, where P is its North Pole. Then, the distance γM measured eastward from γ is called *right ascension* of the star (refer to Fig. 5.6) denoted by the symbol, α. Right ascension is usually measured in units of time, such as 24 hours equals 360°. The distance SM is called the *declination* of the star. It is measured positive north of the equator and negative south of the equator.

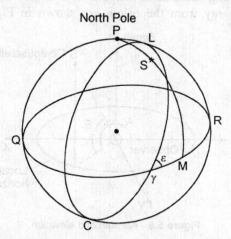

Figure 5.6 Right ascension and declination.

Latitude and longitude of a place Let qr be the Earth's equator and let P be the North Pole, S be a place on the surface of the Earth. Let PS produced meet the equator at M (see Fig. 5.7). The latitude of the place, S, is the angular distance SM and is denoted by the symbol ϕ. A great circle passing through the poles is called *meridian*. A meridian that passes through Greenwich (a place in England) is called *Greenwich meridian* or a *prime meridian*. The angle between Greenwich meridian and the meridian through the place S is called the *longitude* of the place, S and is denoted by the symbol λ. If it is measured east, it is called the *east longitude* λ_E (For Chennai/Madras, $\lambda_E = 80°$, $\phi = 14°$). In general, $-90° \leq \phi \leq 90°$, $0° \leq \lambda_E \leq 360°$.

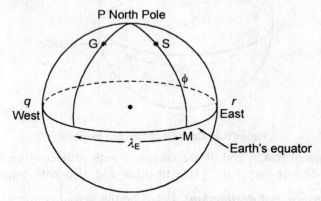

Figure 5.7 Latitude and longitude of a place.

Azimuth and elevation An observer standing on a particular meridian on the surface of the rotating Earth, sees all objects in a rotating coordinate system. In this system, the observer is at the origin of the coordinate system and the fundamental plane is the local horizon. Here, the principal axis is taken pointing towards south. The two angles needed to locate an object along some ray from the origin are shown in Fig. 5.8.

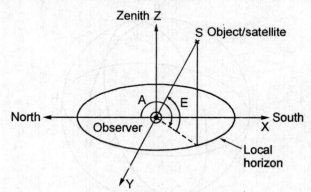

Figure 5.8 Azimuth and elevation.

Elevation (E) The angle measured in the vertical plane between the horizon to the line joining the observer and the object is called the *elevation angle* ($-90° \leq E \leq 90°$).

Azimuth (A) The angle from the north to the line of intersection of vertical plane with the horizon is called the *azimuth*. We shall take azimuth as being measured clockwise from north looking down from zenith ($0° \leq A \leq 360°$).

5.4 CENTRAL FORCE MOTION

When a mass particle possesses an accelerated motion and if the resultant force causing this accelerated motion always passes through a fixed point, such a motion is called *central force* motion. The fixed point is called the *centre of force*. Consider a real-life example: all planets revolve round the Sun in their orbits such that the force of attraction due to the Sun always passes through it. This idea plays an important role in spacecraft dynamics. Therefore, we shall study in this section various properties of a central force motion, which are independent of the form of the force law.

Consider a particle of mass m, whose position vector is $\vec{r}$, relative to a fixed point O. Let the particle trace the curve C under the action of the central force $\vec{F}$, which may be directed towards or away from the origin, O, as shown in Fig. 5.9. Then, $\vec{F} = F\hat{e}_r$. For constant mass, the Newton's second law states

$$F\hat{e}_r = m\dot{\vec{V}},$$

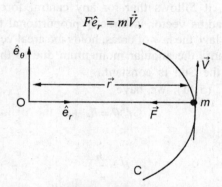

Figure 5.9 Curvilinear central force motion.

and therefore,

$$\vec{r} \times F\hat{e}_r = \vec{r} \times m\dot{\vec{V}}.$$

In view of the collinearity of $\vec{r}$ and $\hat{e}_r$, we get at once

$$\vec{r} \times m\dot{\vec{V}} = 0. \tag{5.8}$$

From Kepler's law, the areal velocity, as defined in Eq. (5.3), is given by

$$\dot{\vec{A}} = \frac{1}{2}\vec{r} \times \vec{V}. \tag{5.9}$$

Differentiating this equation with respect to t, and using Eq. (5.8), we obtain

$$\frac{d\dot{\vec{A}}}{dt} = \frac{1}{2}(\dot{\vec{r}} \times \vec{V} + \vec{r} \times \dot{\vec{V}}) = \frac{1}{2}\vec{r} \times \dot{\vec{V}} = 0,$$

which means that the areal velocity $\dot{\vec{A}}$ is constant. But,

$$\dot{\vec{A}} = \frac{1}{2}\vec{r} \times \vec{V} = \frac{1}{2}r\hat{e}_r \times (\dot{r}\hat{e}_r + r\dot{\theta}\hat{e}_\theta).$$

That is,

$$\dot{\vec{A}} = \frac{1}{2}r^2\dot{\theta}\hat{e}_A = \frac{1}{2}\vec{h}, \qquad (5.10)$$

where $\vec{h}$ is the angular momentum per unit mass and $\hat{e}_A = \hat{e}_r \times \hat{e}_\theta$ is a unit vector in the direction of $\vec{r} \times \vec{V}$, that is, in the direction of the normal to the instantaneous plane containing $\vec{r}$ and $\vec{V}$. Hence, the magnitude of the areal velocity is

$$\dot{A} = \frac{1}{2}r^2\dot{\theta} = \frac{1}{2}h. \qquad (5.11)$$

Therefore,

$$A = \frac{1}{2}ht + \text{constant}. \qquad (5.12)$$

From this analysis, it follows that for any central force motion, the area swept out by the radius vector is directly proportional to the time. Hence, the Kepler's second law, the law of areas, holds the areal velocity of each planet constant. Equivalently, the angular momentum due to the orbital motion of each planet about the Sun is constant.

Also, from Eq. (5.11), we have

$$r^2\dot{\theta} = h.$$

In other words,

$$\dot{\theta} = \frac{h}{r^2}. \qquad (5.13)$$

This result explains the fact that the angular speed of the particle moving under the action of a central force, varies inversely as the square of the distance from the origin to the particle.

5.4.1 Integrals of Energy

We have from Newton's second law, the equation of motion for a particle of mass m is given by

$$m\dot{\vec{V}} = \vec{F} = F\hat{e}_r, \qquad (5.14)$$

where F is the magnitude of the central force and $\hat{e}_r$ is the unit vector in the radial direction.

Now, taking the scalar product of Eq. (5.14) with $\vec{V}$, we get

$$m\dot{\vec{V}} \cdot \vec{V} = F\vec{V} \cdot \hat{e}_r. \qquad (5.15)$$

But,

$$\vec{V} \cdot \dot{\vec{V}} = \frac{d}{dt}\left(\frac{1}{2}\vec{V} \cdot \vec{V}\right) = \frac{d}{dt}\left(\frac{1}{2}v^2\right),$$

and
$$\vec{V} \cdot \hat{e}_r = (\dot{r}\hat{e}_r + r\dot{\theta}\hat{e}_\theta) \cdot \hat{e}_r = \dot{r}.$$
Therefore, Eq. (5.15) becomes
$$\frac{d}{dt}\left(\frac{1}{2}mv^2\right) = F\frac{dr}{dt}. \qquad (5.16)$$

Suppose F depends only on the length of the radius vector, that is, if $F = F(r)$. Then, upon integration of Eq. (5.16) with respect to time t, we get
$$\frac{1}{2}mv^2 = \int F(r)\,dr + E, \qquad (5.17)$$
where E is the constant of integration, which depends upon the initial conditions of the motion. The integral in Eq. (5.17) is the work done by the force F in changing the position of the particle along the orbit. If the analytic form of $F(r)$ is known, the above integral can be evaluated.

Let us consider a conservative force field, that is, $\vec{F} = -\nabla U$, which means that there exists a scalar function $U(r)$ such that
$$F(r) = -\frac{\partial U}{\partial r}. \qquad (5.18)$$
Then, Eq. (5.17) becomes
$$\frac{1}{2}mv^2 + U(r) = E. \qquad (5.19)$$
This equation states that the sum of the kinetic energy and the potential energy of a moving particle under the central force field is constant. This is the law of conservation of energy. Solving Eq. (5.19) for v, we get
$$v = \pm\left[\frac{2}{m}(E - U)\right]^{1/2}. \qquad (5.20)$$
In polar coordinates, Eq. (5.19) can also be rewritten as
$$\frac{1}{2}m(\dot{r}^2 + r^2\dot{\theta}^2) + U(r) = E. \qquad (5.21)$$

5.4.2 Differential Equation of the Orbit

Under the central force motion, both the angular momentum and the total energy of the system are constants and we have
$$r^2\dot{\theta} = h \qquad (5.22)$$
and
$$\frac{1}{2}(\dot{r}^2 + r^2\dot{\theta}^2) + U(r) = E. \qquad (5.23)$$

It can also be noted that
$$\frac{dr}{d\theta} = \frac{\dot{r}}{\dot{\theta}}.$$

Now, let
$$u = \frac{1}{r} \qquad (5.24)$$

then,
$$\frac{du}{d\theta} = -\frac{1}{r^2}\frac{dr}{d\theta} = -\frac{1}{r^2}\frac{dr}{dt}\frac{dt}{d\theta} = -\frac{\dot{r}}{h}$$

and
$$\frac{d^2u}{d\theta^2} = -\frac{d}{d\theta}\left(\frac{\dot{r}}{h}\right)$$
$$= -\frac{d}{dt}\left(\frac{\dot{r}}{h}\right)\frac{dt}{d\theta}$$
$$= -\frac{dt}{d\theta}\frac{d}{dt}\left(\frac{\dot{r}}{h}\right)$$
$$= -\frac{1}{h^2u^2}\frac{d^2r}{dt^2} \qquad \text{[using Eq. (5.22)]}.$$

Noting that the central force $F(r)$ has only radial component, we have from Newton's second law
$$m(\ddot{r} - r\dot{\theta}^2) = F(r).$$

Using Eqs. (5.22) and (5.24), we can rewrite the above equation as
$$m(\ddot{r} - h^2u^3) = F(u) = -m\left(h^2u^2\frac{d^2u}{d\theta^2} + h^2u^3\right).$$

That is,
$$\frac{d^2u}{d\theta^2} + u = -\frac{1}{mh^2u^2}F(u). \qquad (5.25)$$

This is a second order differential equation in u as a function of θ and is known as the polar equation of the orbit. If a proper force law is specified, then the appropriate equation of the orbit can be obtained, for example, if the force is one which varies as an integral power n of the distance. That is, if
$$F(r) = \alpha r^n \qquad (\alpha \text{ is constant}) \qquad (5.26)$$

then

$$F(u) = \alpha u^{-n}$$

and Eq. (5.25) reduces to

$$\frac{d^2 u}{d\theta^2} + u = -\frac{\alpha u^{-(n+2)}}{mh^2}. \tag{5.27}$$

This equation can be directly integrated by multiplying both sides of it with $2(du/d\theta)$. Thus, we have

$$2\frac{du}{d\theta}\frac{d^2 u}{d\theta^2} + 2u\frac{du}{d\theta} = -\left(\frac{\alpha}{mh^2}\right) u^{-(n+2)} \left(2\frac{du}{d\theta}\right)$$

which can be recast as

$$\frac{d}{d\theta}\left[\left(\frac{du}{d\theta}\right)^2 + u^2\right] = -\beta u^{-(n+2)} \frac{du}{d\theta}, \tag{5.28}$$

where

$$\beta = \frac{2\alpha}{mh^2}. \tag{5.29}$$

Now, integrating Eq. (5.28) with respect to θ yields

$$\left(\frac{du}{d\theta}\right)^2 + u^2 = \beta \frac{u^{-(n+1)}}{n+1} + c \qquad (n \neq -1) \tag{5.30}$$

or

$$\frac{du}{d\theta} = \left[c - u^2 + \beta \frac{u^{-(n+1)}}{n+1}\right]^{1/2}.$$

Integrating once again with respect to θ, we get

$$\int_{u_0}^{u} \frac{du}{\left[c - u^2 + \beta u^{-(n+1)}/(n+1)\right]^{1/2}} = \theta - \theta_0, \tag{5.31}$$

where the coordinates (u_0, θ_0) define an initial starting point in the orbit.

5.4.3 The Inverse Square Force

In almost all astronomical applications, the orbits of planets arise from the inverse square force, as postulated by Newton. That is,

$$F(r) = -\frac{GMm}{r^2},$$

where GM is the gravitational coefficient and m is the mass being accelerated. The equation of motion of the orbit (5.25) then becomes

$$\frac{d^2u}{d\theta^2} + u = \frac{GM}{h^2}. \tag{5.32}$$

The general solution of this differential equation is found to be

$$u = \frac{GM}{h^2} + A\cos(\theta - \theta_0), \tag{5.33}$$

where θ_0 and A are constants of integration. Thus,

$$r = \frac{h^2/(GM)}{1 + Ah^2\cos(\theta - \theta_0)/(GM)},$$

which is of the form

$$r = \frac{p}{1 + e\cos(\theta - \theta_0)}, \tag{5.34}$$

a standard equation of a conic, p is the semi-latus rectum, where

$$p = \frac{h^2}{GM}, \qquad e = \frac{Ah^2}{GM}. \tag{5.35}$$

Here, e is the eccentricity of the conic, p/e is the distance of the focus from the directrix. The parameters e and p determine the shape of the conic. From analytical geometry, we know that if

$e < 1$,   the conic is ellipse

$e = 1$,   the conic is parabola

$e > 1$,   the conic is hyperbola

This geometrical parameter obviously depends upon the constant of integration A and the physical constants h, G and M of the system.

Now, multiplying Eq. (5.32) on both sides by $2(du/d\theta)$ and integrating, we get

$$\left(\frac{du}{d\theta}\right)^2 + u^2 = \frac{2GM}{h^2}u + c. \tag{5.36}$$

This equation has a simple physical meaning, seen as follows: If we recall the velocity components in an orbit as

$$\dot{r} = -h\frac{du}{d\theta} \quad \text{and} \quad r\dot{\theta} = hu,$$

then,
$$\vec{V} = \dot{r}\hat{e}_r + r\dot{\theta}\hat{e}_\theta = -h\frac{du}{d\theta}\hat{e}_r + hu\hat{e}_\theta.$$

Hence, the speed of the particle in its obrit is given by

$$v = \left[\left(h\frac{du}{d\theta}\right)^2 + h^2 u^2\right]^{1/2}. \tag{5.37}$$

Also, since
$$F = -\frac{GMm}{r^2} = -\frac{\partial U}{\partial r}$$

Therefore, the potential energy

$$U = -\int F\,dr = GMm\int \frac{dr}{r^2} = -\frac{GMm}{r} = -GMmu.$$

Thus, the total energy of the system is given by

$$\frac{1}{2}mv^2 - GMmu = E. \tag{5.38}$$

Using Eq. (5.37), we get

$$\frac{mh^2}{2}\left[\left(\frac{du}{d\theta}\right)^2 + u^2\right] - GMmu = E$$

or

$$\left(\frac{du}{d\theta}\right)^2 + u^2 = \frac{2GMu}{h^2} + \frac{2E}{mh^2}. \tag{5.39}$$

Now, comparing Eqs. (5.36) and (5.39), we find that

$$c = \frac{2E}{mh^2} \tag{5.40}$$

and Eq. (5.36) represents the total energy of the system, which remains constant.

Noting that at the ends of the transverse axis of the conic, that is, at apses

$$\dot{r} = -h\frac{du}{d\theta} = 0.$$

Equation (5.39) takes the form

$$u^2 - 2\frac{GM}{h^2}u - \frac{2E}{mh^2} = 0, \tag{5.41}$$

which is called an *apsidal quadratic* in u and has two roots given by

$$u_{1,2} = \frac{GM}{h^2} \pm \left(\frac{G^2M^2}{h^4} + \frac{2E}{mh^2}\right)^{1/2}$$

or

$$u = \frac{GM}{h^2}\left[1 \pm \left(1 + \frac{2Eh^2}{mG^2M^2}\right)^{1/2}\right]. \tag{5.42}$$

The values of u, thus obtained correspond to those at the ends of the transverse axis of the conic (see Fig. 5.10).

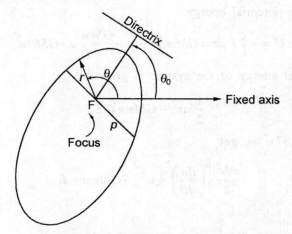

Figure 5.10 Parameters for an elliptic orbit.

Now, recalling Eq. (5.33) and noting that, when $\theta = \theta_0$, $r = r_{min} = u_{max}$ and when $\theta = \pi + \theta_0$, $r = r_{max} = u_{min}$, we have

$$u_{max} = \frac{GM}{h^2}\left[1 + \left(1 + \frac{2Eh^2}{mG^2M^2}\right)^{1/2}\right] = \frac{GM}{h^2} + A$$

and

$$u_{min} = \frac{GM}{h^2}\left[1 - \left(1 + \frac{2Eh^2}{mG^2M^2}\right)^{1/2}\right] = \frac{GM}{h^2} - A,$$

where

$$A = \frac{GM}{h^2}\left(1 + \frac{2Eh^2}{mG^2M^2}\right)^{1/2}$$

But

$$e = \frac{Ah^2}{GM}.$$

Thus, we arrive at a fundamental relation between the eccentricity and the total energy of a particle as

$$e = \left(1 + \frac{2Eh^2}{mG^2M^2}\right)^{1/2}. \tag{5.43}$$

Evidently, we can also classify the orbit in terms of the total energy E of the system as follows:

If $E = 0$, $e = 1$,   the orbit is a parabola

If $E < 0$, $e < 1$,   the orbit is an ellipse

If $E > 0$, $e > 1$,   the orbit is a hyperbola.

5.4.4 Geometry of an Orbit

Parabolic orbit

When $E = 0$ or $e = 1$, the orbit is a parabola. A parabolic orbit rarely occurs in nature, although the orbits of many comets are nearly of the parabolic type. It may be noted that an object or a particle travelling along a parabolic path is on a one-way trip to infinity. There are a few geometrical properties peculiar to a parabola (see Fig. 5.11).

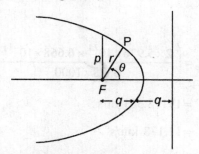

Figure 5.11 Geometry of the parabola.

Firstly, the two arms of the parabola become more and more nearly parallel as we extend them further and further.

Secondly, since its eccentricity is exactly 1,

$$q = \frac{p}{2}, \tag{5.44}$$

where q is the distance from the focus to the vertex and p is the semi-latus rectum. The equation of the parabolic orbit is then given by

$$r = \frac{2q}{1+\cos(\theta-\theta_0)} = \frac{p}{1+\cos(\theta-\theta_0)}. \qquad (5.45)$$

It is known that the strength of the gravitational field of the Sun or a planet decreases rapidly with distance even though it extends to infinity. A finite amount of kinetic energy is sufficient to overcome the effects of gravity and permit an object to coast to an infinite distance without falling back. The speed which is just sufficient to do so is called the *escape speed*. A space probe which is given an escape speed in any direction obviously travels on a parabolic escape trajectory. We can easily compute the escape speed with which a space probe (say a rocket) should be projected if it is to escape the gravitational attraction of the Earth, using the energy Eq. (5.38). That is, per unit mass

$$E = \frac{v^2}{2} - \frac{GM}{r} = 0,$$

which gives

$$v_{escape} = \left(\frac{2GM}{r}\right)^{1/2}. \qquad (5.46)$$

For example, we know that

Mass of the Earth, $M = 5.976 \times 10^{24}$ kg

$G = 6.668 \times 10^{-11}$ m^3/s^2.kg

Radius of the Earth, $r \approx 6378 \times 1000$ m

Using this data,

$$v_{escape} = \left(\frac{2 \times 5.976 \times 10^{24} \times 6.668 \times 10^{-11}}{6378 \times 1000}\right)^{1/2}$$

$$= 11178.301 \text{ m/s}$$

$$= 11.178 \text{ km/s}.$$

This is the minimum velocity with which a space probe should be projected, if it is to escape the gravitational attraction of the Earth.

Elliptic orbit

When $E < 0$ or $e < 1$, the orbit is an ellipse. The orbits of all planets in the solar system as well as the orbits of all Earth-bound satellites are ellipses. Let q denote the radius vector of an object at minimum distance from the focus (perigee) and let q' be the radius vector of it at maximum distance from the focus F (apogee). Let the major axis of the ellipse be $2a$, then (see Fig. 5.12)

Also, from Eq. (5.34)
$$q + q' = 2a.$$

$$r = \frac{p}{1 + e \cos(\theta - \theta_0)}.$$

When $r = q$, $\theta = \theta_0$, we get $q = p/(1+e)$. Similarly when $r = q'$, $\theta = \pi + \theta_0$, we get $q' = p/(1-e)$. Combining the above relations, we arrive at

$$\frac{p}{1+e} + \frac{p}{1-e} = 2a.$$

Therefore, $p = a(1 - e^2)$ and the equation of an ellipse is

$$r = \frac{a(1 - e^2)}{1 + e \cos(\theta - \theta_0)}. \tag{5.47}$$

Since $p = h^2/(GM)$, the angular momentum

$$h = [GMa(1 - e^2)]^{1/2} = \text{constant}. \tag{5.48}$$

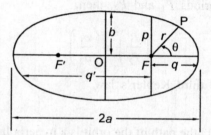

Figure 5.12 Geometry of the ellipse.

In this case, the total energy E of the system can be obtained from Eq. (5.43) after using the result given by Eq. (5.48). Thus,

$$e^2 = 1 + \frac{2Eh^2}{mG^2M^2} = 1 + \frac{2EGMa(1-e^2)}{mG^2M^2}$$

which gives

$$E = -\frac{GMm}{2a}. \tag{5.49}$$

Equating with the total energy E given by the expression (5.38), we get

$$\frac{1}{2}mv^2 - \frac{GMm}{r} = E = -\frac{GMm}{2a},$$

that is,

$$v^2 = GM\left(\frac{2}{r} - \frac{1}{a}\right). \tag{5.50}$$

The time for an object (say satellite) to go once round its orbit is called *period*. To compute the period in an elliptic motion, let A denote the area swept out by the radius vector in time t, then we have

$$A = \frac{1}{2}ht = \frac{1}{2}[GMa(1-e^2)]^{1/2} t.$$

Note that the semi-minor axis b is related to the semi-major axis a by the equation

$$b = a(1-e^2)^{1/2}. \tag{5.51}$$

In a single revolution or in one period P, the area swept is given by

$$\pi ab = \pi a^2 (1-e^2)^{1/2} = \frac{1}{2}[GMa(1-e^2)]^{1/2} P,$$

which gives

$$P = \frac{2\pi}{\sqrt{GM}} a^{3/2}. \tag{5.52}$$

Evidently, if two planets are moving in their orbits with semi-major axes a_1 and a_2 and with periods P_1 and P_2, then,

$$\left(\frac{P_1}{P_2}\right)^2 = \left(\frac{a_1}{a_2}\right)^3, \tag{5.53}$$

which confirms the third Kepler's law.

Hyperbolic orbit

When $E > 0$ or $e > 1$, the path of the orbit is a hyperbola. If $2a$ represents the length of the transverse axis of the hyperbola, then its semi-latus rectum $p = a(e^2 - 1)$ and the equation of hyperbola is

$$r = \frac{a(e^2-1)}{1+e\cos(\theta-\theta_0)}. \tag{5.54}$$

Noting that $p = h^2/(GM)$, the angular momentum is given by

$$h = [GMa(e^2-1)]^{1/2}. \tag{5.55}$$

The total energy is given by

$$E = \frac{GMm}{2a}. \tag{5.56}$$

The speed of the planet at a distance r from the focus is given as

$$v^2 = GM\left(\frac{2}{r} + \frac{1}{a}\right). \tag{5.57}$$

Here follows several examples for illustration.

Example 5.2 A particle of mass m is projected from the point $(r_0, 0)$ with a velocity of magnitude v_0, perpendicular to the line $\theta = 0$. If the particle moves under the inverse cube law, such that it experiences a force directed towards the origin and of magnitude λr^{-3} per unit mass, where λ is a constant and r is the radial distance, find the orbit along which it moves.

Solution The differential equation describing the orbit is given by

$$\frac{d^2 u}{d\theta^2} + u = -\frac{F(u)}{h^2 u^2}. \tag{1}$$

From the given data,

$$F(u) = \frac{-\lambda}{r^3} = -\lambda u^3.$$

Substituting $F(u)$ into Eq. (1), we get

$$\frac{d^2 u}{d\theta^2} + \left(1 - \frac{\lambda}{h^2}\right) u = 0. \tag{2}$$

Clearly, its solution depends on the sign of $(1 - \lambda/h^2)$ and the initial conditions. Given that

$$v = \dot{r}\hat{e}_r + r\dot{\theta}\hat{e}_\theta = v_0 \hat{e}_\theta.$$

we have

$$\dot{r} = 0 \quad \text{and} \quad v_0 = r_0 \dot{\theta},$$

implying

$$\dot{\theta} = \frac{v_0}{r_0}.$$

But

$$h = r_0^2 \dot{\theta} = v_0 r_0.$$

Thus, three cases arise.

Case I: If $v_0 r_0 > \lambda^{1/2}$ or $h^2 > \lambda$, then $(1 - \lambda/h^2) > 0$. Now, let $\alpha^2 = (1 - \lambda/h^2)$, then the solution of Eq. (2) is given by

$$u = A \sin(\alpha\theta + \beta), \tag{3}$$

where A and β are constants to be determined.

Now, the initial conditions state: When $\theta = 0$, (i) $u = 1/r_0$; (ii) $\dot{r} = 0 = -h(du/d\theta)$. Therefore,

$$\frac{du}{d\theta} = 0. \tag{4}$$

We shall use these initial conditions to determine the constants A and β in Eq. (3). Thus, the condition $u = 1/r_0$ when $\theta = 0$ gives

$$\frac{1}{r_0} = A \sin \beta. \tag{5}$$

Using the second condition, $du/d\theta = 0$ when $\theta = 0$, into Eq. (3), we get

$$\frac{du}{d\theta} = A\alpha \cos(\alpha\theta + \beta)\Big|_{\theta=0} = A\alpha \cos \beta = 0$$

implying

$$\beta = \frac{\pi}{2}. \tag{6}$$

Substituting (6) into (5), we obtain

$$A = \frac{1}{r_0}.$$

Hence, the required orbit is given by

$$u = \frac{\sin(\alpha\theta + \pi/2)}{r_0} = \frac{\cos \alpha\theta}{r_0}.$$

That is,

$$r_0 = r \cos \alpha\theta = r \cos\left[\left(1 - \frac{\lambda}{r^2}\right)^{1/2} \theta\right]. \tag{7}$$

Case II: If $\lambda^2 = h$ or $\alpha = 0$, then the differential equation of the orbit (Eq. 2) is given by

$$\frac{d^2u}{d\theta^2} = 0. \tag{8}$$

Integrating twice, its solution can be written as

$$u = A\theta + B.$$

Using the initial condition, $u = 1/r_0$ when $\theta = 0$, we get $B = 1/r_0$. Also, using the initial condition, $du/d\theta = 0$, when $\theta = 0$, we have $A = 0$. Hence, the required orbit is given by

$$u = B \quad \text{or} \quad \frac{1}{r} = \frac{1}{r_0},$$

that is,

$$r = r_0 = \text{constant}. \tag{9}$$

In other words, the orbit is a circle.

Case III: If $h^2 < \lambda$ or $\alpha^2 < 0$, then the differential equation of the orbit is

$$\frac{d^2 u}{d\theta^2} - \alpha^2 u = 0. \qquad (10)$$

Its solution is known to be

$$u = Ae^{-\alpha\theta} + Be^{\alpha\theta}.$$

Using the initial condition, $u = 1/r_0$, when $\theta = 0$, we get

$$\frac{1}{r_0} = A + B. \qquad (11)$$

Using the other initial condition $du/d\theta = 0$, when $\theta = 0$, we obtain

$$(-A\alpha e^{-\alpha\theta} + B\alpha e^{\alpha\theta})\big|_{\theta=0} = 0,$$

which gives $A = B$. Substituting this result into Eq. (11), we get

$$A = B = \frac{1}{2r_0}.$$

Hence, the equation of the required orbit is

$$\frac{1}{r} = \frac{1}{r_0} \frac{e^{\alpha\theta} + e^{-\alpha\theta}}{2} = \frac{\cosh(\alpha\theta)}{r_0}$$

that is,

$$r = \frac{r_0}{\cosh(\alpha\theta)}. \qquad (12)$$

Example 5.3 Communication satellites are placed in a circular orbit in the equatorial plane of the Earth, such that they remain stationary with respect to the surface of the Earth. What is the minimum number of such satellites required for every point on the equator to be in view of at least one satellite?

Solution A communication satellite is a geo-stationary satellite, in the sense that the satellite is at rest relative to the Earth and revolves round the Earth in a circular orbit with periodic time of one day. For a circular orbit of radius $(r_E + x)$ (see Fig. 5.13), the periodic time T is given by

$$T = \frac{2\pi}{\sqrt{GM_E}} a^{3/2} = \frac{2\pi}{\sqrt{GM_E}} (r_E + x)^{3/2}, \qquad (1)$$

where x is the height of the satellite above the Earth. But,

$$T = \text{one day} = 24 \times 60 \times 60 = 8.64 \times 10^4 \text{ seconds}.$$

From Eq. (1), we find that

$$x = -r_E + (GM_E)^{1/3} \left(\frac{8.64 \times 10^4}{2\pi} \right)^{2/3}.$$

170 *Orbital Motion (Ch. 5)*

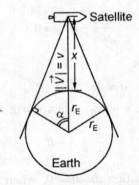

Figure 5.13 Description of Example 5.3.

Taking the value of $GM_E = 398603.6$ km^3/s^2, we get

$$x = -6378 + 42241.2 \approx 35863 \text{ km} = 3.58 \times 10^4 \text{ km}.$$

From the figure, it is clear that a single satellite can cover an angle 2α on the surface of the Earth. Here

$$\cos \alpha = \frac{r_E}{x + r_E} = \frac{6378}{42240},$$

which gives

$$\alpha = 81°.31 \approx 81°. \qquad (2)$$

Hence, the angle covered by one satellite is 162°. Obviously, even two satellites would not be sufficient to cover the Earth's equatorial plane. Hence, at least three satellites are required to cover all points on the equator of the Earth.

Example 5.4 A particle moves in an elliptic orbit about a gravitational force centre at one focus. If the greatest and the least speeds of the particle are v_1 and v_2, find the eccentricity of the orbit.

Solution The greatest and the least speeds of the particle in an elliptic orbit occur at the extremities of a transverse axis, that is, at perigee and apogee (see Fig. 5.14).

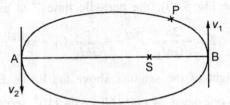

Figure 5.14 Description of Example 5.4.

Since the angular momentum about the centre of force is constant, the angular momentum about perigee and apogee are equal. That is,

$$m \, SB \, v_1 = m \, SA \, v_2,$$

where m is the mass of the particle. Therefore,
$$a(1-e)v_1 = a(1+e)v_2$$
or
$$v_1 - v_2 = e(v_1 + v_2).$$
Hence, we have the eccentricity
$$e = \frac{v_1 - v_2}{v_1 + v_2}.$$

Example 5.5 A particle is projected from a point A at right angles to SA and is acted on by a force, varying inversely as the square of the distance towards S. If the intensity of the force is unity at unit distance SA, and the speed of projection is 1/2, prove that the eccentricity of the orbit is 3/4. Also, find the period of the orbit.

Solution Given the intensity of the force as $GM = 1$, the law of attraction $= 1/r^2$. We know that the energy of the orbit is given by Eq. (5.38) as
$$\frac{1}{2}mv^2 - GMmu = E.$$
That is,
$$E = \frac{v^2}{2} - \frac{1}{r} \quad \text{(per unit mass)}$$
$$= \frac{1}{8} - 1$$
$$= -\frac{7}{8} < 0 \quad \text{(negative)}.$$

Hence, the orbit is an ellipse. Now, using the expression for speed in an elliptic orbit, we have
$$v^2 = GM\left(\frac{2}{r} - \frac{1}{a}\right) = \frac{2}{r} - \frac{1}{a}.$$
That is,
$$\frac{1}{a} = \frac{2}{1} - \frac{1}{4} = \frac{7}{4}.$$
Therefore, $a = 4/7$ units.

The angular momentum of the orbit is $h = 1/2$. That is,
$$h = \sqrt{GMa(1-e^2)} \quad \text{or} \quad h^2 = a(1-e^2),$$
which yields
$$\frac{1}{4} = \frac{4}{7}(1-e^2) \quad \text{or} \quad e^2 = \frac{9}{16} \quad \text{or} \quad e = \frac{3}{4}.$$

Finally, the period of the orbit is given by

$$P = \frac{2\pi}{\sqrt{GM}} a^{3/2} = 2\pi \left(\frac{4}{7}\right)^{3/2} = 2.7141.$$

Definition 5.1 (apses) An *apse* is a point on a satellite orbit where its velocity is perpendicular to its position vector. In Fig. 5.15, we observe that A and B are apses.

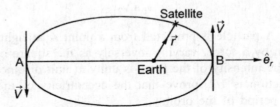

Figure 5.15 Description of apses.

At A and B, it can be seen that the dot product

$$\hat{e}_r \cdot \vec{V} = 0 = \hat{e}_r \cdot (\dot{r}\hat{e}_r + r\dot{\theta}\hat{e}_\theta) = \dot{r},$$

which means that at apses, $\dot{r} = 0$. Here, A and B correspond to maximum and minimum distances of the satellite from the Earth. The closest point is called 'perigee' and the farthest point is called 'apogee'. For planets in their orbits about Sun, the corresponding points are called 'perihelion' and 'aphelion', respectively.

Example 5.6 An artificial Earth satellite is released at an altitude of 400 km with a velocity of 8.85 km/s, in a horizontal direction. Find perigee, apogee, and the period. Also determine the energy required per unit mass of this satellite, by taking the Earth as force centre, radius of the Earth = 6378 km and $GM = 398603.6$ km^3/s^2.

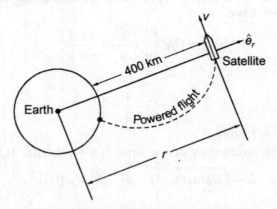

Figure 5.16 Description of Example 5.6.

Solution At the outset, we shall compute the angular momentum and energy per unit mass of the satellite. Taking the radius of the Earth as

6378 km it may be noted that the satellite has been released at an altitude of $r = (6378 + 400) = 6778$ km (see Fig. 5.16).

From the figure, we note that at the completion of the powered flight, the satellite is at a radial distance r and has a velocity v. Now, the angular momentum is given by

$$h = r^2\dot{\theta} = rv = 6778 \times 8.85 = 59985.3 \text{ km}^2/\text{s}.$$

Also, the energy per unit mass E is obtained from

$$E = \frac{1}{2}v^2 - \frac{GM}{r} = \frac{1}{2}(8.85)^2 - \frac{398603.6}{6778}$$

or

$$E = -19.647 \text{ km}^2/\text{s}^2. \tag{1}$$

Since we find that E is negative, the orbit of the satellite is an ellipse.

To compute perigee and apogee, we shall recall the apsidal quadratic given by Eq. (5.41) as

$$u^2 - \frac{2GMu}{h^2} - \frac{2E}{mh^2} = 0.$$

Taking $m = 1$ for unit mass and letting $u = 1/r$, we can rewrite the above equation as

$$r^2 + \frac{GM}{E}r - \frac{h^2}{2E} = 0.$$

Substituting the values of h, E and GM, we obtain

$$r^2 - 20288\, r + 9.157 \times 10^7 = 0,$$

which is a quadratic equation having two roots as

$$\left.\begin{array}{l} r_1 = q = 6777.93 \text{ km (perigee)} \\ r_2 = q' = 13510.3 \text{ km (apogee)}. \end{array}\right\} \tag{2}$$

From the properties of ellipse, we have $2a = q + q'$. Therefore,

$$a = \text{semi-major axis} = \frac{1}{2}(6777.93 + 13510.3)$$

or

$$a = 10144.1 \text{ km}. \tag{3}$$

To compute the period of the orbit, we use the formula

$$P = \frac{2\pi}{\sqrt{GM}} a^{3/2} = \frac{2 \times 3.1415 (10144.1)^{3/2}}{\sqrt{398603.6}}$$

or

$$P \approx 10162.7 \text{ s} = 169.378 \text{ min}. \tag{4}$$

Example 5.7 A satellite is moving in a circular orbit of radius $2r_E$ about the Earth, where r_E is the radius of the Earth. At certain instant of time, the direction of motion of the satellite is changed through an angle α towards the Earth, without changing its speed. Find the angle α, in order that the satellite just touches the Earth.

Solution The satellite is moving in a circular orbit of radius $2r_E$, as shown in Fig. 5.17. We know that the speed of the satellite in an elliptic orbit is given by

$$v^2 = GM\left(\frac{2}{r} - \frac{1}{a}\right).$$

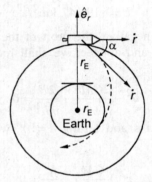

Figure 5.17 Description of Example 5.7.

For a circular orbit, $r = a$. Therefore, the speed of the satellite can be obtained from the equation

$$v^2 = \frac{GM}{a} = \frac{GM}{2r_E}. \tag{1}$$

Now the expression for the total energy of the system is

$$E = \frac{1}{2}mv^2 - \frac{GMm}{r}.$$

In this example, $r = 2r_E$ and thus substituting and using Eq. (1), we obtain

$$E = \frac{1}{2}\frac{GMm}{2r_E} - \frac{GMm}{2r_E} = -\frac{GMm}{4r_E}. \tag{2}$$

Now, at some instant of time, the satellite direction has been changed through an angle α, while the total energy of the system remains unchanged. Hence, the angular momentum $\vec{h}$ is changed. Thus, the magnitude of the new angular momentum is

$$h = |\vec{h}| = m|\vec{r} \times \dot{\vec{r}}|$$

$$= m|\vec{r} \times \vec{V}|$$

$$= m 2 r_E \left(\frac{GM}{2r_E}\right)^{1/2} \sin\left(\frac{\pi}{2} + \alpha\right)$$

or

$$h = m\,(2 r_E\, GM)^{1/2} \cos \alpha = (2 r_E\, GM)^{1/2} \cos \alpha \qquad (3)$$

for unit mass. Now, the condition for the satellite to just touch the Earth is that the apsidal distance, $r_1 = r_E$. Recalling the apsidal equation as

$$u^2 - \frac{2GM}{h^2} u - \frac{2E}{mh^2} = 0$$

that is,

$$r^2 + \frac{GMm}{E} r - \frac{mh^2}{2E} = 0 \quad \left(\text{since } u = \frac{1}{r}\right).$$

Substituting the values of h and E from Eqs. (2) and (3), we get for unit mass, the resulting quadratic as

$$r^2 - 4 r_E r + (4 \cos^2 \alpha) r_E^2 = 0,$$

whose roots are given by

$$r_{1,2} = \frac{[4 r_E \pm \sqrt{16 r_E^2 - 16 r_E^2 \cos^2 \alpha}]}{2} = 2(1 \pm \sin \alpha) r_E.$$

Taking the smaller apsidal distance, we may write

$$\frac{r_1}{r_E} = 2(1 - \sin \alpha).$$

For $r_1 = r_E$, we will have $\sin \alpha = 1/2$ which gives $\alpha = 30°$. Hence the result.

Example 5.8 A satellite is placed into its orbit at an altitude of 300 km above the Earth's surface at a speed of 5 km/s and angle $\theta = 30°$ as depicted in the figure below. Determine the equation for its trajectory and show that its orbit is elliptical. Also, determine the perigee and apogee distances.

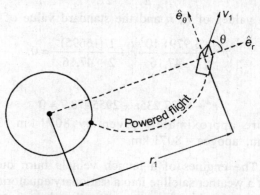

Figure 5.18 Description of Example 5.8.

Solution At the end of the powered flight, let the satellite be at a radial distance r_1 and have a velocity v_1 directed at an angle θ from the radial line (Fig. 5.18). The initial conditions of the given problem can be stated as follows: At $r = r_1$,

$$v = v_1 (\cos \theta \hat{e}_r + \sin \theta \hat{e}_\theta) = \dot{r}\hat{e}_r + r\dot{\theta}\hat{e}_\theta.$$

Therefore,

$$\dot{r}_1 = v_1 \cos \theta, \qquad r_1\dot{\theta} = v_1 \sin \theta \quad \text{or} \quad \dot{\theta} = \frac{v_1 \sin \theta}{r_1}.$$

These initial conditions allow us to compute the total energy E and angular momentum h of the system per unit mass as follows:

$$E = \frac{1}{2}v_1^2 - \frac{GM}{r_1}$$

$$= \frac{5^2}{2} - \frac{3.979 \times 10^5}{6378 + 300}$$

$$= -47.16 \text{ km}^2/\text{s}^2. \qquad (1)$$

E, being negative, the resulting orbit is an ellipse. Now, the angular momentum h is obtained from

$$h = r^2\dot{\theta} = r_1^2 \frac{v_1 \sin \theta}{r_1} = r_1 v_1 \sin \theta$$

or

$$h = (6378 + 300)(5)\left(\frac{1}{2}\right) = 16695 \text{ km}^2/\text{s}. \qquad (2)$$

Finally to compute perigee and apogee, we recall the apsidal quadratic as

$$r^2 + \frac{GM}{E}r - \frac{h^2}{2E} = 0.$$

Substituting the values of E, h and the standard value of GM, we get

$$r^2 + \frac{3.979 \times 10^5}{-47.16}r + \frac{1}{2}\frac{(16695)^2}{47.16} = 0.$$

That is,

$$r^2 - 8437.235r + 2955078.7 = 0$$

whose roots are approximately given by 8071 km, 366 km. Hence perigee = 366 km, apogee = 8071 km.

Example 5.9 The engines of a launch vehicle burn out at the correct altitude to inject a weather satellite into a stationary equatorial orbit about the Earth. The satellite also has the proper speed at burn-out for a stationary orbit. But, due to a malfunction, the velocity at burn-out is only at an angle of 80° with the radial line. Determine the eccentricity and period of the orbit.

Solution At the outset, we shall determine the initial conditions required by studying the desired stationary circular orbit, an orbit whose period is one day. Let the radius be r and velocity be v. Following Newton's law of gravitation, we should have

$$F_r = \frac{GMm}{r^2} = -\frac{mv^2}{r}$$

where for circular orbit, we have

$$v = \sqrt{\frac{GM}{r}}.$$

Further, we have for a circular orbit, $a = r$, and for a stationary orbit, $P = 24$ hours. Hence, taking $GM = 3.979 \times 10^5$ km^3/s^2, we have from the relation

$$P = \frac{2\pi}{\sqrt{GM}} a^{3/2}.$$

that is,

$$r^{3/2} = \frac{P\sqrt{GM}}{2\pi} = \frac{24 \times 60 \times 60 \sqrt{3.979 \times 10^5}}{2\pi}$$

which gives

$$r = 4.22163 \times 10^4 \text{ km} \quad \text{and} \quad v = \sqrt{\frac{GM}{r}} = 3.07 \text{ km/s}.$$

These are the initial conditions, that is, the distance and speed required for the correct orbit. Due to malfunction, the velocity at burn-out makes an angle of 80° with the radial line (launch angle). Actually, it should be 90°, so that its velocity vector is parallel to the local vertical (see Fig. 5.19).

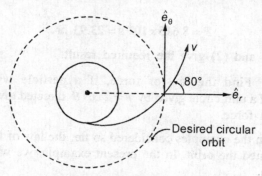

Figure 5.19 Description of Example 5.9.

The initial transverse speed is

$$v_\theta = v \sin 80° = 3.07 \sin 80° = 3.023 \text{ km/s}.$$

Thus, the angular momentum h of the orbit is

$$h = rv_\theta = 4.2216 \times 10^4 \times 3.023 = 1.2762 \times 10^5 \text{ km}^2/\text{s}$$

while the total energy E of the system is given by

$$E = \frac{1}{2}v^2 - \frac{GM}{r} = 4.71245 - 9.4252 = -4.713 \text{ km}^2/\text{s}^2.$$

Due to error in the launch angle, the satellite takes an elliptic orbit whose eccentricity can be computed from Eq. (5.43) as

$$e = \left(1 + \frac{2Eh^2}{mG^2M^2}\right)^{1/2} = \left[1 + 2E\left(\frac{h}{GM}\right)^2\right]^{1/2}$$

for unit mass, or

$$e = \left[1 + 2(-4.713)\left(\frac{1.2762 \times 10^5}{3.979 \times 10^5}\right)^2\right]^{1/2} = 0.174. \tag{1}$$

But Eq. (5.48) gives

$$h^2 = GMa(1 - e^2)$$

which yields

$$a = \frac{h^2}{GM(1 - e^2)} = 4.213 \times 10^4 \text{ km}.$$

Now, the period of the orbit is obtained from

$$P = \frac{2\pi}{\sqrt{GM}} a^{3/2} = \frac{2\pi}{\sqrt{3.979 \times 10^5}} (4.213 \times 10^4)^{3/2}$$

or

$$P = 8.615 \times 10^4 \text{ s} = 23.93 \text{ hr}. \tag{2}$$

Hence, Eqs. (1) and (2) give the required result.

Example 5.10 Find the law of force, if a particle which is on the circumference of a unit circle given by $r = 2\cos\theta$ directed towards the origin, is under central force.

Solution In the examples considered so far, the law of force was given, and we determined the orbit. In the present example, we will work out the inverse problem.

To start with, we shall derive the polar equation of the unit circle in the following steps:

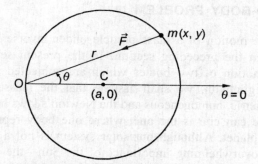

Figure 5.20 Circular motion with origin on the circumference.

In this example, let the equation of the circle be $(x-a)^2 + y^2 = a^2$, but $x = r \cos \theta$, $y = r \sin \theta$. Therefore,

$$x^2 - 2ax + y^2 = 0$$

that is,

$$r^2 \cos^2\theta - 2ar \cos \theta + r^2 \sin^2\theta = 0$$

or

$$r^2 = 2ar \cos \theta.$$

In the case of a unit circle, $a = 1$ and therefore its equation is given by

$$r = 2 \cos \theta \qquad (1)$$

or

$$u = \frac{1}{2 \cos \theta}.$$

Differentiating twice with respect to θ, we get

$$2\frac{du}{d\theta} = \frac{\sin \theta}{\cos^2\theta}$$

and

$$2\frac{d^2u}{d\theta^2} = \frac{\cos^2\theta + 2\sin^2\theta}{\cos^3\theta} = \frac{1+\sin^2\theta}{\cos^3\theta}$$

Now, recalling the general equation of the orbit

$$\frac{d^2u}{d\theta^2} + u = -\frac{F(u)}{mh^2u^2} = \frac{1+\sin^2\theta}{2\cos^3\theta} + \frac{1}{2\cos\theta} = \frac{1}{\cos^3\theta}.$$

Evidently, the law of force can be seen as

$$F(u) = -\frac{mh^2u^3}{\cos^3\theta} = -\frac{mh^2}{r^2(r/2)^3} = -\frac{8mh^2}{r^5}.$$

Hence, the law of force is an inverse fifth power.

5.5 THE TWO-BODY PROBLEM

The central force motion of a mass particle under inverse square law has been discussed in the preceding section. In the present section, we shall investigate the motion of two bodies which are subjected to their mutual attraction. To begin with, we shall assume that the masses involved are spherically symmetric, homogeneous and the Newton's laws hold. The natural example that one can cite is the one, where one body represents the Sun and the other a planet. Although our solar system is not a system of only two bodies, the overwhelming importance of the Sun—the central body of the solar system—the mass of which is 700 times that of all the members of the solar system put together gives rise to one of the characteristic features. To put it differently, every planet moves approximately as if only one central body, i.e. the Sun ruled its motion. Then, the whole solar system can be split into as many two-body problems as there are planets. The gravitational influence exerted by the other planets on each other can be thought of as small perturbations of their orbits about the Sun.

In fact, the motion of an artificial satellite around the Earth is governed primarily by the attraction between the Earth and the satellite. Even for a satellite at an altitude of thousands of kilometres, the effects of gravitational attractions of the Sun, Moon, and other planets on the satellite orbit will be there, though small, and may thus be neglected to a first-order approximation to enable us to consider it as a two-body problem.

5.5.1 The Motion of the Centre of Mass

Let us consider an inertial system having its origin at O, in which the Newton's laws of motion holds true. Let m_1 and m_2 be the masses of the two bodies, whose position vectors relative to O are $\vec{r}_1$ and $\vec{r}_2$, respectively, at a particular instant of time. Let $\vec{R}$ be the position vector of the centre of mass of the pair. Let $\vec{r}$ be the position vector of m_2 relative to m_1 (see Fig. 5.21).

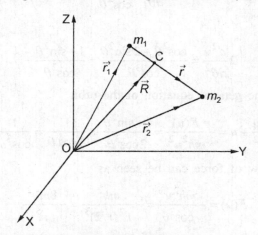

Figure 5.21 Motion of centre of mass in a two-body problem.

By Newton's law of gravitation, the force on m_1 due to m_2 is

$$f_{12} = G \frac{m_1 m_2}{r^2} \hat{e}_r$$

and the force on m_2 due to m_1 is

$$f_{21} = -G \frac{m_1 m_2}{r^2} \hat{e}_r,$$

where G is the constant of gravitation and $\hat{e}_r$ is a unit vector in the direction of $\vec{r}$. The equations of motion can be written as

$$m_1 \ddot{\vec{r}}_1 = G \frac{m_1 m_2}{r^2} \frac{\vec{r}}{r} = G \frac{m_1 m_2}{r^3} (\vec{r}_2 - \vec{r}_1) \qquad (5.58)$$

and

$$m_2 \ddot{\vec{r}}_2 = -G \frac{m_1 m_2}{r^3} (\vec{r}_2 - \vec{r}_1). \qquad (5.59)$$

Now, adding the above two equations, we get at once

$$m_1 \ddot{\vec{r}}_1 + m_2 \ddot{\vec{r}}_2 = 0.$$

Integrating this result twice with respect to t, we have

$$m_1 \vec{r}_1 + m_2 \vec{r}_2 = \vec{c}_1 t + \vec{c}_2, \qquad (5.60)$$

where $\vec{c}_1$ and $\vec{c}_2$ are the constant vectors. Let $M = m_1 + m_2$, the total mass of the pair, then the centre of mass of the system using Eq. (5.60) is given by

$$\vec{R} = \frac{m_1 \vec{r}_1 + m_2 \vec{r}_2}{M} = \frac{\vec{c}_1}{M} t + \frac{\vec{c}_2}{M}, \qquad (5.61)$$

which shows that the centre of mass of the system moves uniformly in a straight line in space.

5.5.2 Relative Motion

Taking the origin at the centre of mass of the two-body system, we can also describe the motions of m_1 and m_2 relative to it. Let $\vec{r}_1'$ and $\vec{r}_2'$ be the position vectors of m_1 and m_2 with respect to the centre of mass C of the system. Suppose the position vector of C with respect to O is $\vec{R}$ and those of m_1 and m_2 are $\vec{r}_1$ and $\vec{r}_2$ respectively. Then, we have (see Fig. 5.22)

$$\vec{r}_1 = \vec{R} + \vec{r}_1', \qquad \vec{r}_2 = \vec{R} + \vec{r}_2', \qquad \vec{r} = \vec{r}_2' - \vec{r}_1'. \qquad (5.62)$$

From Eqs. (5.61) and (5.62), we arrive at once to

$$\ddot{\vec{R}} = 0 \quad \text{and} \quad m_1 \vec{r}_1 = m_1 \vec{R} + m_1 \vec{r}_1',$$

implying

$$m_1 \ddot{\vec{r}}_1 = m_1 \ddot{\vec{r}}_1'.$$

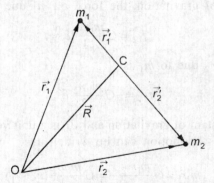

Figure 5.22 Position vectors of two bodies relative to its centre of mass.

Similarly, we have $m_2\ddot{\vec{r}}_2 = m_2\ddot{\vec{r}}'_2$. Now, with the help of Eq. (5.62), Eqs. (5.58) and (5.59) can be rewritten as

$$m_1\ddot{\vec{r}}'_1 = G\frac{m_1 m_2(\vec{r}'_2 - \vec{r}'_1)}{r^3} \tag{5.63}$$

and

$$m_2\ddot{\vec{r}}'_2 = -G\frac{m_1 m_2(\vec{r}'_2 - \vec{r}'_1)}{r^3}. \tag{5.64}$$

Since the origin is taken at the centre of mass, we have

$$m_1\vec{r}'_1 + m_2\vec{r}'_2 = 0$$

or

$$\frac{m_1}{m_2} = -\frac{\vec{r}'_2}{\vec{r}'_1}. \tag{5.65}$$

Eliminating $\vec{r}'_2$ from Eq. (5.63) and $\vec{r}'_1$ from Eq. (5.64), the resulting equations become

$$m_1\ddot{\vec{r}}'_1 = -Gm_1 m_2\left(1 + \frac{m_1}{m_2}\right)\frac{\vec{r}'_1}{r^3}, \tag{5.66}$$

and

$$m_2\ddot{\vec{r}}'_2 = -Gm_1 m_2\left(1 + \frac{m_2}{m_1}\right)\frac{\vec{r}'_2}{r^3}. \tag{5.67}$$

Equations (5.66) and (5.67) can be further simplified as

$$\ddot{\vec{r}}'_1 = -\frac{GM}{r^3}\vec{r}'_1, \tag{5.68}$$

and

$$\ddot{\vec{r}}_2' = -\frac{GM}{r^3}\vec{r}_2'. \qquad (5.69)$$

In principle, we can solve these equations to determine the positions of m_1 and m_2 at any instant. However, there is no way to compute them absolutely, since they are defined with respect to the centre of mass which is fixed in space.

To make this analysis useful, let us fix the origin at m_1, of the two-body system, then subtracting Eq. (5.69) from Eq. (5.68), we get

$$\ddot{\vec{r}} = -\frac{GM}{r^3}\vec{r} = -\frac{G(m_1+m_2)}{r^3}\vec{r}. \qquad (5.70)$$

This differential equation expresses the acceleration of m_2 relative to m_1. In the case of planetary system, we may take m_1 as the Sun and m_2 as a planet. In the case of an artificial satellite, m_1 can be taken as the mass of the Earth and m_2 that of the satellite.

In Cartesian form, Eq. (5.70) is equivalent to three second-order differential equations, requiring six constants of integration for their solution. If we know the position (three coordinates) and the velocity (three components) at any one instant, these six constants of integration can be determined.

Since our efforts will be devoted to studying the motion of artificial satellites, ballistic missiles or space probes orbiting the Earth or the motion of a planet about the Sun, the mass of the orbiting body is very much less than that of the central body. Hence, we may take

$$G(m_1+m_2) \approx Gm_1 \qquad \text{(since } m_1 \gg m_2\text{).}$$

Let $GM = \mu$, then the equation of motion (5.70) of the two-body problem can also be written as

$$\ddot{\vec{r}} = -\frac{\mu\vec{r}}{r^3}.$$

5.5.3 Solution of the Two-body Problem

The equation of motion of the two-body problem in vectorial representation can be written as

$$\ddot{\vec{r}} = -\frac{\mu\vec{r}}{r^3}. \qquad (5.71)$$

Taking the vector product of this equation with $\vec{r}$, we get

$$\vec{r} \times \ddot{\vec{r}} = 0. \qquad (5.72)$$

On integration, we obtain

$$\vec{r} \times \dot{\vec{r}} = \vec{h}, \qquad (5.73)$$

where $\vec{h}$ is a constant vector. Equation (5.73) means that the angular momentum of the system is constant. Now, taking the vector product of Eq. (5.71) with $\vec{h}$, we find

$$\vec{h} \times \ddot{\vec{r}} = -\frac{\mu}{r^3} \vec{h} \times \vec{r} = -\frac{\mu}{r^3} (\vec{r} \times \dot{\vec{r}}) \times \vec{r}$$

$$= -\frac{\mu}{r^3} [(\vec{r} \cdot \vec{r}) \dot{\vec{r}} - (\vec{r} \cdot \dot{\vec{r}}) \vec{r}]$$

$$= -\frac{\mu}{r^3} [r^2 \dot{\vec{r}} - (r \dot{r}) \vec{r}]$$

$$= -\mu \left(\frac{\dot{\vec{r}}}{r} - \frac{\vec{r} \dot{r}}{r^2} \right)$$

$$= -\mu \frac{d}{dt} \left(\frac{\vec{r}}{r} \right) = -\mu \frac{d}{dt} (\hat{e}_r).$$

But,

$$\vec{h} \times \ddot{\vec{r}} = -\frac{d}{dt} (\dot{\vec{r}} \times \vec{h}) \quad \text{(since } \vec{h} \text{ is a constant vector)}.$$

Therefore,

$$\frac{d}{dt} (\dot{\vec{r}} \times \vec{h}) = +\mu \frac{d}{dt} (\hat{e}_r) = \mu \frac{d}{dt} \left(\frac{\vec{r}}{r} \right). \tag{5.74}$$

Integrating both sides with respect to t, we get

$$\dot{\vec{r}} \times \vec{h} = \mu \frac{\vec{r}}{r} + \vec{P}, \tag{5.75}$$

where $\vec{P}$ is the vector constant of integration.

Now, taking the dot product of this equation with $\vec{r}$ we obtain a scalar equation

$$\vec{r} \cdot \dot{\vec{r}} \times \vec{h} = \vec{r} \cdot \mu \frac{\vec{r}}{r} + \vec{r} \cdot \vec{P}.$$

Since, in general,

$$\vec{a} \cdot \vec{b} \times \vec{c} = \vec{a} \times \vec{b} \cdot \vec{c} \quad \text{and} \quad \vec{a} \cdot \vec{a} = a^2$$

we have

$$h^2 = \mu r + rP \cos \nu,$$

where ν is the angle between the constant vector $\vec{P}$ and the radius vector $\vec{r}$. Solving for r, we obtain

$$r = \frac{h^2 / \mu}{1 + (P/\mu) \cos \nu}, \tag{5.76}$$

which has the standard form of the conic $r = p/(1 + e \cos\theta)$ with eccentricity

$$e = \frac{P}{\mu}, \tag{5.77}$$

and semi-latus rectum

$$p = \frac{h^2}{\mu}. \tag{5.78}$$

5.5.4 Earth-bound Satellite Circular Orbits

We may recall that the equation of motion of a satellite relative to the Earth is given by Eq. (5.71), where, we have used the gravitational coefficient μ as

$$\mu = G(m_1 + m_2) = G(m_E + m_S). \tag{5.79}$$

Here, m_E is the mass of the earth and m_S is the mass of the satellite. In the case of artificial satellites it is known that $m_S \ll m_E$ and therefore m_S is negligible and μ can be taken as Gm_E.

Alternatively, μ can also be expressed in terms of acceleration due to gravity at the surface of the Earth as

$$\frac{\mu}{r_E^2} = g \quad \text{or} \quad \mu = gr_E^2, \tag{5.80}$$

where, r_E is the radius of the Earth. Assuming the following standard numerical values

$$r_E = 6378 \text{ km}, \qquad g = 9.806 \text{ m/s}^2,$$

we find

$$\mu = 3.98897 \times 10^{14} \text{ m}^3/\text{s}^2.$$

Let us assume that a satellite moves in a circular orbit on the surface of the Earth (ignoring the presence of the atmosphere). Since the gravitational attraction is equal to the centripetal acceleration, which we may express per unit mass as

$$g = \frac{v_c^2}{r_E} \quad \text{or} \quad v_c = \sqrt{gr_E} = 7908.392 \text{ m/s}, \tag{5.81}$$

where v_c is the speed of a body in a circular orbit on the surface of the Earth. From Eqs. (5.80) and (5.81), we get

$$\mu = r_E v_c^2. \tag{5.82}$$

But, for a general circular orbit about the Earth we equate the magnitudes of gravitational force with that of the centrifugal force per unit mass to get

$$\frac{v^2}{r} = \frac{\mu}{r^2}.$$

Using Eqs. (5.81) and (5.82), we obtain

$$v = v_c \left(\frac{r_E}{r}\right)^{1/2} = r_E \left(\frac{g}{r}\right)^{1/2}, \qquad (5.83)$$

and the total energy can be found to be

$$E = -\frac{r_E v_c^2}{2r} = -\frac{g r_E^2}{2r} \quad \text{[using Eq. (5.81)]}. \qquad (5.84)$$

The period of the circular orbit at the surface of the Earth is obtained from

$$P_c = \frac{2\pi r_E}{v_c} = 5067.295 \text{ s} = 84.455 \text{ min}.$$

For a general circular orbit of radius r, we get from Kepler's third law, the period P as

$$P = \left(\frac{r}{r_E}\right)^{3/2} P_c. \qquad (5.85)$$

5.6 CLASSICAL ORBITAL ELEMENTS

To describe the size, shape and orientation of an orbit, we require five independent elements called *orbital elements*. A sixth element is also required to pinpoint the position of the satellite along the orbit at a particular time. These classical elements are explained below:

1. *a* (*semi-major axis*): a constant which defines the size of the orbit.
2. *e* (*eccentricity*): a constant which defines the shape of the orbit.

In order to explain the orientation of the orbit, consider a geocentric-inertial coordinate system OXYZ, with Earth at O, OX pointing towards the vernal equinox, OZ towards the North Pole of the celestial equator (or the Earth's equator), so that XOY is the fundamental plane/equatorial plane as depicted in Fig. 5.23.

Let $\hat{i}, \hat{j}, \hat{k}$ be the unit vectors along OX, OY and OZ respectively. It may be noted that OXYZ system is not fixed with the Earth and instead rotates with it. Also, the geocentric-inertial frame is non-rotating with respect to stars, while the Earth turns relative to it. Let the plane of the satellite's orbit cut the celestial sphere in a great circle NPA, where N is the point, when the satellite in its orbit rises north of the XOY-plane, called *ascending node*. Now,

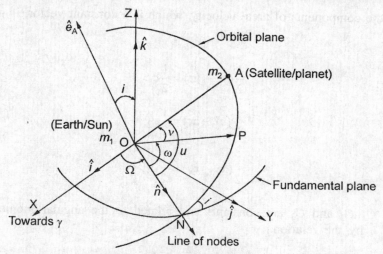

Figure 5.23 Orientation of orbit in space.

3. *i* (*inclination*): The angle between the plane of the orbit and XOY-plane or the angle between $\hat{e}_A$ and $\hat{k}$ is called the *inclination*, *i*, of the orbit.

4. Ω (*longitude of the ascending node*): The angle XON, in the fundamental plane, measured eastwards is called the *longitude* of the ascending node.

5. ω (*argument of perigee*): If OP is the direction of perigee, then the angle NOP, measured in the orbital plane is called the *argument of perigee*.

Here,

$$0 \leq \Omega \leq 360, \qquad 0 \leq i \leq 180°$$

For, $0 \leq i \leq 90°$, the orbit is direct, while for $90° \leq i \leq 180°$, the orbit is retrograde.

If we are considering the orbits of planets about the Sun, we replace the *geocentric inertial* system by the corresponding heliocentric-ecliptic coordinate system, where the ecliptic plane is the fundamental plane. Here ω is called the *argument of perihelion*.

We may recognize that the relative motion of m_2 around m_1 is in fact a central force motion, which we discussed in detail in section 5.4, where we noted that the areal velocity is constant. Mathematically,

$$\dot{\vec{A}} = \frac{1}{2}(\vec{r} \times \dot{\vec{r}}) = \frac{1}{2}h\hat{e}_A, \qquad (5.86)$$

where $\hat{e}_A$ is a unit vector, normal to the orbital plane containing $\vec{r}$ and $\dot{\vec{r}}$. In Cartesian coordinates, we write

$$\dot{\vec{A}} = \frac{1}{2}\begin{vmatrix} \hat{i} & \hat{j} & \hat{k} \\ x & y & z \\ \dot{x} & \dot{y} & \dot{z} \end{vmatrix} = \frac{1}{2}\vec{c} = \frac{1}{2}(c_1\hat{i} + c_2\hat{j} + c_3\hat{k}).$$

Thus, the components of areal velocity, which is a constant vector, is given by

$$\left.\begin{array}{c}\frac{1}{2}(y\dot{z}-\dot{y}z)=\frac{1}{2}c_1\\[6pt]\frac{1}{2}(z\dot{x}-\dot{z}x)=\frac{1}{2}c_2\\[6pt]\frac{1}{2}(x\dot{y}-\dot{x}y)=\frac{1}{2}c_3,\end{array}\right\} \tag{5.87}$$

where C_1, C_2 and C_3 are constants and related to the angular momentum vector $\vec{h}$ by the relation

$$h = (c_1^2 + c_2^2 + c_3^2)^{1/2}. \tag{5.88}$$

Thus, given the initial (injection) position coordinates and velocity components of the satellite m_2 relative to m_1 (Earth), we can determine the constants c_1, c_2 and c_3 and hence h, the absolute value of the angular momentum of the system. In terms of Ω and i, the unit vector $\hat{e}_A$ can be expressed as

$$\hat{e}_A = (\sin i \sin \Omega)\hat{i} + (\sin i \cos \Omega)\hat{j} + (\cos i)\hat{k}. \tag{5.89}$$

Hence, using Eqs. (5.86) and (5.89), the areal velocity can be written as

$$\dot{\vec{A}} = \left(\frac{1}{2}h \sin i \sin \Omega\right)\hat{i} + \left(-\frac{1}{2}h \sin i \cos \Omega\right)\hat{j} + \left(\frac{1}{2}h \cos i\right)\hat{k} \tag{5.90}$$

Comparing Eqs. (5.87) and (5.90), we find at once

$$\left.\begin{array}{c}c_1 = h \sin i \sin \Omega\\[4pt]c_2 = -h \sin i \cos \Omega\\[4pt]c_3 = h \cos i\\[4pt]h = (c_1^2 + c_2^2 + c_3^2)^{1/2}.\end{array}\right\} \tag{5.91}$$

Now, we can compute i and Ω in the following steps:

$$\left.\begin{array}{c}\cos i = \left(\dfrac{c_3}{h}\right) \quad \text{or} \quad i = \cos^{-1}\left(\dfrac{c_3}{h}\right)\\[10pt]\text{and}\\[4pt]\Omega = \tan^{-1}\left(\dfrac{-c_1}{c_2}\right) = \sin^{-1}\left(\dfrac{c_3}{h \sin i}\right)\end{array}\right\} \tag{5.92}$$

Thus, the orientation of the orbit in space can be determined.

5.7 KEPLER'S EQUATION

Consider an elliptic orbit, whose semi-major axis is a and an auxiliary circle of radius a, that has been circumscribed about the ellipse, as shown in Fig. 5.24.

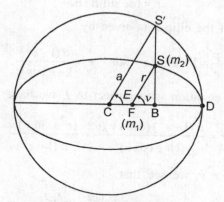

Figure 5.24 Eccentric and true anomaly.

Let the mass m_1 be at the focus F and the mass m_2 move along an elliptic orbit in the anticlockwise direction. Let S be the position of m_2 (say a satellite) at a certain instant and let BS be drawn perpendicular to CD, meeting the auxiliary circle at S'. Join S' with C. Now, we define the angle E (see Fig. 5.24), the eccentric anomaly and v and true anomaly. From the property of an ellipse, we have

$$r \cos v = \text{FB} = \text{CB} - \text{CF} = a \cos E - ae \tag{5.93}$$

and

$$r \sin v = \text{SB} = \text{S'B} - \text{S'S}.$$

But,

$$\frac{\text{BS}}{\text{BS'}} = \frac{b}{a} = \sqrt{1-e^2}.$$

Therefore,

$$\text{SS'} = \text{BS'} - \text{BS} = \text{BS'} - \text{BS'}\sqrt{1-e^2} = \text{BS'}(1 - \sqrt{1-e^2}).$$

Hence,

$$r \sin v = \text{BS'}\sqrt{1-e^2} = a \sin E \sqrt{1-e^2}. \tag{5.94}$$

Now, squaring Eqs. (5.93) and (5.94) and adding, we get

$$r^2 = a^2(1 + e^2 \cos^2 E - 2e \cos E) = a^2(1 - e \cos E)^2$$

or

$$r = a(1 - e \cos E). \tag{5.95}$$

If we know E as a function of time, the corresponding position r can be obtained from Eq. (5.95). To find E as a function of time, we proceed as follows:

Differentiating Eq. (5.95) with respect to time t, we obtain

$$\dot{r} = (ae \sin E)\dot{E}. \tag{5.96}$$

But, the equation of the ellipse is given by

$$\frac{p}{r} = 1 + e \cos v \quad \text{or} \quad r = \frac{a(1-e^2)}{1+e \cos v}.$$

Differentiating this equation with respect to t, we have

$$\dot{r} = \frac{a(1-e^2)(e \sin v)\dot{v}}{(1+e \cos v)^2} = \frac{(r^2 e \sin v)\dot{v}}{a(1-e^2)}. \tag{5.97}$$

However, since $h = r^2 \dot{v}$, we see that

$$p = a(1-e^2) = \frac{h^2}{\mu} = \frac{r^4 \dot{v}^2}{\mu}. \tag{5.98}$$

Using this result, Eq. (5.97) can be rewritten as

$$\dot{r} = \frac{\sqrt{\mu a(1-e^2)}}{a(1-e^2)} e \sin v = \sqrt{\frac{\mu}{a(1-e^2)}} e \sin v. \tag{5.99}$$

Now, equating Eqs. (5.96) and (5.99) and using Eq. (5.94), we get

$$\dot{E} = \sqrt{\frac{\mu}{a^3(1-e^2)}} \frac{\sin v}{\sin E} = \sqrt{\frac{\mu}{a}} \frac{1}{r}. \tag{5.100}$$

Substituting for r from Eq. (5.95), we obtain

$$\dot{E}(1 - e \cos E) = \sqrt{\frac{\mu}{a^3}}.$$

Integrating this result with respect to t, we get at once

$$E - e \sin E = \left(\frac{\mu}{a^3}\right)^{1/2} (t - \tau), \tag{5.101}$$

where τ is the constant of integration. Since $t = \tau$ corresponds to $E = 0$, it can be concluded that τ represents the time of perigee passage.

Frequently, it is convenient to speak of mean motion or mean angular speed, that is, the rate at which the true anomaly v is swept by the radius vector averaged over the whole orbit. Denoting this quantity by 'n', then

$$n = \frac{2\pi}{P}. \tag{5.102}$$

However, the area of the ellipse $= \pi ab = \pi a^2(1-e^2)^{1/2}$. If P is the anomalistic period, we have

$$\frac{dA}{dt} = \frac{1}{2}r^2\dot{v} = \frac{\pi a^2(1-e^2)^{1/2}}{P}.$$

Therefore,

$$r^2\dot{v} = \frac{2\pi a^2(1-e^2)^{1/2}}{P} = na^2(1-e^2)^{1/2}. \tag{5.103}$$

With the help of Eq. (5.98), we can also write

$$r^4\dot{v}^2 = n^2 a^4(1-e^2) = \mu a(1-e^2),$$

which gives

$$n^2 a^3 = \mu. \tag{5.104}$$

Now, substituting Eq. (5.104) into Eq. (5.101), we finally get

$$E - e\sin E = n(t-\tau). \tag{5.105}$$

Very often, the quantity $n(t-\tau)$ is denoted by M, called *mean anomaly*. Thus, Eq. (5.105) takes the form

$$M = E - e\sin E. \tag{5.106}$$

This is known as *Kepler's equation*.

Relation between eccentric and true anomaly

Let us recall the relation connecting the radius vector r and eccentric anomaly E as

$$r = a(1 - e\cos E)$$

and also, we have from Fig. 5.24, that

$$r\cos v = a\cos E - ae. \tag{5.107}$$

But the polar equation of the orbit is given by

$$\frac{p}{r} = \frac{a(1-e^2)}{r} = 1 + e\cos v$$

from which, we get

$$r = \frac{a(1-e^2)}{1 + e\cos v}. \tag{5.108}$$

Now, combining Eqs. (5.107) and (5.108), we obtain

$$\frac{(1-e^2)\cos v}{1 + e\cos v} = \cos E - e,$$

which can be rewritten as
$$\cos E = \frac{e + \cos v}{1 + e \cos v}.$$

Therefore,
$$1 - \cos E = \frac{(1-e)(1-\cos v)}{1+e\cos v} \quad \text{and} \quad 1 + \cos E = \frac{(1+e)(1+\cos v)}{1+e\cos v}.$$

On division, we get
$$\tan^2\left(\frac{E}{2}\right) = \frac{1-e}{1+e}\tan^2\left(\frac{v}{2}\right)$$

or
$$\tan\frac{v}{2} = \left(\frac{1+e}{1-e}\right)^{1/2}\tan\frac{E}{2}. \tag{5.109}$$

The correct quadrant for E is obtained by noting that v and E are always in the same half plane; if $0 \leq v \leq \pi$, so is $0 \leq E \leq \pi$.

5.8 POSITION IN THE ELLIPTIC ORBIT

In Section 5.7, we have established a relation connecting the radius vector r and the eccentric anomaly E as
$$r = a(1 - e\cos E).$$

In the preceding section, we have also obtained an expression for computing the true anomaly v as a function of E given by
$$\tan\frac{v}{2} = \left(\frac{1+e}{1-e}\right)^{1/2}\tan\frac{E}{2}.$$

Given the time after perigee passage, we know the mean anomaly M and therefore E can be computed by solving Kepler's equation
$$M = E - e\sin E$$

either by an analytical technique described in Escobal (1965) or numerically using Newton–Raphson iterative procedure, such as
$$E_{n+1} = E_n - \frac{E_n - e\sin E_n - M}{1 - e\cos E_n},$$

as described in Sankara Rao (2004). Thus, the polar coordinates of a satellite moving round the Earth such as (r, v) can be found, if time after its perigee passage is specified.

Example 5.11 Calculate the eccentric anomaly E of Jupiter, five years after its perihelion passage, given that the Jupiter's period P is 11.8622 years and its eccentricity is 0.04844.

Solution Given the Jupiter's data as
$$P = 11.8622 \text{ yr}, \qquad e = 0.04844.$$
Its mean anomaly M can be computed using the relation
$$M = n(t-\tau) = \frac{2\pi}{P} \times 5 = \frac{10\pi}{P} \text{ rad/yr}$$
or
$$M = \frac{10\pi}{11.8622} \frac{180}{\pi} = \frac{1800}{11.8622} = 151°.743.$$
To determine the eccentric anomaly E, we shall make use of Kepler's equation $M = E - e \sin E$. Taking the first approximation
$$E_0 = M = 151°.743$$
M_0 can be computed as
$$\begin{aligned} M_0 &= E_0 - e \sin E_0 \\ &= 151°.743 - (0.04844)\frac{180}{\pi} \sin 151°.743 \\ &= 150°.429. \end{aligned}$$

Now, we can formulate a Newton's iterative scheme as follows: First we select a trial value for E, call it E_0. Next, we compute the mean anomaly M_0, that results from
$$M_0 = E_0 - e \sin E_0. \tag{1}$$
Now, we shall select a new trial value E_1, from
$$E_1 = E_0 + \frac{M - M_0}{(dM/dE)|_{E=E_0}}, \tag{2}$$
where $(dM/dE)|_{E=E_0}$ is the slope of the M vs E curve at the trial value, E_0. The slope expression is obtained by differentiating the Kepler's equation, thus getting
$$\frac{dM}{dE} = 1 - e \cos E. \tag{3}$$
Therefore, Eq. (2) may be written as
$$E_1 = E_0 + \frac{M - M_0}{1 - e \cos E_0} \tag{4}$$
or, in general, Newton's iteration for solving Kepler's equation is
$$E_{n+1} = E_n + \frac{M - M_n}{1 - e \cos E_n}. \tag{5}$$

Thus, the first approximation E_1 is

$$E_1 = E_0 + \frac{M - M_0}{1 - e \cos E_0} = 151°.743 + \frac{151°.743 - 150°.429}{1 - 0.04844 \cos 151°.743}$$

or

$$E_1 = 151°.743 + 1°.26 = 153°.003.$$

Now, the second approximation M_1 is obtained as

$$M_1 = E_1 - e \sin E_1$$

$$= 153°.003 - 0.04844 \times \frac{180}{\pi} \sin 153°.003$$

$$= 151°.743.$$

Similarly, Eq. (5) gives us the second approximation E_2 as

$$E_2 = E_1 + \frac{M - M_1}{1 - e \cos E_1} = 153°.003 + \frac{151°.743 - 151°.743}{1 - 0.04844 \cos 153°.003}$$

or

$$E_2 = 153°.003 + 0 = 153°.003.$$

As there is no change in the iterations E_1 and E_2, the eccentric anomaly of Jupiter after five years of perihelion passage is found to be $153°.003$.

Example 5.12 On 7 June, 1960, the eccentricity of the orbit of satellite Explorer-I was 0.106. The semi-major axis was 7.52×10^3 km. Calculate (i) the period of the satellite and the mean daily motion and (ii) its position in the orbit, 1 hour after its perigee passage.

Solution The given data is

$$e = 0.106 \quad \text{and} \quad a = 7.52 \times 10^3 \text{ km.}$$

Hence, the period of the orbit can be computed from

$$P = \frac{2\pi}{\sqrt{GM}} a^{3/2} = \frac{2 \times 3.1415}{\sqrt{398603.6}} (\sqrt{10} \times 10^4 \times \sqrt{7.52} \times 7.52)$$

or

$$P = 64.899 \times 10^2 \text{ s} = 1.8027 \text{ hr.} \tag{1}$$

The mean motion

$$n = \frac{2\pi}{P} = \frac{360}{1.8027} = 199°.7/\text{hr}$$

The mean anomaly is

$$M = n(t - \tau) = n \times 1 = 199°.7. \tag{2}$$

For computing eccentric anomaly E, we use Newton's iteration as follows: We take E_0 as a first approximation

$$E_0 = M = 199°.7$$

and compute

$$M_0 = E_0 - e \sin E_0 = 199°.7 - (0.106)\left(\frac{180}{\pi}\right)\sin 199°.7 = 201°.747$$

and

$$E_1 = E_0 + \frac{M - M_0}{1 - e \cos E_0} = 199°.7 + \frac{199°.7 - 201°.747}{1 - 0.106 \cos 199°.7}$$

or

$$E_1 = 199°.7 - 1°.861 = 197°.839.$$

To find the position of the Explorer-I in its orbit, we use

$$r = a(1 - e \cos E)$$

that is,

$$r = 7.52 \times 10^3 (1 - 0.106 \cos 197°.839)$$

$$= 8.2788 \times 10^3 \text{ km}$$

$$\approx 8279 \text{ km}.$$

Recalling the equation of the orbit as

$$\frac{p}{r} = \frac{a(1-e^2)}{r} = 1 + e \cos v,$$

we obtain

$$v = \cos^{-1}\left[\frac{a(1-e^2) - r}{er}\right] = \cos^{-1}(-0.9612) = 163°.9802.$$

Hence, the polar coordinates of the satellite after one hour of perigee passage are found to be

$$(8279, 163°.9802). \tag{3}$$

Example 5.13 An Earth satellite in an orbit whose plane coincides with that of the Earth's equator is 1000 km above the Earth's surface at perigee and 2000 km at apogee. What are the values of a and e for its orbit and what will be its polar coordinates 1½ hours after its perigee passage?

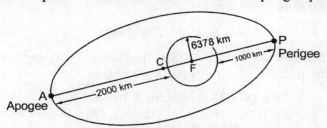

Figure 5.25 Description of Example 5.13.

Solution Let F be the centre of the Earth which is also a focus of the elliptic orbit, as shown in Fig. 5.25. From the given data, we have (see Fig. 5.25)

$$FP = q = a(1-e) = 6378 + 1000 = 7378 \text{ km}$$

$$FA = q' = a(1+e) = 6378 + 2000 = 8378 \text{ km},$$

which results in two equations in two unknowns a and e. On division, we get

$$\frac{1-e}{1+e} = \frac{7378}{8378},$$

which gives

$$e = 0.0635. \tag{1}$$

But, $a(1-e) = 7378$. Therefore,

$$a = \frac{7378}{1 - 0.0635} = 7877.9365 \approx 7878 \text{ km}. \tag{2}$$

From the known relation for period of the orbit, we have

$$P = \frac{2\pi}{\sqrt{GM}} a^{3/2} = \frac{2\pi(7878)^{3/2}}{\sqrt{398603}} = 6958.789 \approx 6959 \text{ s}$$

or

$$P = 0.08054 \text{ day}.$$

The mean motion n is defined by

$$n = \frac{2\pi}{P} = \frac{2\pi}{0.08054} = 78.01 \text{ rad/day}.$$

1½ hours after perigee passage, the mean anomaly M is given by

$$M = n(t-\tau) = 78.01 \times \frac{1.5}{24} = 4.8758 \text{ rad}$$

or

$$M = 279°.3643. \tag{3}$$

To find the polar coordinates (r, v) of the satellite at the desired time, where v is true anomaly, we solve Kepler's equation, using Newton's iterative scheme, as follows: We assume

$$E_0 = M = 279°.3643.$$

After one iteration, we find $E_1 = 275.7371$. Taking $E = E_1$ approximately and recalling the relation $r = a(1-e \cos E)$, we obtain

$$r = 7878(1 - 0.0635 \cos E_1) = 7827.99 \text{ km}. \tag{4}$$

Now, using the relation

$$\tan \frac{v}{2} = \left(\frac{1+e}{1-e}\right)^{1/2} \tan \frac{E}{2}$$

we find

$$\frac{v}{2} = 136°.05 \quad \text{or} \quad v = 272°.1. \tag{5}$$

Hence, the polar coordinates of the satellite in its orbit at the required time are (7827.99 km, 272°.1).

Example 5.14 Consider the Earth to be a point mass revolving round the Sun at a speed of 30 km/s. The XY-plane is the plane of its orbit. The mean distance (Earth to Sun) is one astronomical unit (AU) which is 1.5×10^8 km. At $x = 0.707$, $y = 0.707$ and $z = 0$ (AU), let a mass particle m be projected perpendicular to the XY-plane with a speed of 15 km/s. What will be the orientation of the resulting orbit, given that:

$$\dot{x} = -21 \text{ km/s}, \qquad \dot{y} = 21 \text{ km/s}, \qquad \dot{z} = 15 \text{ km/s}.$$

Solution The orientation of the orbit is determined if the longitude of the ascending node Ω and inclination i, of the orbit are known. The description of the example will be clear through the following diagram (see Fig. 5.26).

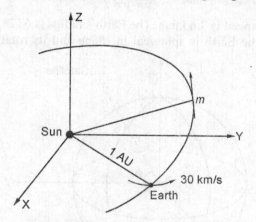

Figure 5.26 Description of Example 5.14.

The fact that the areal velocity is constant (Kepler's law) gives the areal velocity components as

$$c_1 = y\dot{z} - \dot{y}z = (0.707 \times 1.5 \times 10^8)\,15 - 0 = 1.59 \times 10^9$$

$$c_2 = z\dot{x} - \dot{z}x = -1.59 \times 10^9$$

$$c_3 = x\dot{y} - \dot{x}y = 4.454 \times 10^9$$

and therefore, the magnitude of the angular momentum of the system is given by

$$h = \sqrt{c_1^2 + c_2^2 + c_3^2} = 5.0 \times 10^9 \text{ km}^2/\text{s}.$$

Now, the longitude of the ascending node Ω and the inclination of the orbit i is obtained from Eq. (5.92), as shown below:

$$\tan \Omega = -\frac{c_1}{c_2} = 1.$$

Therefore,

$$\Omega = 45° \tag{1}$$

and

$$\cos i = \frac{c_3}{h} = \frac{4.46 \times 10^9}{5 \times 10^9} = 0.892,$$

which yields

$$i = 26°.87. \tag{2}$$

Hence, Eqs. (1) and (2) constitute the required result.

Example 5.15 A satellite is launched 500 km above the Earth's surface at longitude 45°E (λ_E) and latitude (ϕ_N) 30°N in such a way that

$$\dot{x} = \dot{y} = \dot{z}$$

and its resultant speed is 7.6 km/s. The Earth's radius is 6378 km. Find Ω and i. Assume that the Earth is spherical in shape and its rotation is neglected.

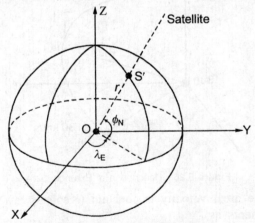

Figure 5.27 Description of Example 5.15.

Solution From Fig. 5.27, the position of the satellite at launch is given by

$$x = r \cos \phi_N \cos \lambda_E,$$
$$y = r \cos \phi_N \sin \lambda_E,$$
$$z = r \sin \phi_N.$$

From the given data, we have
$$r = 6378 + 500 = 6878 \text{ km}, \quad \lambda_E = 45°, \quad \phi_N = 30°.$$
Thus, the position of the satellite at launch is found to be
$$x = 6878 \cos 30° \cos 45° = 4211.8975 \text{ km},$$
$$y = 6878 \cos 30° \sin 45° = 4211.8975 \text{ km},$$
$$z = 6878 \sin 30° = 3439 \text{ km}.$$
It is also given that
$$\dot{x}^2 + \dot{y}^2 + \dot{z}^2 = 3\dot{x}^2 = 7.6^2$$
or
$$\dot{x}^2 = 19.253383.$$
Thus, we can write
$$\dot{x} = -4.387862, \quad \dot{y} = 4.387862, \quad \dot{z} = 4.387862.$$
Now, we can compute the areal velocity components as
$$c_1 = y\dot{z} - \dot{y}z = 3391.476 \text{ km}^2/\text{s},$$
$$c_2 = z\dot{x} - \dot{z}x = -33572.14 \text{ km}^2/\text{s},$$
$$c_3 = x\dot{y} - \dot{x}y = 36962.449 \text{ km}^2/\text{s}.$$
The magnitude of the angular momentum is
$$h = \sqrt{c_1^2 + c_2^2 + c_3^2} = 50048.975 \text{ km}^2/\text{s}.$$
Finally,
$$i = \cos^{-1}\left(\frac{c_3}{h}\right) = \cos^{-1}\left(\frac{36962.449}{50048.975}\right) = 42°39'$$
and
$$\Omega = \tan^{-1}\left(-\frac{c_1}{c_2}\right) = \tan^{-1}\left(\frac{3391.476}{33572.14}\right) = 5°.769.$$

Example 5.16 Three masses m_1, m_2 and m_3 at the corners of an equilateral triangle of side s, attract one another according to Newton's law of gravitation. Determine the rotational motion which leaves the relative separation of each mass unchanged.

Solution Let the positions of the three masses in the Cartesian frame at the corners of an equilateral triangle be as shown in Fig. 5.28; then, the three masses have coordinates $(0, 0)$, $(s, 0)$, $(s/2, \sqrt{3}s/2)$, respectively, so that the position vector of each mass can be written as

$$\vec{r}_1 = \vec{0}, \qquad \vec{r}_2 = (s\hat{i}, 0), \qquad \vec{r}_3 = \left(\frac{s}{2}\hat{i} + \frac{\sqrt{3}}{2}s\hat{j}\right).$$

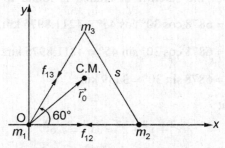

Figure 5.28 Description of Example 5.16.

Then, the position vector of the centre of mass (C.M.) of the system is

$$\vec{r}_0 = \frac{\Sigma m_i \vec{r}_i}{\Sigma m_i}$$

or

$$\vec{r}_0 = \frac{s}{m_1 + m_2 + m_3} \left[\left(m_2 + \frac{m_3}{2}\right)\hat{i} + \frac{\sqrt{3}}{2} m_3 \hat{j}\right]. \tag{1}$$

Now, consider the forces acting on m_1. There are two attractive forces due to m_2 and m_3 and are given by

$$\vec{f}_{12} = G\frac{m_1 m_2}{s^2}\hat{i}, \qquad \vec{f}_{13} = G\frac{m_1 m_3}{s^2}\left(\frac{\hat{i}}{2} + \frac{\sqrt{3}}{2}\hat{j}\right). \tag{2}$$

Let their resultant be denoted by

$$\vec{f}_1 = \vec{f}_{12} + \vec{f}_{13} = G\frac{m_1}{s^2}\left[\left(m_2 + \frac{m_3}{2}\right)\hat{i} + \frac{\sqrt{3}}{2}m_3\hat{j}\right] \tag{3}$$

which indicates that $\vec{f}_1$ is parallel to $\vec{r}_0$, of course both originate from the same point O. The force $\vec{f}_1$ passes through the centre of mass of the system. Thus, m_1 is acted upon by a central force with centre at the C.M. of the system. Therefore, m_1 moves in a circle about the centre of mass. The radius of this circular orbit is

$$R = |\vec{r}_0| = \frac{s}{m_1 + m_2 + m_3}(m_2^2 + m_3^2 + m_2 m_3)^{1/2}. \tag{4}$$

Let the linear velocity of m_1 be v_1, which can be computed from the fact $|\vec{f}_1| = m_1 v_1^2 / r_0$. Thus,

$$v_1^2 = \frac{R}{m_1}|\vec{f}_1| = \frac{G}{s}\frac{m_2^2 + m_3^2 + m_2 m_3}{m_1 + m_2 + m_3} \tag{5}$$

In fact, the results of Eqs. (4) and (5) hold true for all the three masses, which can be seen by permuting the choices of the position of three masses. The period of the orbit of each mass can be computed as

$$P = \frac{2\pi R}{v_1} = 2\pi s \left[\frac{s}{G(m_1 + m_2 + m_3)} \right]^{1/2}, \tag{6}$$

which determines the rotational motion of each mass when they are located at the vertices of an equilateral triangle.

5.9 POSITION IN A PARABOLIC ORBIT

The parabolic orbit is rarely found in nature. However, the orbits of some comets are nearly parabolic. An object travelling in a parabolic path is on a one-way trip to infinity and will never retrace the same path again. To find the position (r, v) of an object in a parabolic path as a function of time, we may recall that its absolute angular momentum is constant. That is,

$$r^2 \frac{dv}{dt} = h = \sqrt{p\mu}. \tag{5.110}$$

However, the general equation of the conic is written as

$$\frac{p}{r} = 1 + e \cos v. \tag{5.111}$$

Eliminating r between Eqs. (5.110) and (5.111), we get

$$\frac{dv}{(1 + e \cos v)^2} = \sqrt{\frac{\mu}{p^3}} \, dt. \tag{5.112}$$

This expression can be integrated for all e. But for the cases $e = 0$ and 1, an additional substitution will put the result in a useful form. $e = 0$ corresponds to the circular orbit and the theory developed for elliptic orbit is applicable for circular orbits, too, merely by setting $e = 0$. The perigee distance for an elliptic orbit is

$$q = a(1 - e) \quad \text{and} \quad p = a(1 - e^2) = q(1 + e).$$

Thus, in the case of parabolic orbit, when $e = 1$, $p = 2q$, Eq. (5.111) can be written as

$$\frac{2q}{r} = 1 + \cos v = 2 \cos^2 \left(\frac{v}{2} \right)$$

or

$$r = q \sec^2 \left(\frac{v}{2} \right). \tag{5.113}$$

Here, q is the distance of the focus from the vertex (see Fig. 5.29).

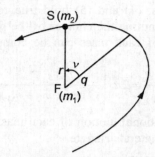

Figure 5.29 Parameters in a parabolic orbit.

Since the parabolic motion is applicable to the motion of comets, let us briefly describe comets. A comet is a loose aggregation of rocks moving around the Sun. When observed, it is surrounded by a haze of gases which give it fuzzy appearance. When close to the Sun, it may develop a tail which will point away from the Sun. The orbits of comets are nearly parabolic and travel in orbits that remain close to the Sun. Taking M as the mass of the Sun itself, so that $\mu = GM$, Eq. (5.112) reduces to

$$\frac{1}{4}\sec^4\left(\frac{v}{2}\right)dv = \sqrt{\frac{GM}{p^3}}\,dt = \sqrt{\frac{GM}{2^3 q^3}}\,dt$$

or

$$\sec^4\left(\frac{v}{2}\right)dv = \sqrt{\frac{2GM}{q^3}}\,dt. \qquad (5.114)$$

That is,

$$\left[1+\tan^2\left(\frac{v}{2}\right)\right]\sec^2\left(\frac{v}{2}\right)d\left(\frac{v}{2}\right) = \sqrt{\frac{GM}{2q^3}}\,dt.$$

Integrating this expression, we get

$$\frac{1}{3}\tan^3\left(\frac{v}{2}\right) + \tan\left(\frac{v}{2}\right) = \sqrt{\frac{GM}{2q^3}}\,t + \text{constant}. \qquad (5.115)$$

At $t = T$, when the comet passes through perihelion, that is, when $v = 0$, we obtain

$$\text{constant} = -\sqrt{\frac{GM}{2q^3}}\,T.$$

Thus, we arrive at a cubic equation

$$\frac{1}{3}\tan^3\left(\frac{v}{2}\right) + \tan\left(\frac{v}{2}\right) = \sqrt{\frac{GM}{2q^3}}\,(t-T). \qquad (5.116)$$

This equation is known as *Baker's equation*, which gives us a relation between true anomaly v and time for parabolic orbits. This cubic equation can be solved numerically to get v as a function of t, with the help of a digital computer. Finally, the radius vector can be readily obtained from Eq. (5.113) as

$$r = q\sec^2\left(\frac{v}{2}\right) = q\left[1 + \tan^2\left(\frac{v}{2}\right)\right]. \quad (5.117)$$

Some of the useful astronomical constants are given in Table 5.1, in the SI system of units.

Table 5.1 Some Useful Astronomical Constants

Description (symbol)	Numerical value
Gravitational constant (G)	6.668×10^{-11} N·m^2/kg^2
Mass of the Earth (M_E)	5.98×10^{24} kg
Gaussian gravitational constant (GM_E)	398603.6 km^3/s^2
π	3.1415926535897
Equatorial radius of the Earth (r_E)	6378.165 km
Polar radius of the Earth (r_P)	6356.912 km
Astronomical unit (AU)	1.49598845×10^8 km
Angular velocity of the Earth about its axis ($d\theta/dt$)	4.178074×10^{-3} deg/s $= 7.292115156 \times 10^{-5}$ rad/s
GMs	1.34×10^{11} km^3/s^2
Mass of the Moon	7.35×10^{22} kg
Radius of the Moon	1738 km
Surface gravity of the Moon	$0.16g$
Mass of the Sun	1.99×10^{30} kg
Surface gravity of the Sun	$27.9g$
Radius of the Sun	696000 km
Earth's gravity (g)	9.806 m/s^2
Velocity of light (c)	3×10^8 m/s

5.10 THE HYPERBOLIC ORBIT

Meteors, which generally strike the Earth and interplanetary space probes sent from the Earth, travel in hyperbolic paths relative to the Earth. The hyperbola is very interesting in the sense that it has two branches. The geometry of the hyperbolic orbit is shown in Fig. 5.30.

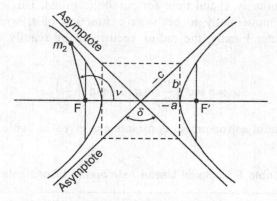

Figure 5.30 Geometry of the hyperbolic orbit.

The arms of a hyperbola are asymptotic to the two intersecting straight lines (the asymptotes). Considering the left-hand focus, F, as the prime focus, where the centre of our gravitating Earth is located, then, the left-hand branch of the hyperbola alone represents the possible orbit. If, instead, we assume a force of repulsion between our satellite (m_2) and the body located at F (such as the force between two like-charged particles), then the right-hand branch represents the orbit. The parameters a, b and c, shown in Fig. 5.30, satisfy the relation

$$c^2 = a^2 + b^2 \qquad (5.118)$$

for the hyperbola. The angle between the asymptotes, that is, the angle through which, the path of a space probe is turned by its encounter with a planet denoted by δ is related to the geometry of the hyperbola by the formula

$$\sin \frac{\delta}{2} = \frac{a}{c}. \qquad (5.119)$$

But we know that $e = c/a$, thus the above equation becomes

$$\sin \frac{\delta}{2} = \frac{1}{e}, \qquad (5.120)$$

from which, we can observe that the greater the eccentricity of the hyperbola, the smaller will be the turning angle.

The equation of the hyperbolic orbit is

$$r = \frac{a(e^2 - 1)}{1 + e \cos v} \qquad (e > 1). \qquad (5.121)$$

When, v, the true anomaly is known, r can be found from the above relation.

EXERCISES

5.1 State the Kepler's laws of planetary motion.

5.2 Obtain the expression for the escape velocity of a satellite.

5.3 Define true anomaly, mean anomaly and eccentric anomaly in the case of an elliptic orbit. Derive the Kepler's equation.

5.4 Define a central force motion. Derive the differential equation of the orbit of a planet which is moving under a central force $F(r)$.

5.5 Verify that the Keplerian motion of a planet is governed by the inverse square law.

5.6 Derive the following expression in an orbital motion:
$$\tan\frac{v}{2} = \left(\frac{1+e}{1-e}\right)^{1/2} \tan\frac{E}{2}.$$

5.7 A particle moves in a circular orbit of radius a, under an attraction to a point inside the circle. The greatest and least speeds of the particle in its orbit are v_1 and v_2 respectively. Prove that the period of the orbit is
$$\pi a \frac{v_1 + v_2}{v_1 v_2}.$$

5.8 A satellite was launched with a velocity of 28000 km/h at an altitude of 640 km. If the burn-out velocity of the last stage is parallel to the Earth's surface, compute the maximum altitude from the Earth's surface that the satellite will reach.

5.9 Calculate the semi-major axis, the eccentricity, and the time period of RS-1 satellite projected to a distance of 300 km with a velocity of 7.864 km/s.

5.10 The orbit of an artificial satellite Explorer-VI is given as follows:
$$a = 27411.8 \text{ km}, \quad e = 0.758, \quad P = 752.77 \text{ min}.$$
Compute its velocities at perigee and apogee.

5.11 A particle moves under a central force which varies as some integral power of its distance from the centre of force. Show that there are almost two apsidal distances.

5.12 Find the law of force, if a particle under central force moves along a curve cardioide
$$r = a(1 + \cos\theta).$$

5.13 Show that the solution of the Kepler's equation can be written in the form
$$E = M + 2\sum_{k=1}^{\infty} \frac{1}{k} J_k(kl) \sin kM.$$

5.14 From the Kepler's equation, find E for a given $M = 62°$ and $e = 0.1$ using the successive approximation method or the Newton's iterative method.

5.15 Starting from the Kepler's equation, find the value of E using the method of successive approximation when $e = 0.2$ and $M = 45°$.

5.16 A comet probe has the following orbital elements:

$$a = 3.45482 \text{ AU}$$

$$e = 0.552735$$

$$P = 6.42154 \text{ yr}$$

$$T = 1948 \text{ September } 16.194 \text{ (perifocal passage time)}$$

What is the radius vector and true anomaly on November 1.00, 1949?

5.17 A satellite is in an elliptic orbit about the Earth. Show that its velocity away from the Earth is greatest when the radius vector is perpendicular to the major axis of the orbit, and find the magnitude of this velocity.

5.18 A satellite is moving in an elliptic orbit round the Earth. With the usual notation, show that the rate of change of true anomaly is given by

$$\frac{dv}{dt} = \frac{\sqrt{GM}}{a^{3/2}} \frac{(1+e\cos v)^2}{(1-e^2)^{3/2}}, \qquad 0 \le e \le 1.$$

Hence show that the time of flight from perigee is given by

$$t = \frac{a^{3/2}}{\sqrt{GM}} \left[2 \tan^{-1} \left(\sqrt{\frac{1-e}{1+e}} \tan \frac{v}{2} \right) - \frac{e\sqrt{1-e^2} \sin v}{1 + e \cos v} \right].$$

5.19 When free of the Earth's gravitational field, a space vehicle moves around the Sun under the sole influence of the latter. At $t = 0$, the vehicle is located at

$$x_0 = 0.707 \text{ AU}, \qquad y_0 = 0.707 \text{ AU}, \qquad z_0 = 0$$

and has a velocity given by

$$\dot{x}_0 = -21 \text{ km/s}, \qquad \dot{y}_0 = 21 \text{ km/s}, \qquad \dot{z}_0 = 15 \text{ km/s}.$$

Find the elements of the orbit.

5.20 From the radius and mass ratios

$$R(\text{Moon})/R(\text{Earth}) \approx 1/3.66$$

$$M(\text{Moon})/M(\text{Earth}) \approx 1/81.6$$

Show that the gravitational accelerations on the Moon and Earth are related by

$$g(\text{Moon})/g(\text{Earth}) \approx \frac{1}{6}.$$

Also, find the escape velocity from the surface of the Moon.

6

Lagrange and Hamilton Equations

Equations are more important to me, because politics is for the present, but an equation is something for eternity.

—Einstein

6.1 INTRODUCTION

The study of any dynamical system essentially involves two basic important steps: (i) mathematical formulation of the problem, that is, writing down the governing differential equations of motion and (ii) finding its solution. We have so far used the Newton's laws of motion for writing down the governing differential equations and analyzed the motion with the help of concepts like force, momentum and angular momentum, torque, and so on. As one encounters more difficult problems, it has been realized that the direct application of Newton's laws of motion becomes increasingly difficult. A different approach to study mechanics is attributed to Lagrange and Hamilton. Joseph Louis Lagrange (1736–1813) reformulated Newton's laws in a way that avoids the need to calculate forces acting on individual parts of a mechanical system and considers the system as a whole. They formulated the problems in mechanics in terms of fundamental quantities such as kinetic energy and work, which are scalars. In fact, the methods of Lagrange and Hamilton gave us a systematic procedure to write down the equations of motion for any dynamical system. We shall study, in what follows some of the basic concepts required for the study of Lagrangian description of a dynamical system.

Constraints and degrees of freedom

A constrained motion is one which takes place in a restricted manner. For example, the motion of a sphere along an inclined plane. The motion along a specified path is also a constrained motion as in the case of a guided missile or a rocket. As another example, we may consider the motion of a particle, which is constrained to move on the surface of a sphere of radius R. Suppose, the position of the particle is specified in terms of Cartesian coordinates (x, y, z). Evidently, we observe that these coordinates are not independent. In fact, they are connected by the equation of constraint

$$(x-x_0)^2 + (y-y_0)^2 + (z-z_0)^2 = R^2,$$

where the centre of the sphere is located at (x_0, y_0, z_0). Thus, there are three coordinates and one constraint resulting in two degrees of freedom. Instead, if we choose the origin of the coordinate system at the centre of the sphere and use the spherical coordinates θ and ϕ to locate the particle, then, no additional coordinate such as r, is required, since the motion of the particle is confined to the surface of the sphere, implying $r = R$ is not a variable. Hence, the coordinates θ and ϕ are independent and conform to the constraint of the problem.

Thus, the number of independent ways in which a mechanical system can move without violating the prescribed constraints on the system is known as the *number of degrees of freedom*. In other words, the degrees of freedom of a dynamical system form the minimum number of independent coordinates required to specify the system compatible with the constraints. For example, the motion of a particle moving freely in space can be described by a set of three coordinates, say (x, y, z). Hence, the number of degrees of freedom is three. If the system consists of two particles moving freely in space, such a system can be described by two sets of coordinates, say (x_1, y_1, z_1) and (x_2, y_2, z_2). Thus, its degrees of freedom is six. Similarly, for a system of n particles moving independently of each other in three-dimensional space, the number of degrees of freedom is obviously $3n$, because, there exists a set of $3n$ coordinates $(x_1, y_1, z_1, ..., x_n, y_n, z_n)$ that uniquely specifies the configuration of the system. If this system of n particles is subjected to k independent constraints expressed as

$$f_1(x_1, y_1, z_1, ..., x_n, y_n, z_n, t) = c_1$$
$$f_2(x_1, y_1, z_1, ..., x_n, y_n, z_n, t) = c_2$$
$$\vdots$$
$$f_k(x_1, y_1, z_1, ..., x_n, y_n, z_n, t) = c_k$$

then, the number of degrees of freedom of this system is $(3n - k)$.

Generalized coordinates

To describe the configuration of a system, we naturally select the smallest possible number of variables, which we call *generalized coordinates* of the system. It may be noted that we need not restrict our choice to Cartesian coordinates alone to describe a system. Sometimes, it may not be convenient to introduce all coordinates with respect to a fixed origin. For instance, when we deal with a rigid body motion, we specify three Cartesian coordinates to locate the centre of mass with respect to an external fixed origin and three Eulerian angles. In general, the number of degrees of freedom coincides with the minimum number of independent coordinates necessary to describe a system uniquely. While selecting generalized coordinates, we should bear in mind the following important points:

(i) They should determine the system completely.
(ii) They may be varied arbitrarily and independently of each other without violating the constraints on the system.
(iii) The generalized coordinates are conventionally denoted by the symbols $q_1, q_2, q_3, \ldots, q_n$. They may not have physical meaning, nor are they unique. They should give us a reasonable mathematical simplification.

For illustration, we consider the following examples:

Example 6.1 When a particle is moving in a plane, we may describe its motion in Cartesian coordinates or, more conveniently, in polar coordinates (r, θ). We may then write,

$$q_1 = x, \quad q_2 = y$$

or

$$q_1 = r = (x^2 + y^2)^{1/2}, \quad q_2 = \theta = \tan^{-1}\left(\frac{y}{x}\right).$$

Alternatively, we may also write as

$$x = q_1 \cos q_2, \quad y = q_1 \sin q_2$$

or

$$x_i = x_i(q_1, q_2).$$

If the problem involves spherical symmetry, we may conveniently use spherical polar coordinates (r, θ, ϕ) as (q_1, q_2, q_3), such that

$$x = r \sin \theta \cos \phi, \quad y = r \sin \theta \sin \phi, \quad z = r \cos \theta$$

or

$$x = q_1 \sin q_2 \cos q_3, \quad y = q_1 \sin q_2 \sin q_3, \quad z = q_1 \cos q_2$$

or

$$x_i = x_i(q_1, q_2, q_3).$$

Thus, in general, the relation between the old Cartesian coordinates and the generalized coordinates can be written in the form

$$x_i = x_i(q_1, q_2, q_3, \ldots, q_n).$$

Example 6.2 If a particle is constrained to move on a spherical surface, its position can be specified by two angles θ and ϕ. The Cartesian coordinates of the particle will be functions of two variables θ and ϕ. Of course, time is an independent variable on which both θ and ϕ depend. The variables θ and ϕ are called 'generalized coordinates' usually denoted by q_1 and q_2. We may then write

$$x_i = x_i(q_1, q_2).$$

With the help of coordinates, θ and ϕ, the position of the particle can be specified uniquely. Then the natural question that arises is, why the pair (x, y) cannot be treated as generalized coordinates. The reason being that for every pair (x, y), there are two possible values for z, consistent with the constraint equation

$$x^2 + y^2 + z^2 = r^2,$$

and hence not suitable to become a member of generalized coordinates.

Example 6.3 A rigid body moving freely in space has six degrees of freedom.

Example 6.4 A rod of mass m and length l which is free to move in any direction on a frictionless inclined plane has only one degree of freedom, as it can slide down vertically.

6.2 CLASSIFICATION OF A DYNAMICAL SYSTEM

In general, there is an eight-fold classification of all known dynamical systems. They are:

(1, 2): Scleronomic and rheonomic system

A configuration of the scleronomic system is given, when the values of the generalized coordinates $q_1, q_2, ..., q_n$ are given by the equations of the form

$$x_i = x_i(q_1, q_2, ..., q_n),$$

which do not depend explicitly on time.

In a rheonomic system, it is necessary to specify the time t as well. Then the equations will be of the form

$$x_i = x_i(q_1, q_2, ..., q_n, t).$$

Thus in a *scleronomic* system, there will be only fixed constraints, whereas in a *rheonomic* system there will be moving constraints.

(3, 4): Conservative and non-conservative system

In a *conservative* system, the generalized forces are derivable from a potential energy $V = V(q_1, q_2, ..., q_n)$. That is, if $Q_1, Q_2, ..., Q_n$ represent generalized forces, then

$$Q_1 = -\frac{\partial V}{\partial q_1}, \quad Q_2 = -\frac{\partial V}{\partial q_2}, ..., \quad Q_n = -\frac{\partial V}{\partial q_n}$$

or

$$\vec{Q} = -\nabla \vec{V}$$

otherwise, the system is *non-conservative*.

(5, 6): Holonomic and non-holonomic systems

Let $q_1, q_2, ..., q_n$ denote the generalized coordinates, describing a system and let t denote the time and if all the constraints of the system can be expressed as equations of the form

$$f_i(q_1, q_2, ..., q_n, t) = 0,$$

which is independent of velocities $\dot{q}_1, \dot{q}_2, ..., \dot{q}_n$, where i is the ith component, then the system is called *holonomic*.

However, if the constraints cannot be expressed as relations among the generalized coordinates, then the system is referred as *non-holonomic*.

(7, 8): Bilateral and unilateral systems

If at any point on the constraint surface, both the forward and backward motions are possible, such a system is called *bilateral*. In this system, the constraint relations are in the form of equations but not in the form of inequalities.

However, if at some point on the constraint surface, the forward motion is not possible, while the constraint relations are expressed in the form of inequalities, then such a system is called *unilateral*.

Systems which are scleronomic, conservative and holonomic are called *simple*, while the others are termed *general* systems.

6.3 LAGRANGE'S EQUATIONS FOR SIMPLE SYSTEMS

Consider a simple dynamical system consisting of n-particles. Let m_i be the mass of a typical particle and $\vec{r}_i$ be its position vector. Suppose our simple system has n degrees of freedom and n generalized coordinates q_ρ ($\rho = 1, 2, ..., n$). Then, the position vector $\vec{r}_i$ is a function of q_ρ, which can be expressed as (scleronomic system)

$$\vec{r}_i = \vec{r}_i(q_1, q_2, ..., q_n), \qquad (6.1)$$

where q_ρ explicitly depends on t. As the system moves in a specified manner, we have by using the chain rule of partial differentiation that

$$\frac{d\vec{r}_i}{dt} = \sum_{\rho=1}^{n} \frac{\partial \vec{r}_i}{\partial q_\rho} \frac{dq_\rho}{dt}$$

or

$$\dot{\vec{r}}_i = \sum_{\rho=1}^{n} \frac{\partial \vec{r}_i}{\partial q_\rho} \dot{q}_\rho. \qquad (6.2)$$

Here, the quantities $\dot{q}_\rho$ are called *generalized velocities* and, therefore, we may write

$$\dot{\vec{r}}_i = \vec{V}_i(q_1, q_2, ..., q_n, \dot{q}_1, \dot{q}_2, ..., \dot{q}_n), \qquad (6.3)$$

which means that we have a function of $2n$ quantities $q_\rho, \dot{q}_\rho$. Differentiating Eq. (6.2) partially with respect to $\dot{q}_\rho$, we obtain

$$\frac{\partial \dot{\vec{r}}_i}{\partial \dot{q}_\rho} = \frac{\partial \vec{r}_i}{\partial q_\rho}, \qquad (6.4)$$

since all other quantities are treated as constants.

This important result is called *cancellation of dots*. Similarly, from Eq. (6.2), we get

$$\frac{\partial \dot{\vec{r}}_i}{\partial q_s} = \sum_{\rho=1}^{n} \frac{\partial^2 \vec{r}_i}{\partial q_s \, \partial q_\rho} \dot{q}_\rho. \qquad (6.5)$$

At the same time, it may be noted that

$$\frac{d}{dt}\left(\frac{\partial \vec{r}_i}{\partial q_s}\right) = \sum_{\rho=1}^{n} \frac{\partial}{\partial q_\rho}\left(\frac{\partial \vec{r}_i}{\partial q_s}\right) \frac{dq_\rho}{dt} = \sum_{\rho=1}^{n} \frac{\partial^2 \vec{r}_i}{\partial q_\rho \, \partial q_s} \dot{q}_\rho. \qquad (6.6)$$

Now, comparing Eqs. (6.5) and (6.6), we obtain

$$\frac{d}{dt}\left(\frac{\partial \vec{r}_i}{\partial q_s}\right) = \frac{\partial \dot{\vec{r}}_i}{\partial q_s}. \qquad (6.7)$$

This result, which is equally important is called *interchange of d and ∂*. These two mathematical tricks, that is, cancellation of dots and interchange of d and ∂ are widely used in Lagrangian method.

The kinetic energy of the system considered can be written as

$$T = \sum_{i=1}^{n} \frac{1}{2} m_i \, (\dot{\vec{r}}_i \cdot \dot{\vec{r}}_i) = T(\vec{q}, \dot{\vec{q}}). \qquad (6.8)$$

Differentiating this expression, partially with respect to q_ρ, we get

$$\frac{\partial T}{\partial q_f} = \sum_{i=1}^{n} m_i \dot{\vec{r}}_i \cdot \frac{\partial \dot{\vec{r}}_i}{\partial q_\rho}. \qquad (6.9)$$

Similarly, differentiating Eq. (6.8) with respect to $\dot{q}_\rho$, we obtain, on using cancellation of dots that

$$\frac{\partial T}{\partial \dot{q}_\rho} = \sum_{i=1}^{n} m_i \dot{\vec{r}}_i \cdot \frac{\partial \dot{\vec{r}}_i}{\partial \dot{q}_\rho} = \sum_{i=1}^{n} m_i \dot{\vec{r}}_i \cdot \frac{\partial \vec{r}_i}{\partial q_\rho}. \qquad (6.10)$$

Now, differentiating the above equation with respect to t, we get

$$\frac{d}{dt}\left(\frac{\partial T}{\partial \dot{q}_\rho}\right) = \sum_{i=1}^{n} m_i \ddot{\vec{r}}_i \cdot \frac{\partial \vec{r}_i}{\partial q_\rho} + \sum_{i=1}^{n} m_i \dot{\vec{r}}_i \cdot \frac{d}{dt}\left(\frac{\partial \vec{r}_i}{\partial q_\rho}\right).$$

Interchanging d and ∂, we have

$$\frac{d}{dt}\left(\frac{\partial T}{\partial \dot{q}_\rho}\right) = \sum_{i=1}^{n} m_i \ddot{\vec{r}}_i \cdot \frac{\partial \vec{r}_i}{\partial q_\rho} + \sum_{i=1}^{n} m_i \dot{\vec{r}}_i \cdot \frac{\partial \dot{\vec{r}}_i}{\partial q_\rho}. \tag{6.11}$$

Subtracting Eq. (6.9) from Eq. (6.11), we get

$$\frac{d}{dt}\left(\frac{\partial T}{\partial \dot{q}_\rho}\right) - \frac{\partial T}{\partial q_\rho} = \sum_{i=1}^{n} m_i \ddot{\vec{r}}_i \cdot \frac{\partial \vec{r}_i}{\partial q_\rho}. \tag{6.12}$$

It may be observed that, this result is obtained from kinematics without using Newton's laws of motion.

Suppose $\vec{F}_i$ is the total force (applied force + internal force) acting on an ith particle, then, the Newton's second law gives

$$m_i \ddot{\vec{r}}_i = m_i \vec{a}_i = \vec{F}_i. \tag{6.13}$$

Now, Eq. (6.12) can be rewritten as

$$\frac{d}{dt}\left(\frac{\partial T}{\partial \dot{q}_\rho}\right) - \frac{\partial T}{\partial q_\rho} = \sum_{i=1}^{n} \vec{F}_i \cdot \frac{\partial \vec{r}_i}{\partial q_\rho} = Q_\rho \text{ (say)}. \tag{6.14}$$

This is one form of Lagrange's equation, where Q_ρ are generalized forces. Now, consider a set of arbitrary infinitesimal increments δq_ρ. In view of Eq. (6.1), they give corresponding displacements to the particles as specified below:

$$\delta \vec{r}_i = \sum_{\rho=1}^{n} \frac{\partial \vec{r}_i}{\partial q_\rho} \delta q_\rho.$$

Then, the work done due to these displacements can be expressed as

$$\delta W = \sum_{i=1}^{n} \vec{F}_i \cdot \delta \vec{r}_i = \sum_{\rho=1}^{n} \left[\sum_{i=1}^{n} \vec{F}_i \cdot \frac{\partial \vec{r}_i}{\partial q_\rho}\right] \delta q_\rho. \tag{6.15a}$$

It may also be noted that the system has a potential energy V, which is a function of q's only, which we may write as

$$V = V(\vec{q}).$$

Hence, the work done in the above displacements can be expressed as

$$\delta W = -\delta V = \sum_{\rho=1}^{n} \frac{\partial V}{\partial q_\rho} \delta q_\rho. \tag{6.15b}$$

Comparing Eqs. (6.15a) and (6.15b) and noting that δq_ρ are arbitrary, we obtain

$$\sum_{i=1}^{n} \vec{F}_i \cdot \frac{\partial \vec{r}_i}{\partial q_\rho} = -\frac{\partial V}{\partial q_\rho}. \qquad (6.16)$$

Consequently, Eq. (6.14) can be recast as

$$\frac{d}{dt}\left(\frac{\partial T}{\partial \dot{q}_\rho}\right) - \frac{\partial T}{\partial q_\rho} = -\frac{\partial V}{\partial q_\rho}. \qquad (6.17)$$

Now, defining the Lagrangian (L) as

$$L = T - V = L(\vec{q}, \dot{\vec{q}}). \qquad (6.18)$$

Noting that V is a function of $\vec{q}$ only, we can write

$$\frac{\partial L}{\partial \dot{q}_\rho} = \frac{\partial T}{\partial \dot{q}_\rho} \quad \text{and} \quad \frac{\partial L}{\partial q_\rho} = \frac{\partial T}{\partial q_\rho} - \frac{\partial V}{\partial q_\rho}.$$

Substituting these expressions into Eq. (6.17), we arrive at the following final form:

$$\frac{d}{dt}\left(\frac{\partial L}{\partial \dot{q}_\rho}\right) - \frac{\partial L}{\partial q_\rho} = 0, \qquad (6.19)$$

which is known as the Lagrange's equation of motion for a conservative, scleronomic (as there are no moving constraints) and holonomic (as we can vary the generalized coordinates arbitrarily without violating the constraints) system. In other words, Eq. (6.19) represents the Lagrange's equation of motion for a simple system. These are n ordinary differential equations of second order and hence their solution contains $2n$ arbitrary constants, which can be determined by considering the initial conditions at $t = 0$ on both the generalized coordinates q_ρ and generalized velocities $\dot{q}_\rho$.

In case of a non-conservative system, the Lagrange's equation of motion assumes the form

$$\frac{d}{dt}\left(\frac{\partial T}{\partial \dot{q}_\rho}\right) - \frac{\partial T}{\partial q_\rho} = Q_\rho, \qquad (6.20)$$

where Q_ρ are generalized forces.

For illustration of these concepts, we shall consider the following well-known examples:

Example 6.5 Use the Lagrangian method to obtain the equation of motion for one-dimensional harmonic oscillator.

Solution The harmonic oscillator consists of a spring and a particle which can move on a straight line, which we shall take as x-axis (see Fig. 1.10). The particle is attracted towards the origin by a controlling force, $-kx\hat{i}$, where k is the spring constant. The harmonic oscillator is a simple system, that is scleronomic, conservative and holonomic system. Hence, the kinetic energy is given by

$$T = \frac{1}{2} m\dot{x}^2. \tag{1}$$

The potential energy is

$$V = -\int F(x)\, dx = -\int -kx\, dx = \frac{1}{2} kx^2 + \text{constant}.$$

If we choose the horizontal plane passing through the position of equilibrium, then $V = 0$, $x = 0$, which gives constant of integration as zero. Thus,

$$V = \frac{1}{2} kx^2. \tag{2}$$

Now, we define the Lagrangian (L) as

$$L = T - V = \frac{1}{2} m\dot{x}^2 - \frac{1}{2} kx^2$$

and get

$$\frac{\partial L}{\partial \dot{x}} = m\dot{x}, \qquad \frac{\partial L}{\partial x} = -kx. \tag{3}$$

Hence, the Lagrangian equation of motion for a harmonic oscillator is

$$\frac{d}{dt}\left(\frac{\partial L}{\partial \dot{x}}\right) - \frac{\partial L}{\partial x} = 0. \tag{4}$$

Substituting (3) into (4), we get

$$\frac{d}{dt}(m\dot{x}) + kx = 0$$

or

$$m\ddot{x} + kx = 0. \tag{5}$$

This is the Lagrangian equation of motion for a harmonic oscillator.

Example 6.6 Use the Lagrangian method and obtain the equation of motion for planetary motion (Kepler's problem).

Solution Here, we consider a particle of mass m, moving in a plane and attracted towards the origin O of coordinates with a force proportional to the inverse square of the distance from it (Fig. 6.1). That is,

$$\vec{F} = \frac{-m\mu}{r^2}$$

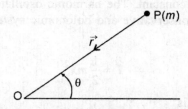

Figure 6.1 Description of Example 6.6.

If (r, θ) be the plane polar coordinates of the particle at a given instant, then its kinetic energy is given by

$$T = \frac{1}{2} mv^2 = \frac{1}{2} m (\dot{r}^2 + r^2 \dot{\theta}^2) \tag{1}$$

while its potential energy is

$$V = -\int F(r)\, dr = \int \frac{m\mu}{r^2} dr = -\frac{m\mu}{r}. \tag{2}$$

Taking (r, θ) as generalized coordinates, we define the Lagrangian

$$L = T - V = \frac{1}{2} m (\dot{r}^2 + r^2 \dot{\theta}^2) + \frac{m\mu}{r}, \tag{3}$$

which gives

$$\frac{\partial L}{\partial \dot{r}} = m\dot{r}, \qquad \frac{\partial L}{\partial r} = mr\dot{\theta}^2 - \frac{m\mu}{r^2},$$
$$\frac{\partial L}{\partial \dot{\theta}} = mr^2 \dot{\theta}, \qquad \frac{\partial L}{\partial \theta} = 0. \tag{4}$$

Thus, the Lagrange's equations for the planetary or central force motion can be written as follows:

(i) r-equation is

$$\frac{d}{dt} \left(\frac{\partial L}{\partial \dot{r}} \right) - \frac{\partial L}{\partial r} = 0,$$

that is,

$$m (\ddot{r} - r\dot{\theta}^2) + \frac{m\mu}{r^2} = 0. \tag{5}$$

(ii) θ-equation is

$$\frac{d}{dt} \left(\frac{\partial L}{\partial \dot{\theta}} \right) - \frac{\partial L}{\partial \theta} = 0,$$

that is,
$$\frac{d}{dt}(mr^2\dot{\theta}) = 0.$$

On integration, we have

$$mr^2\dot{\theta} = \text{constant} \quad \text{or} \quad r^2\dot{\theta} = \text{constant}. \tag{6}$$

Equations (5) and (6) constitute the required result. Equation (6) can be interpreted to state that the angular momentum of the particle about the centre of attraction is a constant.

Example 6.7 Use the Lagrangian method and obtain the equations of motion for a spherical pendulum.

Solution In the case of a spherical pendulum, the bob moves under gravity but is constrained to move on a smooth sphere of radius r. The position of the bob P is located by spherical coordinates (r, θ, ϕ). The distance r of the bob from the centre of the sphere on which the bob moves is the radius of the sphere (see Fig. 6.2).

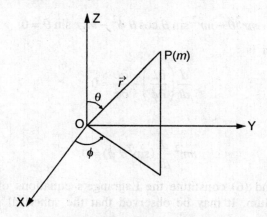

Figure 6.2 Description of Example 6.7.

The velocity of the point P in spherical polar coordinates is given by

$$\vec{V} = \dot{r}\hat{e}_r + r\dot{\theta}\hat{e}_\theta + r\dot{\phi}\sin\theta\,\hat{e}_\phi.$$

Since the bob is constrained to move on the surface of the sphere, $\dot{r} = 0$. Therefore, the kinetic energy T is given by

$$T = \frac{1}{2}mr^2(\dot{\theta}^2 + \sin^2\theta\,\dot{\phi}^2), \tag{1}$$

while the potential energy V is

$$V = mgr\cos\theta. \tag{2}$$

Taking θ and ϕ as generalized coordinates, the Lagrangian of the system is defined by

$$L = T - V = \frac{1}{2} mr^2 (\dot{\theta}^2 + \sin^2\theta \, \dot{\phi}^2) - mgr \cos\theta \qquad (3)$$

from which, we have

$$\left. \begin{array}{ll} \dfrac{\partial L}{\partial \dot{\theta}} = mr^2 \dot{\theta}, & \dfrac{\partial L}{\partial \theta} = mr^2 \sin\theta \cos\theta \, \dot{\phi}^2 + mgr \sin\theta \\[2mm] \dfrac{\partial L}{\partial \dot{\phi}} = mr^2 \sin^2\theta \, \dot{\phi}, & \dfrac{\partial L}{\partial \phi} = 0. \end{array} \right\} \qquad (4)$$

Thus, the Lagrange's equations of motion for the given system can be written as follows:

(i) θ-equation is

$$\frac{d}{dt}\left(\frac{\partial L}{\partial \dot{\theta}}\right) - \frac{\partial L}{\partial \theta} = 0.$$

that is,

$$mr^2 \ddot{\theta} - mr^2 \sin\theta \cos\theta \, \dot{\phi}^2 - mgr \sin\theta = 0. \qquad (5)$$

(ii) ϕ-equation is

$$\frac{d}{dt}\left(\frac{\partial L}{\partial \dot{\phi}}\right) - \frac{\partial L}{\partial \phi} = 0.$$

that is,

$$mr^2 \frac{d}{dt}(\sin^2\theta \, \dot{\phi}) = 0. \qquad (6)$$

Equations (5) and (6) constitute the Lagrange's equations of motion for a spherical pendulum. It may be observed that the spherical pendulum is a simple system.

Example 6.8 Consider a dynamical system with

$$T = \frac{1}{2}(\dot{q}_1^2 + \dot{q}_2^2), \qquad V = f(q_1 - q_2),$$

where f is a given function. By choosing suitable new coordinates Q_1 and Q_2, reduce the problem of determining the motion to the evaluation of an integral involving the function f. Determine q_1, q_2 as functions of t if $f(x) = x^2$.

Solution Let us choose

$$Q_1 = q_1 - q_2 \qquad (1)$$

$$Q_2 = q_1 + q_2 \qquad (2)$$

so that

$$q_1 = \frac{1}{2}(Q_1 + Q_2), \qquad q_2 = -\frac{1}{2}(Q_1 - Q_2),$$

$$\dot{q}_1 = \frac{1}{2}(\dot{Q}_1 + \dot{Q}_2), \qquad \dot{q}_2 = -\frac{1}{2}(\dot{Q}_1 - \dot{Q}_2).$$

Now, the kinetic energy T is given by

$$T = \frac{1}{2}(\dot{q}_1^2 + \dot{q}_2^2) = \frac{1}{4}(\dot{Q}_1^2 + \dot{Q}_2^2) \qquad (3)$$

while the potential energy V is

$$V = f(Q_1). \qquad (4)$$

We define the Lagrangian, $L = T - V$, that is,

$$L = \frac{1}{4}(\dot{Q}_1^2 + \dot{Q}_2^2) - f(Q_1) \qquad (5)$$

from which we have

$$\left.\begin{array}{ll} \dfrac{\partial L}{\partial \dot{Q}_1} = \dfrac{1}{2}\dot{Q}_1, & \dfrac{\partial L}{\partial \dot{Q}_2} = \dfrac{1}{2}\dot{Q}_2, \\[2mm] \dfrac{\partial L}{\partial Q_1} = -f'(Q_1), & \dfrac{\partial L}{\partial Q_2} = 0. \end{array}\right\} \qquad (6)$$

Thus, the Lagrange's equation for Q_1 is

$$\frac{d}{dt}\left(\frac{\partial L}{\partial \dot{Q}_1}\right) - \frac{\partial L}{\partial Q_1} = 0$$

that is,

$$\frac{1}{2}\ddot{Q}_1 = -f'(Q_1), \qquad (7)$$

while the Lagrange's equation for Q_2 is

$$\frac{d}{dt}\left(\frac{\partial L}{\partial \dot{Q}_2}\right) = 0,$$

that is,

$$\frac{1}{2}\ddot{Q}_2 = 0. \qquad (8)$$

On integrating twice, we get

and
$$\dot{Q}_2 = A \quad \text{(a constant)}$$
$$Q_2 = At + B, \tag{9}$$
where A and B are constants of integration. Let
$$\dot{Q}_1 = v. \tag{10}$$
Then Eq. (7) becomes
$$\frac{dv}{dt} = -2f'(Q_1) = \frac{dv}{dQ_1}\frac{dQ_1}{dt}$$
or
$$\frac{dQ_1}{dt} dv = -2f'(Q_1) dQ_1 = v\, dv.$$
On integration, we get at once
$$\frac{v^2}{2} = -2\int f'(Q_1)\, dQ_1$$
or
$$\dot{Q}_1^2 = -4\int f'(Q_1)\, dQ_1$$
that is,
$$\frac{dQ_1}{dt} = i\sqrt{\text{function of } Q_1}$$
or
$$\int \frac{dQ_1}{\sqrt{\text{function of } Q_1}} = i\int dt. \tag{11}$$
Given $f(x) = x^2$, in this problem $Q_1 = x$, then, Eq. (11) gives us
$$\frac{dx}{\sqrt{x^2}} = i\, dt \quad \text{or} \quad \int \frac{dx}{x} = i\int dt,$$
which gives
$$\log x + \log c = \log cx = it.$$
That is,
$$x = ke^{it} = Q_1 = q_1 - q_2. \tag{12}$$
Finally, Eqs. (9) and (12) give
$$q_1 + q_2 = At + B, \qquad q_1 - q_2 = ke^{it}. \tag{13}$$

The solution of which gives

$$q_1 = \frac{1}{2}(At + B + ke^{it}) = at + b + k'e^{it}$$

and

$$q_2 = \frac{1}{2}(At + B - ke^{it}) = at + b - k'e^{it},$$

which is the required result.

Example 6.9 Obtain the equations of motion for a symmetrical spinning top using Lagrange's method.

Solution In Section 4.2, we have seen that by using Euler angles as generalized coordinates and noting that $I_{xx} = I_{yy}$ for a symmetric top, the kinetic energy T as derived in Eq. (4.22) is given by

$$T = \frac{1}{2}I_{xx}(\dot{\psi}^2 \sin^2\theta + \dot{\theta}^2) + \frac{1}{2}I_{zz}(\dot{\psi}\cos\theta + \dot{\phi})^2 \tag{1}$$

and the potential energy as obtained in Eq. (4.23) is

$$V = mgl\cos\theta, \tag{2}$$

where l is the height of the centre of mass of the top above the XY-plane.

Thus, the Lagrangian (L) is

$$L = \frac{1}{2}I_{xx}(\dot{\psi}^2 \sin^2\theta + \dot{\theta}^2) + \frac{1}{2}I_{zz}(\dot{\psi}\cos\theta + \dot{\phi})^2 - mgl\cos\theta. \tag{3}$$

Then,

$$\left.\begin{aligned}
\frac{\partial L}{\partial \dot{\theta}} &= I_{xx}\dot{\theta} \\
\frac{\partial L}{\partial \theta} &= I_{xx}(\dot{\psi}^2 \sin\theta \cos\theta) + I_{zz}(\dot{\psi}\cos\theta + \dot{\phi})(-\dot{\psi}\sin\theta) + mgl\sin\theta, \\
\frac{\partial L}{\partial \dot{\psi}} &= I_{xx}\dot{\psi}\sin^2\theta + I_{zz}(\dot{\psi}\cos\theta + \dot{\phi})\cos\theta, \\
\frac{\partial L}{\partial \psi} &= 0, \\
\frac{\partial L}{\partial \dot{\phi}} &= I_{zz}(\dot{\psi}\cos\theta + \dot{\phi}), \\
\frac{\partial L}{\partial \phi} &= 0.
\end{aligned}\right\} \tag{4}$$

Here, θ represents nutational motion and ψ that of precession of the top. Now, the Lagrange's equations are given by

222 *Lagrange and Hamilton Equations (Ch. 6)*

$$\left.\begin{array}{l}\dfrac{d}{dt}\left(\dfrac{\partial L}{\partial \dot\theta}\right)-\dfrac{\partial L}{\partial \theta}=0,\\[6pt]\dfrac{d}{dt}\left(\dfrac{\partial L}{\partial \dot\psi}\right)-\dfrac{\partial L}{\partial \psi}=0,\\[6pt]\dfrac{d}{dt}\left(\dfrac{\partial L}{\partial \dot\phi}\right)-\dfrac{\partial L}{\partial \phi}=0.\end{array}\right\} \quad (5)$$

Substituting Eqs. (4) into (5), we obtain the required equations of motion of the spinning top as

$$I_{xx}\ddot\theta - I_{xx}\dot\psi^2 \sin\theta\cos\theta + I_{zz}(\dot\psi\cos\theta+\dot\phi)\dot\psi\sin\theta - mgl\sin\theta = 0, \quad (6)$$

where

$$\frac{d}{dt}[I_{xx}\dot\psi\sin^2\theta + I_{zz}(\dot\psi\cos\theta+\dot\phi)\cos\theta]=0, \quad (7)$$

and

$$\frac{d}{dt}[I_{zz}(\dot\psi\cos\theta+\dot\phi)]=0. \quad (8)$$

Example 6.10 A particle of mass m_1 is suspended from a fixed point by a string of length l_1. A second particle of mass m_2 is suspended from the first particle by a string of length l_2. The system is constrained to move in a vertical plane. Choosing the independent generalized angles θ and ϕ, which the two strings make respectively with the vertical, calculate the generalized forces and obtain the equations of motion.

Solution The given system is described as shown in Figure 6.3.

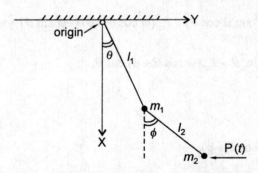

Figure 6.3 Description of Example 6.10.

We have defined the generalized forces as

$$Q_\rho = \sum \vec{F_i}\cdot\frac{\partial \vec{r_i}}{\partial q_\rho}. \quad (1)$$

Taking $\theta = q_1, \phi = q_2$, let the position vectors of the two particles be $\vec{r}_1$ and $\vec{r}_2$, then

and
$$\left. \begin{array}{l} \vec{r}_1 = (l_1 \cos \theta)\hat{i} + (l_1 \sin \theta)\hat{j} \\ \\ \vec{r}_2 = (l_1 \cos \theta + l_2 \cos \phi)\hat{i} + (l_1 \sin \theta + l_2 \sin \phi)\hat{j}. \end{array} \right\} \quad (2)$$

Also,
$$\vec{F}_1 = m_1 g \hat{i}, \qquad \vec{F}_2 = m_2 g \hat{i} - P(t)\hat{j} \quad (3)$$

and
$$\left. \begin{array}{l} \dfrac{\partial \vec{r}_1}{\partial \theta} = -l_1 \sin \theta \hat{i} + l_1 \cos \theta \hat{j}, \\ \\ \dfrac{\partial \vec{r}_2}{\partial \theta} = -l_1 \sin \theta \hat{i} + l_1 \cos \theta \hat{j}, \\ \\ \dfrac{\partial \vec{r}_1}{\partial \phi} = 0, \\ \\ \dfrac{\partial \vec{r}_2}{\partial \phi} = -l_2 \sin \phi \hat{i} + l_2 \cos \phi \hat{j}. \end{array} \right\} \quad (4)$$

Substituting Eqs. (3) and (4) into Eq. (1), we find the generalized forces as

$$Q_\theta = \vec{F}_1 \cdot \frac{\partial \vec{r}_1}{\partial \theta} + \vec{F}_2 \cdot \frac{\partial \vec{r}_2}{\partial \theta}$$

which gives
$$Q_\theta = -m_1 g \, l_1 \sin \theta - m_2 g \, l_1 \sin \theta - P(t) \, l_1 \cos \theta. \quad (5)$$

Similarly,
$$Q_\phi = \vec{F}_1 \cdot \frac{\partial \vec{r}_1}{\partial \phi} + \vec{F}_2 \cdot \frac{\partial \vec{r}_2}{\partial \phi}$$

which gives
$$Q_\phi = -m_2 g \, l_2 \sin \phi - P(t) \, l_2 \cos \phi. \quad (6)$$

To obtain the equations of motion, let (x_1, y_1) and (x_2, y_2) be the coordinates of m_1 and m_2. Then,

$$\left. \begin{array}{ll} x_1 = l_1 \cos \theta, & y_1 = l_1 \sin \theta, \\ x_2 = l_1 \cos \theta + l_2 \cos \phi, & y_2 = l_1 \sin \theta + l_2 \sin \phi. \end{array} \right\} \quad (7)$$

Thus,

$$\begin{aligned}\dot{x}_1 &= -l_1\dot{\theta}\sin\theta, & \dot{y}_1 &= l_1\dot{\theta}\cos\theta, \\ \dot{x}_2 &= -l_1\dot{\theta}\sin\theta - l_2\dot{\phi}\sin\phi, & \dot{y}_2 &= l_1\dot{\theta}\cos\theta + l_2\dot{\phi}\cos\phi.\end{aligned} \quad (8)$$

The total kinetic energy of the system is

$$T = \frac{1}{2}m_1(\dot{x}_1^2 + \dot{y}_1^2) + \frac{1}{2}m_2(\dot{x}_2^2 + \dot{y}_2^2)$$

$$= \frac{1}{2}m_1(l_1^2\dot{\theta}^2\sin^2\theta + l_1^2\dot{\theta}^2\cos^2\theta) + \frac{1}{2}m_2(l_1^2\dot{\theta}^2\sin^2\theta + l_2^2\dot{\phi}^2\sin^2\phi$$

$$+ 2l_1l_2\dot{\theta}\dot{\phi}\sin\theta\sin\phi + l_1^2\dot{\theta}^2\cos^2\theta + l_2^2\dot{\phi}^2\cos^2\phi + 2l_1l_2\dot{\theta}\dot{\phi}\cos\theta\cos\phi)$$

or

$$T = \frac{1}{2}m_1 l_1^2\dot{\theta}^2 + \frac{1}{2}m_2[l_1^2\dot{\theta}^2 + l_2^2\dot{\phi}^2 + 2l_1l_2\dot{\theta}\dot{\phi}\cos(\theta-\phi)]$$

$$= \frac{1}{2}(m_1+m_2)l_1^2\dot{\theta}^2 + \frac{1}{2}m_2 l_2^2\dot{\phi}^2 + m_2 l_1 l_2 \dot{\theta}\dot{\phi}\cos(\theta-\phi). \quad (9)$$

The potential energy of the given system is

$$V = -m_1 g l_1\cos\theta - m_2 g(l_1\cos\theta + l_2\cos\phi)$$

$$= -(m_1+m_2)g l_1\cos\theta - m_2 g l_2\cos\phi. \quad (10)$$

Recalling the definition of Lagrangian

$$L = T - V$$

$$= \frac{1}{2}(m_1+m_2)l_1^2\dot{\theta}^2 + \frac{1}{2}m_2 l_2^2\dot{\phi}^2 + m_2 l_1 l_2\dot{\theta}\dot{\phi}\cos(\theta-\phi)$$

$$+ (m_1+m_2)g l_1\cos\theta + m_2 g l_2\cos\phi \quad (11)$$

The required Lagrange's equations of motion

$$\frac{d}{dt}\left(\frac{\partial L}{\partial \dot{q}_\rho}\right) - \frac{\partial L}{\partial q_\rho} = 0,$$

on taking $q_\rho = \theta, \phi$ and $\alpha = \theta - \phi$, becomes

$$(m_1+m_2)l_1\ddot{\theta} + m_2 l_2(\ddot{\phi}\cos\alpha + \dot{\phi}^2\sin\alpha) + (m_1+m_2)g\sin\theta = 0 \quad (12)$$

and

$$l_2\ddot{\phi} + l_1\ddot{\theta}\cos\alpha - l_1\dot{\theta}^2\sin\alpha - g\sin\phi = 0. \quad (13)$$

Thus, Eqs. (5), (6), (12) and (13) constitute the desired results.

6.4 PRINCIPLE OF VIRTUAL WORK—D'ALEMBERT'S PRINCIPLE

The adjective 'virtual' means possible or permissible. According to Goldstein, "a virtual displacement (infinitesimal) of a system refers to a change in the configuration of the system as a result of any arbitrary infinitesimal change of coordinates δr_i, consistent with the forces and constraints imposed on the system at a given instant, t, and does not involve a change in time. The displacement δr_i is called *virtual* to distinguish it from an *actual* displacement dr_i of the system occurring in a time interval dt, during which the forces and constraints may be changing." Let us consider a system and any motion of the system during which it remains in equilibrium. Suppose, we apply the work energy principle, $\Delta W = \Delta T$, to the system. Since, each part of the system moves with at most a constant velocity, the change in kinetic energy, that is, ΔT is zero. Thus, for a system in equilibrium, the work done by all forces on the system is zero. In other words, if the system is in equilibrium the total force $\vec{F}_i$ acting on each particle of the system is zero. Hence, the virtual work of the force $\vec{F}_i$ in the displacement $\delta \vec{r}_i$ is given by the dot product

$$\vec{F}_i \cdot \delta \vec{r}_i = 0.$$

Therefore, the sum over all particles is zero. That is,

$$\sum \vec{F}_i \cdot \delta \vec{r}_i = 0. \tag{6.21}$$

Suppose, we decompose $\vec{F}_i$ into a sum of applied force $\vec{F}_i^{(a)}$ and the force of constraint f_i, then $\vec{F}_i = \vec{F}_i^{(a)} + \vec{f}_i$, and Eq. (6.21) becomes

$$\sum \vec{F}_i^{(a)} \cdot \delta \vec{r}_i + \sum \vec{f}_i \cdot \delta \vec{r}_i = 0. \tag{6.22}$$

Assume that the net virtual work of the forces of constraint is zero, which in fact is true. For, if a rigid body is constrained to move on a surface, the force of constraint is perpendicular to the surface, while the virtual displacement is tangential to it and hence, the virtual work vanishes. For the condition of equilibrium, we therefore have

$$\sum \vec{F}_i^{(a)} \cdot \delta \vec{r}_i = 0. \tag{6.23}$$

This is called the *principle of virtual work*. This is the approach to statics and indeed to mechanics. But, we want to have a principle involving general motion of the system. This was developed by D'Alembert, which is explained as follows: Newton's second law states that the rate of change of linear momentum is equal to the applied force. Therefore,

$$\vec{F}_i = \dot{\vec{p}}_i.$$

Thus, Eq. (6.21) can be written in the form

$$\sum (\vec{F}_i - \dot{\vec{p}}_i) \cdot \delta \vec{r}_i = 0. \tag{6.24}$$

If $\vec{F}_i$ is the applied force acting on the ith particle of mass m_i, then the above equation becomes

$$\sum (\vec{F}_i^{(a)} - \dot{\vec{p}}_i) \cdot \delta \vec{r}_i = 0. \qquad (6.25)$$

This is often called *D'Alembert's principle*.

6.5 LAGRANGE'S EQUATIONS FOR GENERAL SYSTEMS

Consider a general system of n particles in the sense that it is rheonomic, non-conservative and non-holonomic. Let m_i be the mass of a typical particle, whose position vector is $\vec{r}_i$. Also, let F_i be the total force acting on a typical particle (the reactions of the constraints being included). Suppose r_i is a function of $q_1, q_2, \ldots, q_n, t$. That is,

$$\vec{r}_i = \vec{r}_i (q_1, q_2, \ldots, q_n, t) \qquad (6.26)$$

or

$$\vec{r}_i = \vec{r}_i (\vec{q}, t).$$

Here, the time t is included to account for moving constraints (rheonomic system). Using the chain rule of partial differentiation, we have

$$\frac{d\vec{r}_i}{dt} = \dot{\vec{r}}_i = \sum_{\rho=1}^{n} \frac{\partial \vec{r}_i}{\partial q_\rho} \frac{dq_\rho}{dt} + \frac{\partial \vec{r}_i}{\partial t}$$

or

$$\frac{d\vec{r}_i}{dt} = \dot{\vec{r}}_i = \sum_{\rho=1}^{n} \frac{\partial \vec{r}_i}{\partial q_\rho} \dot{q}_\rho + \frac{\partial \vec{r}_i}{\partial t}. \qquad (6.27)$$

Differentiating the above equation partially with respect to q_β, we get

$$\frac{\partial \dot{\vec{r}}_i}{\partial q_\beta} = \sum_{\rho=1}^{n} \frac{\partial^2 \vec{r}_i}{\partial q_\beta \partial q_\rho} \dot{q}_\rho + \frac{\partial}{\partial q_\beta} \left(\frac{\partial \vec{r}_i}{\partial t} \right)$$

Changing the order of differentiation, we get

$$\frac{\partial \dot{\vec{r}}_i}{\partial q_\beta} = \sum_{\rho=1}^{n} \frac{\partial}{\partial q_\rho} \left(\frac{\partial \vec{r}_i}{\partial q_\beta} \right) \dot{q}_\rho + \frac{\partial}{\partial t} \left(\frac{\partial \vec{r}_i}{\partial q_\beta} \right) = \frac{d}{dt} \left(\frac{\partial \vec{r}_i}{\partial q_\beta} \right).$$

Therefore,

$$\frac{\partial \dot{\vec{r}}_i}{\partial q_\beta} = \frac{d}{dt} \left(\frac{\partial \vec{r}_i}{\partial q_\beta} \right). \qquad (6.28)$$

Now, differentiating partially Eq. (6.27) with respect to $\dot{q}_\rho$, we have at once

$$\frac{\partial \dot{\vec{r}}_i}{\partial \dot{q}_\rho} = \frac{\partial \vec{r}_i}{\partial q_\rho}. \qquad (6.29)$$

The kinetic energy of the system T is given by

$$T = \frac{1}{2} \sum m_i (\dot{\vec{r}}_i \cdot \dot{\vec{r}}_i). \qquad (6.30)$$

Differentiating this result partially with respect to $\dot{q}_\rho$, we obtain after using cancellation of dot

$$\frac{\partial T}{\partial \dot{q}_\rho} = \sum m_i \dot{\vec{r}}_i \cdot \frac{\partial \dot{\vec{r}}_i}{\partial \dot{q}_\rho} = \sum m_i \dot{\vec{r}}_i \cdot \frac{\partial \vec{r}_i}{\partial q_\rho}.$$

Therefore,

$$\frac{d}{dt}\left(\frac{\partial T}{\partial \dot{q}_\rho}\right) = \sum m_i \ddot{\vec{r}}_i \cdot \frac{\partial \vec{r}_i}{\partial q_\rho} + \sum m_i \dot{\vec{r}}_i \cdot \frac{\partial \dot{\vec{r}}_i}{\partial q_\rho} \qquad (6.31)$$

and

$$\frac{\partial T}{\partial q_\rho} = \sum m_i \dot{\vec{r}}_i \cdot \frac{\partial \dot{\vec{r}}_i}{\partial q_\rho}. \qquad (6.32)$$

Let us now define S_ρ by writing

$$S_\rho = \frac{d}{dt}\left(\frac{\partial T}{\partial \dot{q}_\rho}\right) - \frac{\partial T}{\partial q_\rho}. \qquad (6.33)$$

Then, substituting Eqs. (6.31) and (6.32) into Eq. (6.33), we get

$$S_\rho = \sum m_i \ddot{\vec{r}}_i \cdot \frac{\partial \vec{r}_i}{\partial q_\rho}.$$

If $\vec{F}_i$ is the force acting on m_i, then

$$m_i \ddot{\vec{r}}_i = \vec{F}_i$$

or

$$\sum (m_i \ddot{\vec{r}}_i - \vec{F}_i) = 0.$$

Since the system is in equilibrium, using the principle of virtual work, we have

$$\sum (m_i \ddot{\vec{r}}_i - \vec{F}_i) \cdot \delta \vec{r}_i = 0, \qquad (6.34)$$

where $\delta \vec{r}_i$ is the virtual displacement. From Eq. (6.26), it follows that

$$\delta \vec{r}_i = \sum \frac{\partial \vec{r}_i}{\partial q_\rho} \delta q_\rho. \qquad (6.35)$$

Here, a virtual displacement is interpreted as a first variation of the function r_i. Thus, Eq. (6.34) becomes

$$\sum_{\rho=1}^n \left(\sum_i m_i \ddot{\vec{r}}_i \cdot \frac{\partial \vec{r}_i}{\partial q_\rho} \right) \delta q_\rho = \sum_{\rho=1}^n \left(\sum_i \vec{F}_i \cdot \frac{\partial \vec{r}_i}{\partial q_\rho} \right) \delta q_\rho. \qquad (6.36)$$

Now, if we define the generalized forces Q_ρ by

$$\sum_i \vec{F}_i \cdot \frac{\partial \vec{r}_i}{\partial q_\rho} = Q_\rho, \quad (6.37)$$

and using Eq. (6.33), Eq. (6.36) reduces to

$$\sum_{\rho=1}^{n} (S_\rho - Q_\rho)\delta q_\rho = 0. \quad (6.38)$$

If δq_ρ were arbitrary, we can conclude that

$$S_\rho = Q_\rho, \quad (6.39)$$

which is same as Lagrange's equation, given by Eq. (6.14), in Section 6.3, for a simple system.

Now, we shall consider a new class of constraints, which does not restrict the motion of the particle to a surface, but to the domain bounded by the surface. These are known as *inequality constraints* of the form,

$$f(q_1, q_2, \ldots, q_n, t) \geq 0 \quad (6.40)$$

or

$$f(\vec{q}, t) \geq 0.$$

If the bounding surface is in motion, it corresponds to non-holonomic constraints. These constraints can be generalized by defining a system of n dependent generalized coordinates $q_1, q_2, \ldots, q_n$ but related by m constraint equations of the form,

$$\sum_{\rho=1}^{n} A_{\alpha\rho} dq_\rho + A_\alpha dt = 0, \quad \alpha = 1, 2, \ldots, m < n, \quad (6.41)$$

where A's are functions of q's and t. We assume that Eqs. (6.41) are independent which implies that the matrix of the coefficients is of rank m. Further, the possible displacements dq_ρ satisfy Eqs. (6.41), while the virtual displacements satisfy

$$\sum_{\rho=1}^{n} A_{\alpha\rho} \delta q_\rho = 0, \quad \alpha = 1, 2, \ldots, m. \quad (6.42)$$

Now, multiplying each of these equations by $\lambda_1, \lambda_2, \ldots, \lambda_m$ respectively and subtracting from Eqs. (6.39) and then defining F by

$$F = (S_1 - Q_1 - \lambda_1 A_{11} - \lambda_2 A_{21} - \cdots - \lambda_m A_{m1})\delta q_1 + (S_2 - Q_2 - \lambda_1 A_{12} - \lambda_2 A_{22}$$
$$- \cdots - \lambda_m A_{m2})\delta q_2 + \cdots + (S_n - Q_n - \lambda_1 A_{1n} - \lambda_2 A_{2n} - \cdots - \lambda_m A_{mn})\delta q_n, \quad (6.43)$$

where we have chosen λ's for the moment arbitrarily. It may be observed that $F = 0$ for all δq_ρ satisfying Eqs. (6.42), since Eq. (6.42) implies Eq. (6.39). Now, we choose λ's to satisfy the m equations

$$\left.\begin{aligned} S_1 - Q_1 &= \lambda_1 A_{11} + \lambda_2 A_{21} + \cdots + \lambda_m A_{m1} \\ S_2 - Q_2 &= \lambda_1 A_{12} + \lambda_2 A_{22} + \cdots + \lambda_m A_{m2} \\ &\vdots \\ S_m - Q_m &= \lambda_1 A_{1m} + \lambda_2 A_{2m} + \cdots + \lambda_m A_{mm} \end{aligned}\right\} \quad (6.44)$$

So that Eq. (6.43) becomes

$$\begin{aligned} F = &(S_{m+1} - Q_{m+1} - \lambda_1 A_{1,\,m+1} - \lambda_2 A_{2,\,m+1} - \cdots - \lambda_m A_{m,\,m+1})\delta q_{m+1} \\ &+ (S_{m+2} - Q_{m+2} - \lambda_1 A_{1,\,m+2} - \lambda_2 A_{2,\,m+2} - \cdots - \lambda_m A_{m,\,m+2})\delta q_{m+2} \\ &+ \cdots + (S_n - Q_n - \lambda_1 A_{1,\,n} - \lambda_2 A_{2,\,n} - \cdots - \lambda_m A_{m,\,n})\delta q_n. \end{aligned} \quad (6.45)$$

Equation (6.45) must vanish for arbitrary values of $\delta q_{m+1}, \delta q_{m+2}, \ldots, \delta q_n$, since when these values are specified, it is always possible to choose $\delta q_1, \delta q_2, \ldots, \delta q_m$ so that the m equations (6.42) are satisfied, which means $F = 0$. Thus, we have

$$\left.\begin{aligned} S_{m+1} - Q_{m+1} &= \lambda_1 A_{1,\,m+1} + \lambda_2 A_{2,\,m+1} + \cdots + \lambda_m A_{m,\,m+1} \\ S_{m+2} - Q_{m+2} &= \lambda_1 A_{1,\,m+2} + \lambda_2 A_{2,\,m+2} + \cdots + \lambda_m A_{m,\,m+2} \\ &\vdots \\ S_n - Q_n &= \lambda_1 A_{1,\,n} + \lambda_2 A_{2,\,n} + \cdots + \lambda_m A_{m,\,n} \end{aligned}\right\} \quad (6.46)$$

Now, combining Eqs. (6.44) and (6.46), we find that

$$S_\rho - Q_\rho = \sum_{\alpha=1}^{m} \lambda_\alpha A_{\alpha\rho}, \qquad \rho = 1, 2, \ldots, n \quad (6.47)$$

or

$$S_\rho = Q_\rho + \sum_{\alpha=1}^{m} \lambda_\alpha A_{\alpha\rho}, \qquad \rho = 1, 2, \ldots, n.$$

By recalling the definition of S_ρ, from Eq. (6.33), we have

$$\frac{d}{dt}\left(\frac{\partial T}{\partial \dot{q}_\rho}\right) - \frac{\partial T}{\partial q_\rho} = Q_\rho + \sum_{\alpha=1}^{m} \lambda_\alpha A_{\alpha\rho}, \qquad \rho = 1, 2, \ldots, n. \quad (6.48)$$

These are the Lagrange's equations of motion for rheonomic, non-conservative and non-holonomic systems or for a general system. Here, T is the kinetic energy, Q_ρ are the generalized forces and λ's are called Lagrange multipliers. Equations (6.41) and (6.48) together constitute $(n + m)$ equations, whose solutions give us $(n + m)$ quantities, such as generalized coordinates $q_1, q_2, \ldots, q_n$ and the Lagrange multipliers $\lambda_1, \lambda_2, \ldots, \lambda_m$. For illustration, we consider the following example:

Example 6.11 A particle of mass m moves under gravity on a smooth sphere of radius a. Obtain the Lagrange's equations of motion, taking x, y, z as generalized coordinates with origin at the centre of the sphere and z being measured vertically upwards.

Solution Since the particle is constrained to move on the sphere, the equation of the constraint is given by

$$x^2 + y^2 + z^2 = a^2$$

or

$$x\dot{x} + y\dot{y} + z\dot{z} = 0. \tag{1}$$

The general equation of the constraint is of the form,

$$A_{\alpha\rho} dq_\rho + A_\alpha dt = 0$$

or

$$A_{\alpha\rho} \dot{q}_\rho + A_\alpha = 0. \tag{2}$$

Comparing Eqs. (1) and (2), we have

$$A_\alpha = 0, \quad A_{11} = x, \quad A_{12} = y, \quad A_{13} = z. \tag{3}$$

Since the particle is moving on the sphere under gravity, the generalized forces are

$$Q_1 = 0, \quad Q_2 = 0, \quad Q_3 = -mg. \tag{4}$$

The kinetic energy T of the system is given by

$$T = \frac{1}{2} m (\dot{x}^2 + \dot{y}^2 + \dot{z}^2). \tag{5}$$

Since there is only one constraint, the Lagrange's equation for a general system is

$$\frac{d}{dt}\left(\frac{\partial T}{\partial \dot{x}}\right) - \frac{\partial T}{\partial x} = Q_1 + \lambda_1 A_{11} = \lambda_1 x,$$

$$\frac{d}{dt}\left(\frac{\partial T}{\partial \dot{y}}\right) - \frac{\partial T}{\partial y} = Q_2 + \lambda_1 A_{12} = \lambda_1 y,$$

$$\frac{d}{dt}\left(\frac{\partial T}{\partial \dot{z}}\right) - \frac{\partial T}{\partial z} = Q_3 + \lambda_1 A_{13} = -mg + \lambda_1 z.$$

Thus, the Lagrange's equations of motion for the given problem are found to be

$$\left.\begin{array}{l} m\ddot{x} = \lambda_1 x, \\ m\ddot{y} = \lambda_1 y, \\ m\ddot{z} = -mg + \lambda_1 z. \end{array}\right\} \qquad (6)$$

6.6 HAMILTON'S EQUATIONS

One way of solving dynamical problems is through Lagrangian. Yet another important and powerful way of solving them is due to Hamilton. Consider a dynamical system with n degrees of freedom whose behaviour obeys the Lagrange's equations of motion of the form,

$$\frac{d}{dt}\left(\frac{\partial L}{\partial \dot{q}_\rho}\right) - \frac{\partial L}{\partial q_\rho} = 0, \qquad (6.49)$$

where L is a Lagrangian and is a function of n quantities q_ρ. Their derivatives $\dot{q}_\rho$ are with respect to t, and t itself. Thus, we write

$$L = L(\vec{q}, \dot{\vec{q}}, t). \qquad (6.50)$$

We shall continue to call this function Lagrangian, though we have inserted an additional argument t, for convenience, in the definition of L as described in Eq. (6.18), just to include in this way some typical systems with moving constraints.

In Hamiltonian formulation, we introduce a new independent variable called *generalized momenta* p_ρ, defined as

$$p_\rho = \frac{\partial L(\vec{q}, \dot{\vec{q}}, t)}{\partial \dot{q}_\rho} = \frac{\partial L}{\partial \dot{q}_\rho}. \qquad (6.51)$$

Like Lagrangian $L(\vec{q}, \dot{\vec{q}}, t)$, a new function of this formalism is Hamiltonian $H(\vec{q}, \vec{p}, t)$. It means that there is a change of basis from $(\vec{q}, \dot{\vec{q}}, t)$ set to $(\vec{q}, \vec{p}, t)$ set. In Hamiltonian mechanics, q_ρ, p_ρ and t are independent variables and $\dot{q}_\rho$ is a dependent quantity. That is,

$$\dot{q}_\rho = \dot{q}_\rho(\vec{q}, \vec{p}, t). \qquad (6.52)$$

The Hamiltonian function H is defined as

$$H = \sum_{\rho=1}^{n} \dot{q}_\rho \frac{\partial L}{\partial \dot{q}_\rho} - L = \sum_{\rho=1}^{n} \dot{q}_\rho p_\rho - L. \qquad (6.53)$$

To establish the physical significance of Hamiltonian, let us consider a simple dynamical system, with $T = T(\vec{q}, \dot{\vec{q}})$, which does not contain time, t, explicitly. We can then show that

$$T = \frac{1}{2}\sum_{i=1}^{n} m_i \left(\sum_{j=1}^{n} \sum_{k=1}^{n} \frac{\partial \vec{r}_i}{\partial q_j} \cdot \frac{\partial \vec{r}_i}{\partial q_k} \dot{q}_j \dot{q}_k \right),$$

$$= \frac{1}{2}\sum_{j=1}^{n} \sum_{k=1}^{n} \alpha_{jk} \dot{q}_j \dot{q}_k \qquad \text{(see Example 6.1 under Exercises)} \qquad (6.54)$$

is a homogeneous quadratic function in the generalized velocities. But, from Eq. (6.53), we may write

$$H = \sum_{\rho=1}^{n} \dot{q}_\rho \frac{\partial L}{\partial \dot{q}_\rho} - L.$$

Since

$$L = T - V = T(\vec{q}, \dot{\vec{q}}) - V(\vec{q}),$$

we have

$$\frac{\partial L}{\partial \dot{\vec{q}}} = \frac{\partial T}{\partial \dot{\vec{q}}} \qquad \text{(since } V \text{ is not a function of } \dot{\vec{q}}\text{)}.$$

Therefore, using Euler's theorem for homogeneous functions,

$$H = \sum \dot{q}_\rho \frac{\partial T}{\partial \dot{\vec{q}}} - (T - V) = 2T - (T - V)$$

or

$$H = T + V, \qquad (6.55)$$

which is the sum of kinetic and potential energies.

Hamilton's canonical equations of motion

Recalling that Hamiltonian is a function of generalized coordinates, generalized momenta and time, that is

$$H = H(\vec{q}, \vec{p}, t).$$

The total differential of H is given by

$$dH = \sum_{\rho=1}^{n} \frac{\partial H}{\partial q_\rho} dq_\rho + \sum_{\rho=1}^{n} \frac{\partial H}{\partial p_\rho} dp_\rho + \frac{\partial H}{\partial t} dt. \qquad (6.56)$$

From the definition of Hamiltonian function H, given by Eq. (6.53), we have

$$H = \sum_{\rho=1}^{n} \dot{q}_\rho p_\rho - L(\vec{q}, \dot{\vec{q}}, t),$$

then we can also write its total differential in the form,

$$dH = \sum_{\rho=1}^{n} \dot{q}_\rho \, dp_\rho + \sum_{\rho=1}^{n} p_\rho \, d\dot{q}_\rho - \sum_{\rho=1}^{n} \frac{\partial L}{\partial q_\rho} dq_\rho - \sum_{\rho=1}^{n} \frac{\partial L}{\partial \dot{q}_\rho} d\dot{q}_\rho - \frac{\partial L}{\partial t} dt. \quad (6.57)$$

Following the definition of generalized momenta, given by Eq. (6.51), the second and fourth terms on the right-hand side of Eq. (6.57) cancel and we are left with

$$dH = \sum_{\rho=1}^{n} \dot{q}_\rho \, dp_\rho - \sum_{\rho=1}^{n} \frac{\partial L}{\partial q_\rho} dq_\rho - \frac{\partial L}{\partial t} dt. \quad (6.58)$$

Let us now define the generalized conjugate momenta as

$$\dot{p}_\rho = \frac{\partial L}{\partial q_\rho}. \quad (6.59)$$

Then Eq. (6.58) becomes

$$dH = \sum \dot{q}_\rho \, dp_\rho - \sum \dot{p}_\rho \, dq_\rho - \frac{\partial L}{\partial t} dt. \quad (6.60)$$

Now, comparing Eqs. (6.56) and (6.60), we find that

$$\left. \begin{array}{l} \dot{p}_\rho = \dfrac{\partial L}{\partial q_\rho} = -\dfrac{\partial H}{\partial q_\rho} \\[6pt] \dot{q}_\rho = \dfrac{\partial H}{\partial p_\rho} \end{array} \right\} \quad (6.61)$$

and

$$\frac{\partial L}{\partial t} = -\frac{\partial H}{\partial t}. \quad (6.62)$$

Equations (6.61) are called *Hamilton's canonical equations* of motion. These are $2n$ differential equations of first order. Equations (6.61) can also be rewritten as

$$\frac{dq_\rho}{\partial H/\partial p_\rho} = \frac{dp_\rho}{-\partial H/\partial q_\rho} = dt. \quad (6.63)$$

For illustration, we consider the following examples:

Example 6.12 Use Hamiltonian method and obtain the equation of motion for one-dimensional harmonic oscillator.

Solution From Example 6.5, we find that

$$T = \frac{1}{2} m\dot{x}^2, \tag{1}$$

$$V = \frac{1}{2} kx^2 \tag{2}$$

$$L = T - V = \frac{1}{2} m\dot{x}^2 - \frac{1}{2} kx^2. \tag{3}$$

Taking $q = x$, the generalized momenta p is found to be

$$p = \frac{\partial L}{\partial \dot{q}} = \frac{\partial L}{\partial \dot{x}} = m\dot{x}. \tag{4}$$

The Hamiltonian

$$H = \sum \dot{q}_\rho p_\rho - L = p\dot{x} - \left(\frac{1}{2} m\dot{x}^2 - \frac{1}{2} kx^2 \right).$$

Using Eq. (4), we have

$$H = \frac{p^2}{m} - \frac{p^2}{2m} + \frac{1}{2} kx^2 = \frac{p^2}{2m} + \frac{1}{2} kx^2. \tag{5}$$

Recalling the Hamilton canonical equations of motion

$$\frac{\partial H}{\partial q_\rho} = -\dot{p}_\rho, \qquad \frac{\partial H}{\partial p_\rho} = \dot{q}_\rho.$$

Therefore,

$$\frac{\partial H}{\partial x} = -\dot{p}, \qquad \frac{\partial H}{\partial p} = \dot{x}. \tag{6}$$

Using Eq. (5), we get

$$kx = -\dot{p}, \qquad \frac{p}{m} = \dot{x}. \tag{7}$$

These are two first-order differential equations. Eliminating p between them, we obtain

$$-m\ddot{x} = kx \quad \text{or} \quad m\ddot{x} + kx = 0. \tag{8}$$

This is the required equation of motion.

Example 6.13 A mass m is placed on a frictionless plane, which is tangential to the surface of the Earth as shown in Fig. 6.4. Determine the equation of motion using Hamilton's method, taking x as the independent generalized coordinate.

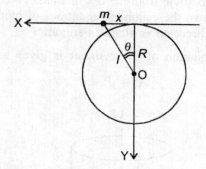

Figure 6.4 Description of Example 6.13.

Solution The kinetic energy T of the system is

$$T = \frac{1}{2}m\dot{x}^2 \tag{1}$$

while the potential energy V is given by

$$V = lmg - Rmg = \left(\sqrt{x^2 + R^2} - R\right)mg. \tag{2}$$

Since $q = x$, the generalized momenta is

$$p_x = \frac{\partial L}{\partial \dot{x}} = \frac{\partial}{\partial \dot{x}}(T-V) = m\dot{x}. \tag{3}$$

But,

$$H = \sum \dot{q}_\rho p_\rho - L = m\dot{x}^2 - \frac{1}{2}m\dot{x}^2 + V = \frac{m\dot{x}^2}{2} + V.$$

Therefore, using Eq. (3), we find

$$H = \frac{p_x^2}{2m} + \left(\sqrt{x^2 + R^2} - R\right)mg.$$

Now, applying Hamilton's canonical equations, we have

$$\left.\begin{array}{l} \dot{x} = \dfrac{\partial H}{\partial p_x} = \dfrac{p_x}{m}, \\[2mm] \dot{p}_x = -\dfrac{\partial H}{\partial x} = \dfrac{-mgx}{\sqrt{x^2 + R^2}}. \end{array}\right\} \tag{4}$$

which is the required result.

Example 6.14 Consider a particle of mass m that is constrained to move on the surface of a cylinder defined by the equation $x^2 + y^2 = R^2$. The particle is subjected to a force directed towards the origin and proportional to the distance of the particle from the origin and is given by $\vec{F} = -kr$. Using Hamiltonian, write down the canonical equations of motion. Also show that the motion in the Z-direction is simple harmonic.

Solution The equation of the cylinder is given as (see Fig. 6.5)

$$x^2 + y^2 = R^2. \tag{1}$$

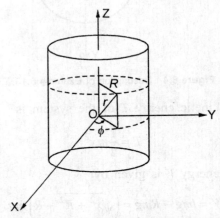

Figure 6.5 Description of Example 6.14.

Since the system is conservative, we have

$$\vec{F} = -\nabla V = -kr. \tag{2}$$

On integration, we get

$$V = \frac{k}{2} r^2 = \frac{k}{2}(x^2 + y^2 + z^2) = \frac{k}{2}(R^2 + z^2). \tag{3}$$

In cylindrical coordinates,

$$v^2 = \dot{r}^2 + r^2 \dot{\phi}^2 + \dot{z}^2. \tag{4}$$

But, due to the given constraint

$$r = R = \text{constant}.$$

Therefore, the kinetic energy T of the system is

$$T = \frac{1}{2} m (R^2 \dot{\phi}^2 + \dot{z}^2). \tag{5}$$

Thus, the Lagrangian L is given by

$$L = T - V = \frac{1}{2} m (R^2 \dot{\phi}^2 + \dot{z}^2) - \frac{k}{2}(R^2 + z^2). \tag{6}$$

Taking ϕ and z as generalized coordinates q_1 and q_2, the corresponding generalized momenta are

$$p_\phi = \frac{\partial L}{\partial \dot\phi} = mR^2\dot\phi,$$
$$p_z = \frac{\partial L}{\partial \dot z} = m\dot z.$$
(7)

Therefore, the Hamiltonian H of the system is

$$H = \sum \dot q_\rho p_\rho - L = mR^2\dot\phi^2 + m\dot z^2 - \frac{mR^2\dot\phi^2}{2} - \frac{m\dot z^2}{2} + \frac{kR^2}{2} + \frac{kz^2}{2}$$

that is,

$$H = \frac{mR^2\dot\phi^2}{2} + \frac{m\dot z^2}{2} + \frac{kR^2}{2} + \frac{kz^2}{2}$$
$$= \frac{p_\phi^2}{2mR^2} + \frac{p_z^2}{2m} + \frac{k(z^2 + R^2)}{2}. \tag{8}$$

Here, ϕ does not occur explicitly in H. Hence, the Hamiltonian canonical equations are

$$\dot p_\phi = -\frac{\partial H}{\partial \phi} = 0 \tag{9}$$

$$\dot p_z = -\frac{\partial H}{\partial z} = -kz \tag{10}$$

$$\dot\phi = \frac{\partial H}{\partial p_\phi} = \frac{p_\phi}{mR^2} \tag{11}$$

$$\dot z = \frac{\partial H}{\partial p_z} = \frac{p_z}{m} \tag{12}$$

Now, Eqs. (9) and (11) give

$$p_\phi = mR^2\dot\phi = \text{constant}, \tag{13}$$

which indicates that the angular momentum about the Z-axis is constant.

Combining Eqs. (10) and (12), we find that

$$\ddot z = \frac{\dot p_z}{m} = -\frac{k}{m}z = -\omega^2 z \quad \text{(say)}.$$

That is,

$$\ddot z + \omega^2 z = 0, \tag{14}$$

which suggests that the motion in the Z-direction is simple harmonic.

6.7 IGNORABLE COORDINATES

There exists certain systems in which a particular coordinate, say q_ρ, is abscent from the Lagrangian L, although its time derivative $\dot{q}_\rho$ is present in L. If such a system is holonomic, then the corresponding Lagrange's equations take the form

$$\frac{d}{dt}\left(\frac{\partial L}{\partial \dot{q}_\rho}\right) = 0. \tag{6.64}$$

Now, let us recall the definition of generalized momenta

$$p_\rho = \frac{\partial L}{\partial \dot{q}_\rho}. \tag{6.65}$$

Then Eq. (6.64), together with Eq. (6.65), gives

$$\frac{dp_\rho}{dt} = 0. \tag{6.66}$$

Integrating Eq. (6.66) with respect to time t, we get

$$p_\rho = \text{constant}. \tag{6.67}$$

Hence, we conclude that the generalized momenta associated with an ignorable coordinate is conserved.

If there is an ignorable coordinate, the number of degrees of freedom is reduced by unity. Similarly if there are m ignorable coordinates, then the number of degrees of freedom of the system is at once reduced by m.

In case we start with Hamiltonian instead of Lagrangian, the above argument also holds true. For illustration, suppose, q_1 is absent in H, so that

$$\frac{\partial H}{\partial q_1} = 0. \tag{6.68}$$

Then, the Hamilton's canonical equations of motion (6.61) give

$$\dot{p}_1 = -\frac{\partial H}{\partial q_1} = 0. \tag{6.69}$$

Therefore, $p_1 = \text{constant} = c_1$ (say). Now, replacing p_1 by c_1 in the right-hand sides of the Eqs. (6.61) and excluding the two equations corresponding to $\rho = 1$ from the $2n$ equations (6.61), we are left with $(2n-2)$ canonical equations for the $(2n-2)$ quantities $q_2, q_3, \ldots, q_n, p_2, p_3, \ldots, p_n$. Under these conditions, we say that q_1 is an ignorable coordinate. For illustration of this concept, we consider the following examples:

Example 6.15 Consider the Kepler's problem or planetary motion as discussed in Example 6.6 and show that the generalized momenta p_θ is constant.

Solution From Example 6.6, we find that the Lagrangian

$$L = T - V = \frac{1}{2}mv^2 + \frac{m\mu}{r}. \tag{1}$$

In terms of plane rectangular coordinates,

$$L = \frac{1}{2}m(\dot{x}^2 + \dot{y}^2) + \frac{m\mu}{(x^2 + y^2)^{1/2}}. \tag{2}$$

This gives no indication that an ignorable coordinate actually exists. But the same Lagrangian when expressed in polar coordinates, we have

$$L = \frac{1}{2}m(\dot{r}^2 + r^2\dot{\theta}^2) + \frac{m\mu}{r}. \tag{3}$$

From this expression, it is evident that L does not contain θ. Thus,

$$\frac{\partial L}{\partial \theta} = 0.$$

Therefore, θ is an ignorable coordinate. The generalized momenta, from Eq. (6.65), gives

$$p_\theta = \frac{\partial L}{\partial \dot{\theta}} = mr^2\dot{\theta} = \text{constant} \qquad \text{[from Eq. (6.67)]}.$$

Example 6.16 Use the Hamiltonian method and obtain the equation of motion of the non-ignorable coordinate for the spherical pendulum as described in Example 6.7 and show that the generalized momenta p_ϕ is constant.

Solution In Example 6.7, we obtained the kinetic and potential energies T and V, respectively, as

$$\left. \begin{array}{l} T = \dfrac{1}{2}mr^2(\dot{\theta}^2 + \sin^2\theta\,\dot{\phi}^2) \\[4pt] V = mgr\cos\theta \end{array} \right\} \tag{1}$$

The Hamiltonian H is defined as

$$H = T + V = \frac{1}{2}mr^2(\dot{\theta}^2 + \sin^2\theta\,\dot{\phi}^2) + mgr\cos\theta. \tag{2}$$

Taking θ and ϕ as generalized coordinates q_1 and q_2 we have the corresponding generalized momenta p_θ and p_ϕ as

$$p_\theta = \frac{\partial L}{\partial \dot{\theta}} = mr^2\dot{\theta}, \quad p_\phi = \frac{\partial L}{\partial \dot{\phi}} = mr^2\sin\theta\,\dot{\phi}. \tag{3}$$

Substituting Eq. (3) into Eq. (2), H can be rewritten as

$$H = \frac{mr^2}{2}\left(\frac{p_\theta^2}{m^2 r^4} + \frac{p_\phi^2}{m^2 r^4 \sin^2\theta}\right) + mgr\cos\theta.$$

That is,

$$H = \frac{1}{2mr^2}(p_\theta^2 + \csc^2\theta\, p_\phi^2) + mgr\cos\theta. \qquad (4)$$

We observe from Eqs. (2) and (4) that ϕ is absent in H and, therefore, ϕ is an ignorable coordinate as hence

$$\frac{\partial H}{\partial \phi} = 0. \qquad (5)$$

Thus, the generalized momenta $\dot{p}_\phi = 0$ and so

$$p_\phi = \text{constant} = c. \qquad (6)$$

Now, we are left with only two canonical equations:

$$\dot{p}_\theta = -\frac{\partial H}{\partial \theta}, \qquad \dot{\theta} = \frac{\partial H}{\partial p_\theta}. \qquad (7)$$

With the help of Eq. (4), we get from Eq. (7) that

$$\dot{p}_\theta = \frac{p_\phi^2}{mr^2}\csc^2\theta\cot\theta - mgr\sin\theta \qquad (8)$$

and

$$\dot{\theta} = \frac{p_\theta}{mr^2}. \qquad (9)$$

Eliminating p_θ between Eqs. (8) and (9), we find at once that

$$mr^2\ddot{\theta} = \frac{p_\phi^2}{mr^2}\frac{\cos\theta}{\sin^3\theta} - mgr\sin\theta, \qquad (10)$$

which is a second-order differential equation in θ, the non-ignorable coordinate. Hence Eqs. (6) and (10) constitute the required solution.

6.8 THE ROUTHIAN FUNCTION

In the Lagrangian description of a dynamical system, we noted that, if there are any ignorable coordinates, their elimination will reduce the degree of freedom of the system. Such an elimination is carried out through the introduction of Routhian function. To illustrate this concept, let us consider a simple holonomic system whose configuration is described through n independent generalized coordinates $q_1, q_2, \ldots, q_n$. Suppose the first k generalized coordinates are ignorable. This means that the Lagrangian L is a function of $q_{k+1}, q_{k+2}, \ldots, q_n, \dot{q}_1, \dot{q}_2, \ldots, \dot{q}_n$ and t. In the previous section, we showed that the generalized momenta associated with an ignorable coordinate is constant and therefore, it follows at once that

$$p_1 = c_1, \qquad p_2 = c_2, \ldots, p_k = c_k.$$

The Routhian Function (Sec. 6.8)

Now, let us construct a Routhian function

$$R(q_{k+1}, q_{k+2}, \ldots, q_n, \dot{q}_1, \dot{q}_2, \ldots, \dot{q}_n, c_1, c_2, \ldots, c_k, t)$$

defined by the relation

$$R = L - \sum_{\rho=1}^{k} c_\rho \dot{q}_\rho. \tag{6.70}$$

Now, let us form an arbitrary variation of all the variables in the Routhian function, then we will have

$$\delta R = \delta L - \sum_{\rho=1}^{k} \delta c_\rho \dot{q}_\rho - \sum_{\rho=1}^{k} c_\rho \delta \dot{q}_\rho, \tag{6.71}$$

where we should note that c's are regarded as variables. That is,

$$\delta R = \sum_{\rho=k+1}^{n} \frac{\partial L}{\partial q_\rho} \delta q_\rho + \sum_{\rho=1}^{n} \frac{\partial L}{\partial \dot{q}_\rho} \delta \dot{q}_\rho - \sum_{\rho=1}^{k} \delta c_\rho \dot{q}_\rho - \sum_{\rho=1}^{k} c_\rho \delta \dot{q}_\rho. \tag{6.72}$$

But,

$$p_\rho = c_\rho = \frac{\partial L}{\partial \dot{q}_\rho}. \tag{6.73}$$

Using this expression for c_ρ into Eq. (6.72) it becomes

$$\delta R = \sum_{\rho=k+1}^{n} \frac{\partial L}{\partial q_\rho} \delta q_\rho + \sum_{\rho=1}^{n} \frac{\partial L}{\partial \dot{q}_\rho} \delta \dot{q}_\rho - \sum_{\rho=1}^{k} \delta c_\rho \dot{q}_\rho - \sum_{\rho=1}^{k} \frac{\partial L}{\partial \dot{q}_\rho} \delta \dot{q}_\rho.$$

That is,

$$\delta R = \sum_{\rho=k+1}^{n} \frac{\partial L}{\partial q_\rho} \delta q_\rho + \sum_{\rho=k+1}^{n} \frac{\partial L}{\partial \dot{q}_\rho} \delta \dot{q}_\rho - \sum_{\rho=1}^{k} \delta c_\rho \dot{q}_\rho. \tag{6.74}$$

On the other hand,

$$\delta R = \sum_{\rho=k+1}^{n} \frac{\partial R}{\partial q_\rho} \delta q_\rho + \sum_{\rho=k+1}^{n} \frac{\partial R}{\partial \dot{q}_\rho} \delta \dot{q}_\rho + \sum_{\rho=1}^{k} \frac{\partial R}{\partial c_\rho} \delta c_\rho. \tag{6.75}$$

Comparing Eqs. (6.74) and (6.75), we get at once

$$\left.\begin{array}{l} \dfrac{\partial L}{\partial q_\rho} = \dfrac{\partial R}{\partial q_\rho}, \\[2mm] \dfrac{\partial L}{\partial \dot{q}_\rho} = \dfrac{\partial R}{\partial \dot{q}_\rho}, \quad \rho = k+1, \ldots, n \end{array}\right\} \tag{6.76}$$

and

$$-\frac{\partial R}{\partial c_\rho} = \dot{q}_\rho, \qquad \rho = 1, 2, \ldots, k. \qquad (6.77)$$

Substituting Eq. (6.76) into Lagrange's equations of motion for a simple system,

$$\frac{d}{dt}\left(\frac{\partial L}{\partial \dot{q}_\rho}\right) - \frac{\partial L}{\partial q_\rho} = 0, \qquad \rho = 1, 2, \ldots, n \qquad (6.78)$$

we obtain

$$\frac{d}{dt}\left(\frac{\partial R}{\partial \dot{q}_\rho}\right) - \frac{\partial R}{\partial q_\rho} = 0, \qquad \rho = k+1, \ldots, n \qquad (6.79)$$

which can be regarded as Routhian equations of motion for a system with $(n-k)$ degrees of freedom. These are $(n-k)$ second-order differential equations in the non-ignorable variables/coordinates. Thus, the Routhian procedure eliminates the ignorable coordinates from the equations of motion. If the solution of Eqs. (6.79) can be found for $(n-k)$ non-ignorable coordinates, then we can integrate Eq. (6.77) and get expressions for ignorable coordinates in the form

$$q_\rho = -\int \frac{\partial R}{\partial c_\rho} dt, \qquad \rho = 1, 2, \ldots, k. \qquad (6.80)$$

We shall illustrate this procedure through the following examples:

Example 6.17 Use the Routhian function method and find the equation of motion in the case of planetary motion (Kepler's problem) for the non-ignorable coordinate r.

Solution In Example 6.6, we described the so-called Kepler's problem, where we showed that the Lagrangian for planetary motion is given as

$$L = T - V = \frac{1}{2}m(\dot{r}^2 + r^2\dot{\theta}^2) + \frac{m\mu}{r}. \qquad (1)$$

We can notice that θ is the ignorable coordinate and the corresponding generalized momenta p_θ is constant. That is,

$$p_\theta = \frac{\partial L}{\partial \dot{\theta}} = mr^2\dot{\theta} = c. \qquad (2)$$

Therefore,

$$\dot{\theta} = \frac{c}{mr^2}. \qquad (3)$$

Now, we construct the Routhian function R as

$$R = L - c\dot{\theta} = \frac{1}{2}m(\dot{r}^2 + r^2\dot{\theta}^2) + \frac{m\mu}{r} - c\dot{\theta}.$$

Using Eq. (3), we have

$$R = \frac{1}{2}m\left(\dot{r}^2 + \frac{c^2}{m^2 r^2}\right) + \frac{m\mu}{r} - \frac{c^2}{mr^2}$$

or

$$R = \frac{1}{2}m\left(\dot{r}^2 - \frac{c^2}{m^2 r^2}\right) + \frac{m\mu}{r}. \tag{4}$$

Differentiating Eq. (4), we obtain

$$\left.\begin{array}{l} \dfrac{\partial R}{\partial r} = \dfrac{c^2}{mr^3} - \dfrac{m\mu}{r^2}, \\[1em] \dfrac{\partial R}{\partial \dot{r}} = m\dot{r}, \\[1em] \dfrac{d}{dt}\left(\dfrac{\partial R}{\partial \dot{r}}\right) = m\ddot{r}. \end{array}\right\} \tag{5}$$

Hence, the resulting Routhian equation for the non-ignorable coordinate r, is

$$\frac{d}{dt}\left(\frac{\partial R}{\partial \dot{r}}\right) - \frac{\partial R}{\partial r} = m\ddot{r} - \frac{c^2}{mr^3} + \frac{m\mu}{r^2} = 0, \tag{6}$$

which is a second-order differential equation involving only r, and can be integrated.

Example 6.18 Using the Routhian function procedure for eliminating the ignorable coordinates in the case of the symmetric spinning top, under uniform gravitational field, obtain the equation of motion for the non-ignorable coordinate.

Solution Let M be the mass of the top and I_{xx}, I_{yy}, I_{zz} denote the principal moments of inertia of the top with respect to the inertial frame having origin at the bottom tip of the top. Let l be the distance of the centre of mass of the top from the tip. Taking ψ, θ and ϕ as Euler angles, the kinetic energy T and the potential energy V, described by Eqs. (4.16), (4.20), (4.22) and (4.23), can be written as

$$T = \frac{1}{2}[I_{xx}(\dot{\theta}^2 + \dot{\psi}^2 \sin^2\theta) + I_{zz}(\dot{\psi}\cos\theta + \dot{\phi})^2] \tag{1}$$

and

$$V = Mgl\cos\theta. \tag{2}$$

Therefore, the Lagrangian $L = T - V$ is

$$L = \frac{1}{2}[I_{xx}(\dot{\theta}^2 + \dot{\psi}^2 \sin^2\theta) + I_{zz}(\dot{\psi}\cos\theta + \dot{\phi})^2] - Mgl\cos\theta. \qquad (3)$$

It is evident from L that both ψ and ϕ are ignorable coordinates and the corresponding generalized momenta are constants. Thus,

$$p_\phi = \frac{\partial L}{\partial \dot{\phi}} = I_{zz}(\dot{\psi}\cos\theta + \dot{\phi}) = c_1, \qquad (4)$$

and

$$p_\psi = \frac{\partial L}{\partial \dot{\psi}} = I_{xx}\dot{\psi}\sin^2\theta + I_{zz}(\dot{\psi}\cos\theta + \dot{\phi})\cos\theta = c_2. \qquad (5)$$

Here, c_1 and c_2 are constants of motion.

Now, we construct the Routhian function R as

$$R = L - (p_\phi \dot{\phi} + p_\psi \dot{\psi}). \qquad (6)$$

Using Eqs. (3), (4) and (5), the above equation becomes

$$R = \frac{1}{2} I_{xx}\dot{\theta}^2 + \frac{1}{2} I_{xx}\dot{\psi}^2 \sin^2\theta + \frac{1}{2} I_{zz}(\dot{\psi}\cos\theta + \dot{\phi})^2 - Mgl\cos\theta$$

$$- I_{zz}\dot{\phi}\dot{\psi}\cos\theta - I_{zz}\dot{\phi}^2 - I_{xx}\dot{\psi}^2\sin^2\theta - I_{zz}\dot{\psi}^2\cos^2\theta - I_{zz}\dot{\phi}\dot{\psi}\cos\theta.$$

or

$$R = \frac{1}{2} I_{xx}\dot{\theta}^2 - \frac{1}{2} I_{xx}\dot{\psi}^2 \sin^2\theta - \frac{1}{2} I_{zz}(\dot{\psi}\cos\theta + \dot{\phi})^2 - Mgl\cos\theta. \qquad (7)$$

But,

$$(p_\psi - p_\phi \cos\theta) = I_{xx}\dot{\psi}\sin^2\theta + I_{zz}(\dot{\psi}\cos\theta + \dot{\phi})\cos\theta - I_{zz}\dot{\psi}\cos^2\theta - I_{zz}\dot{\phi}\cos\theta$$

$$= I_{xx}\dot{\psi}\sin^2\theta.$$

Therefore,

$$(p_\psi - p_\phi \cos\theta)^2 = I_{xx}^2 \dot{\psi}^2 \sin^4\theta, \qquad (8)$$

and

$$p_\phi^2 = I_{zz}^2 (\dot{\psi}\cos\theta + \dot{\phi})^2. \qquad (9)$$

Introducing Eqs. (8) and (9), Eq. (7) can be rewritten as

$$R = \frac{1}{2} I_{xx}\dot{\theta}^2 - \frac{(p_\psi - p_\phi \cos\theta)^2}{2 I_{xx} \sin^2\theta} - \frac{p_\phi^2}{2 I_{zz}} - Mgl\cos\theta. \qquad (10)$$

Here, θ is the non-ignorable coordinate, which represents nutation of the top. Thus, having constructed Routhian function, as in Eq. (10), the Routhian equation of motion for the non-ignorable coordinate θ can be obtained from

$$\frac{d}{dt}\left(\frac{\partial R}{\partial \dot{\theta}}\right) - \frac{\partial R}{\partial \theta} = 0. \qquad (11)$$

From Eq. (10), we have

$$\frac{\partial R}{\partial \dot{\theta}} = I_{xx}\dot{\theta},$$

$$\frac{\partial R}{\partial \theta} = \frac{(p_\psi - p_\phi \cos\theta)^2 \cos\theta}{\sin^3\theta} + \frac{p_\phi(p_\psi - p_\phi \cos\theta)}{\sin\theta} + Mgl\sin\theta.$$

Substituting these expressions into Eq. (11), we get

$$I_{xx}\ddot{\theta} - \frac{(p_\psi - p_\phi \cos\theta)^2 \cos\theta}{\sin^3\theta} + \frac{p_\phi(p_\psi - p_\phi \cos\theta)}{\sin\theta} + Mgl\sin\theta = 0, \qquad (12)$$

which is the equation of motion involving non-ignorable coordinate θ, the nutation.

EXERCISES

6.1 In a holonomic system, the dependent variables r_i are expressed in terms of generalized coordinates q_ρ and time t in the form

$$r_i = r_i(q_1, q_2, \ldots, q_n, t), \qquad i = 1, 2, \ldots, n.$$

Express the kinetic energy T in the form of the following functional dependence

$$T = T(q_1, q_2, \ldots, q_n, \dot{q}_1, \dot{q}_2, \ldots, \dot{q}_n, t).$$

6.2 For a rigid body turning about a fixed point under no forces, we have

$$T = \frac{1}{2}(I_{xx}\omega_x^2 + I_{yy}\omega_y^2 + I_{zz}\omega_z^2), \qquad V = 0.$$

Taking Euler angles as the generalized coordinates, find the Lagrange's equations of motion.

6.3 A double pendulum consists of two weightless rods connected to each other and a point of support, as shown in Figure 6.6. The masses m_1 and m_2 are not equal, while the lengths of the rods are equal. The pendulums are free to swing only in one vertical plane. Taking θ and ϕ as generalized coordinates, find the Lagrange's equations of motion.

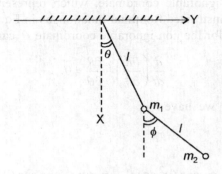

Figure 6.6 Description of Problem 6.3.

6.4 Consider the motion of a spinning top in a uniform gravitational field, which has its kinetic energy T and potential energy V given as

$$T = \frac{1}{2}[I_{xx}(\dot{\theta}^2 + \dot{\psi}^2 \sin^2\theta) + I_{zz}(\dot{\psi}\cos\theta + \dot{\phi})^2]$$

and

$$V = Mgl\cos\theta.$$

Obtain the expressions for generalized momenta and hence construct the Hamiltonian function.

6.5 Explain the concept of an ignorable coordinate and show that the generalized momenta corresponding to each ignorable coordinate is constant.

6.6 State and explain in detail the principle of virtual work with reference to a rigid body constrained to move on a surface.

6.7 Show that in a conservative system, the Hamiltonian function is a constant and that in a simple dynamical system it represents the sum of kinetic energy and potential energy.

6.8 Explain D'Alembert's principle.

6.9 Derive the Lagrange's equations of motion for a simple dynamical system.

6.10 Define Routhian function and show how the Routhian procedure succeeds in eliminating the ignorable coordinates from the equations of motion of a standard simple dynamical system.

6.11 A particle of mass m moves in a frictionless thin circular tube of radius r (Fig. 6.7). If the tube rotates with angular velocity ω about a vertical diameter, obtain the equation of motion of the particle.

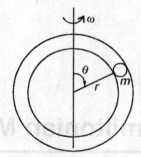

Figure 6.7 Description of Problem 6.11.

6.12 A particle is constrained to move along the inner surface of a fixed hemispherical bowl. What is the number of degrees of freedom of the particle?

6.13 An artificial satellite moves in the gravitational field of the Earth of mass M.

(a) Assuming the satellite as a point particle of mass m, write down the Lagrangian L of the satellite in terms of spherical polar coordinates with the origin at the centre of Earth.

(b) Obtain the Lagrange's equations of motion for the satellite.

(c) Find the expressions for the canonical momenta p_r, p_θ, p_ϕ.

7

Hamiltonian Methods

Truths beautifully arranged together in their proper places are like a fine garland.
—Swami Vivekananda

7.1 INTRODUCTION

We observed in Chapter 6 that a dynamical system can be described elegantly through Lagrangian formulation. Another formulation of a dynamical system is through Hamiltonian. For a dynamical system with n degrees of freedom, the Lagrangian formulation gives n, second-order differential equations, while the Hamiltonian formulation gives $2n$ first-order differential equations. The Hamiltonian formulation has several advantages over that of Lagrangian. They are:

1. Hamiltonian formulation provides deeper insight into the behaviour of a dynamical system.
2. Hamilton's formulation proves superior in the indirect integration of the equations of motion by means of suitable transformations.
3. The transformations used in Hamiltonian mechanics are known as *contact transformation*, which is characterized by a single function, namely the generating function. If a generating function can be found, any dynamical problem can be reduced to one of differentiations and elimination, which is discussed in detail in the following sections:

7.2 HAMILTON'S PRINCIPLE

Before we discuss in detail the Hamilton's principle, we should be familiar with certain terminology:

Natural motion Consider a dynamical system with n degrees of freedom and a Hamiltonian $H(q_\rho, p_\rho, t)$, $\rho = 1, 2, \ldots, n$. A motion is said to be *natural*, if the Hamiltonian of the physical system considered satisfies the following:

$$\dot{q}_\rho = \frac{\partial H}{\partial p_\rho}, \qquad \dot{p}_\rho = -\frac{\partial H}{\partial q_\rho} \tag{7.1}$$

and

$$\dot{H} = \frac{\partial H}{\partial t}. \tag{7.2}$$

Space of events A point in a representative space of $(n+1)$ dimensions, that is, in E_{n+1}, is denoted by a set of numbers $(q_1, q_2, \ldots, q_n, t)$. Such a point corresponds to a configuration of the system at certain time t. We may refer to this point as an event and the space E_{n+1} is called *space* of events. Any motion of the system, whether it is natural or not can be described by taking q's as functions of time t. For example, if $n = 2$, the geometrical motion in E_3 is a curve C given by the set of equations

$$q_1 = q_1(t), \qquad q_2 = q_2(t),$$

as shown in Fig. 7.1.

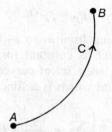

Figure 7.1 Motion represented by a curve C in E_3 space of events.

Action for an arbitrary motion We can describe a motion from an event A to an event B by writing

$$q_\rho = q_\rho(u), \qquad p_\rho = p_\rho(u), \qquad t = t(u), \tag{7.3}$$

where u is a parameter which runs from $u = u_1$ at A to $u = u_2$ at B. These are $(2n+1)$ functions of a parameter u. Imagine a curve C in E_{n+1} with p's assigned, that is, we may think of them as defining a momentum vector p_ρ attached at each point of C, as shown in Fig. 7.2.

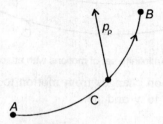

Figure 7.2 Curve of motion with attached momentum.

Thus, we may say that Eq. (7.3) defines a curve C with momentum. Now, we define the action along C as an integral

$$S = \int_{u_1}^{u_2} \left(\sum_{\rho=1}^{n} p_\rho \frac{dq_\rho}{du} - H \frac{dt}{du} \right) du \qquad (7.4)$$

or briefly

$$S = \int_A^B \left(\sum_{\rho=1}^n p_\rho \, dq_\rho - H \, dt \right) = \int_A^B (p \, dq - H \, dt), \qquad (7.4a)$$

where the integration is carried out along C. In Eq. (7.4a), we have suppressed summation finally, of course with the understanding that it actually exists.

The variation of action Let us consider an infinity of motions, each with attached momentum described as

$$q_\rho = q_\rho(u, v), \qquad p_\rho = p_\rho(u, v), \qquad t = t(u, v), \qquad (7.5)$$

where u is a parameter running from u_1 to u_2 for each of the motions and v is a second parameter which is constant for each motion. Thus, the set of motions appearing in E_{n+1} are a set of curves as depicted in Fig. 7.3. For each motion, there is an action which is a function of v, which we write as

$$S(v) = \int_{u_1}^{u_2} \left(\sum_{\rho=1}^n p_\rho \frac{\partial q_\rho}{\partial u} - H \frac{\partial t}{\partial u} \right) du. \qquad (7.6)$$

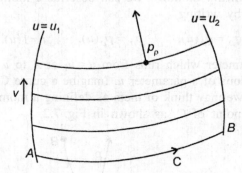

Figure 7.3 An infinite set of motions with attached momenta.

In order to see how action changes from motion to motion, we differentiate the action with respect to v and get

$$\frac{dS}{dv} = \int_{u_1}^{u_2} \left(\sum_{\rho=1}^n \frac{\partial p_\rho}{\partial v} \frac{\partial q_\rho}{\partial u} + \sum_{\rho=1}^n p_\rho \frac{\partial^2 q_\rho}{\partial v \partial u} - \frac{\partial H}{\partial v} \frac{\partial t}{\partial u} - H \frac{\partial^2 t}{\partial v \partial u} \right) du. \qquad (7.7)$$

Since

$$\frac{\partial^2 q_\rho}{\partial v \partial u} = \frac{\partial^2 q_\rho}{\partial u \partial v} = \frac{\partial}{\partial u}\left(\frac{\partial q_\rho}{\partial v}\right).$$

Using the integration by parts rule, we have

$$\int_{u_1}^{u_2} p_\rho \frac{\partial^2 q_\rho}{\partial v \partial u} du = \int_{u_1}^{u_2} p_\rho \frac{\partial}{\partial u}\left(\frac{\partial q_\rho}{\partial v}\right) du$$

$$= \left[p_\rho \frac{\partial q_\rho}{\partial v}\right]_{u_1}^{u_2} - \int_{u_1}^{u_2} \frac{\partial q_\rho}{\partial v} \frac{\partial p_\rho}{\partial u} du. \quad (7.8)$$

Similarly, integrating the last term in Eq. (7.7), we can recast Eq. (7.7) in the form

$$\frac{dS}{dv} = \left[\sum_{\rho=1}^{n} p_\rho \frac{\partial q_\rho}{\partial v} - H \frac{\partial t}{\partial v}\right]_{u_1}^{u_2}$$

$$+ \int_{u_1}^{u_2} \left(\sum_{\rho=1}^{n} \frac{\partial p_\rho}{\partial v} \frac{\partial q_\rho}{\partial u} - \sum_{\rho=1}^{n} \frac{\partial p_\rho}{\partial u} \frac{\partial q_\rho}{\partial v} - \frac{\partial H}{\partial v} \frac{\partial t}{\partial u} + \frac{\partial H}{\partial u} \frac{\partial t}{\partial v}\right) du. \quad (7.9)$$

If we consider an infinitesimal change in S due to an infinitesimal change δv in v, that is, in passing from a curve C to the neighbouring curve, we write

$$\left.\begin{aligned}\frac{\partial q_\rho}{\partial u} du = dq_\rho, & \qquad \frac{\partial p_\rho}{\partial u} du = dp_\rho \\ \frac{\partial t}{\partial u} du = dt, & \qquad \frac{\partial H}{\partial u} du = dH,\end{aligned}\right\} \quad (7.10)$$

which represents increments in passing along a curve C and also write

$$\left.\begin{aligned}\frac{dS}{dv}\delta v = \delta S, & \qquad \frac{\partial q_\rho}{\partial v}\delta v = \delta q_\rho \\ \frac{\partial p_\rho}{\partial v}\delta v = \delta p_\rho, & \qquad \frac{\partial t}{\partial v}\delta v = \delta t \\ \frac{\partial H}{\partial v}\delta v = \delta H. & \end{aligned}\right\} \quad (7.11)$$

which represents variations as a result of change in v. Now, multiplying Eq. (7.9) by δv and using the variations, as given in Eqs. (7.10) and (7.11), we obtain

$$\delta S = \left[\sum_{\rho=1}^{n} p_\rho \frac{\partial q_\rho}{\partial v} \delta v - H \frac{\partial t}{\partial v} \delta v \right]_{u_1}^{u_2}$$

$$+ \int_{u_1}^{u_2} \left[\sum_{\rho=1}^{n} \left(\frac{\partial p_\rho}{\partial v} \delta v \right) \frac{\partial q_\rho}{\partial u} - \sum \frac{\partial p_\rho}{\partial u} \left(\frac{\partial q_\rho}{\partial v} \delta v \right) \right.$$

$$\left. - \left(\frac{\partial H}{\partial v} \delta v \right) \frac{\partial t}{\partial u} + \frac{\partial H}{\partial u} \left(\frac{\partial t}{\partial v} \right) \delta v \right] du$$

or

$$\delta S = \left[\sum_{\rho=1}^{n} p_\rho \, \delta q_\rho - H \delta t \right]_{u_1}^{u_2}$$

$$+ \int_{u_1}^{u_2} \left(\sum_{\rho=1}^{n} \delta p_\rho \, dq_\rho - \sum_{\rho=1}^{n} \delta q_\rho \, dp_\rho - \delta H \, dt + \delta t \, dH \right). \quad (7.12)$$

Here the integration is to be carried out with respect to d, not δ. Further, it may be noted that if f is any function of u and v, then we have

$$\frac{\partial^2 f}{\partial u \partial v} = \frac{\partial^2 f}{\partial v \partial u}. \quad (7.13)$$

Multiplying this equation with $du \delta v$ and using the fact that these infinitesimals are constants, we obtain

$$du \frac{\partial}{\partial u} \left(\frac{\partial f}{\partial v} \delta v \right) = \delta v \frac{\partial}{\partial v} \left(\frac{\partial f}{\partial u} du \right).$$

Now, using the results from Eqs. (7.10) and (7.11), the above equation can be written as

$$du \frac{\partial}{\partial u} (\delta f) = \delta v \frac{\partial}{\partial v} (df),$$

that is,

$$d\delta f = \delta df \quad (7.14)$$

or

$$d\delta = \delta d. \quad (7.15)$$

This rule, that is, interchange of d and δ is very useful in variational calculus.

We can also find the variation of action δS from S using the above rule of interchange of d and δ, as follows:

$$S = \int_A^B (p\,dq - H\,dt)$$

$$\delta S = \int_A^B [\delta p\,dq + p\delta(dq) - \delta H\,dt - H\delta(dt)]$$

$$= \int_A^B [\delta p\,dq + p\,d(\delta q) - \delta H\,dt - H\,d(\delta t)].$$

On integrating by parts the second and the last terms, we get

$$\delta S = [p\delta q - H\delta t]_A^B + \int_A^B (\delta p\,dq - \delta q\,dp - \delta H\,dt + \delta t\,dH), \tag{7.16}$$

which is same as Eq. (7.12).

Hamilton's principle The integral of action

$$S = \int_A^B (p\,dq - H\,dt)$$

has a stationary value for the natural motion when compared with the adjacent motions having the same end events. (Here q denotes the set of n quantities q_ρ and p denotes p_ρ.)

Proof The integral of action is defined as

$$S = \int_A^B (p\,dq - H\,dt). \tag{7.17}$$

Its variation is

$$\delta S = \int_A^B [\delta p\,dq + p\delta(dq) - \delta H\,dt - H\delta(dt)]$$

Interchanging d and δ, we have

$$\delta S = \int_A^B [\delta p\,dq + p\,d(\delta q) - \delta H\,dt - H\,d(\delta t)].$$

Integrating the second and last terms using the parts rule, we get after rearrangement

$$\delta S = [p\delta q - H\delta t]_A^B + \int_A^B (\delta p\,dq - \delta q\,dp - \delta H\,dt + \delta t\,dH). \tag{7.18}$$

Now, consider the two curves in E_{n+1} with the same end-events A and B with attached momenta, as shown in Fig. 7.4, that is, at A and B, $\delta t = 0$, $\delta q = 0$. Using this fact, Eq. (7.18) gives

$$\delta S = \int_A^B (\delta p\,dq - \delta q\,dp - \delta H\,dt + \delta t\,dH). \tag{7.19}$$

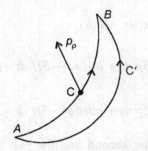

Figure 7.4 Hamilton's principle.

But

$$\delta H(q, p, t) = \frac{\partial H}{\partial q}\delta q + \frac{\partial H}{\partial p}\delta p + \frac{\partial H}{\partial t}\delta t. \qquad (7.20)$$

Substituting Eq. (7.20) into Eq. (7.19), we arrive at

$$\delta S = \int_A^B \left[\delta p\left(dq - \frac{\partial H}{\partial p}dt\right) - \delta q\left(dp + \frac{\partial H}{\partial q}dt\right) + \delta t\left(dH - \frac{\partial H}{\partial t}dt\right)\right]$$

$$= \int_A^B \left[\delta p\left(\dot{q} - \frac{\partial H}{\partial p}\right) - \delta q\left(\dot{p} + \frac{\partial H}{\partial q}\right) + \delta t\left(\dot{H} - \frac{\partial H}{\partial t}\right)\right] dt. \qquad (7.21)$$

Now, using Hamilton's canonical equations of motion for a natural motion described by Eqs. (7.1) and (7.2), the above equation at once becomes

$$\delta S = \delta \int_A^B (p\, dq - H\, dt) = 0 \qquad (7.22)$$

whatever may be the variations $\delta p, \delta q$ and δt.

Thus, S has a stationary value for a natural motion when compared with arbitrary adjacent motions for the same end-events.

Conversely if S has a stationary value for arbitrary variations $\delta p, \delta q, \delta t$ except for end-events, then C represents a natural motion.

Proof It is given that $\delta S = 0$ for any arbitrary variations $\delta p, \delta q, \delta t$. Suppose, we choose these variations as

$$\left.\begin{array}{l} \delta q = -\left(\dot{p} + \dfrac{\partial H}{\partial q}\right)F\delta v, \\[2mm] \delta p = \left(\dot{q} - \dfrac{\partial H}{\partial p}\right)F\delta v, \\[2mm] \delta t = \left(\dot{H} - \dfrac{\partial H}{\partial t}\right)F\delta v, \end{array}\right\} \qquad (7.23)$$

where $\delta v > 0$, and F is any arbitrary function along the curve C such that $F \geq 0$ and $F = 0$ at the end-events. Substituting (7.23) into Eq. (7.21), we find that

$$\delta S = \int_A^B \left[\left(\dot{p} + \frac{\partial H}{\partial q} \right)^2 F + \left(\dot{q} - \frac{\partial H}{\partial p} \right)^2 F + \left(\dot{H} - \frac{\partial H}{\partial t} \right)^2 F \right] \delta v \, dt. \quad (7.24)$$

However, $\delta S = 0$, only if

$$\dot{p} + \frac{\partial H}{\partial q} = 0, \qquad \dot{q} - \frac{\partial H}{\partial p} = 0, \qquad \dot{H} - \frac{\partial H}{\partial t} = 0. \quad (7.25)$$

on C, which of course, are Hamilton's canonical equations. Therefore, C is a natural path and hence the converse is proved. It is also possible to bring out a connection between the Hamiltonian H and the Lagrangian L through Hamilton's principle, which can be seen in the following steps:

Hamilton's principle states

$$\delta S = \delta \int_A^B (p \, dq - H \, dt) = 0. \quad (7.26)$$

Recalling the definition of Hamiltonian, that is,

$$H = \dot{q} p - L, \quad (7.27)$$

the Hamilton's principle can also be stated in the form

$$\delta S = \delta \int_A^B (\dot{q} p - H) \, dt = \delta \int L \, dt = 0. \quad (7.28)$$

Thus, alternatively, Hamilton's principle can also be stated as

$$\delta \int_A^B L \, dt = 0. \quad (7.29)$$

Noting that

$$L = L(q, \dot{q}), \quad (7.30)$$

$$\delta \int_A^B L \, dt = \int_A^B \delta L \, dt = \int_A^B \left(\frac{\partial L}{\partial \dot{q}} \delta \dot{q} + \frac{\partial L}{\partial q} \delta q \right) dt \quad (7.31)$$

$$= \int_A^B \frac{\partial L}{\partial \dot{q}} \delta \left(\frac{dq}{dt} \right) dt + \int_A^B \frac{\partial L}{\partial q} \delta q \, dt$$

$$= \int_A^B \frac{\partial L}{\partial \dot{q}} d(\delta q) + \int_A^B \frac{\partial L}{\partial q} \delta q \, dt.$$

Evaluating the first integral by parts rule, we get

$$\delta \int_A^B L \, dt = \left[\frac{\partial L}{\partial \dot{q}} \delta q \right]_A^B - \int_A^B \frac{d}{dt} \left(\frac{\partial L}{\partial \dot{q}} \right) \delta q \, dt + \int \frac{\partial L}{\partial q} \delta q \, dt = 0.$$

Since $\delta q = 0$ at A and B, the first term vanishes and we are left with

$$\int_A^B \left[\frac{\partial L}{\partial q} - \frac{d}{dt}\left(\frac{\partial L}{\partial \dot{q}}\right) \right] \delta q \, dt = 0. \tag{7.32}$$

This expression is true for all δq and dt. Consequently, we have

$$\frac{\partial L}{\partial q} - \frac{d}{dt}\left(\frac{\partial L}{\partial \dot{q}}\right) = 0. \tag{7.33}$$

This equation is called *Euler–Lagrange equations* associated with the variational Eq. (7.29).

Theorem 7.1 If the Hamiltonian $H(q_\rho, p_\rho, t)$ is not an explicit function of time, then H is a constant of the motion.

Proof Given $H = H(q_\rho, p_\rho, t)$, its time rate of change is

$$\frac{dH}{dt} = \sum_{\rho=1}^n \frac{\partial H}{\partial q_\rho} \dot{q}_\rho + \sum_{\rho=1}^n \frac{\partial H}{\partial p_\rho} \dot{p}_\rho + \frac{\partial H}{\partial t}.$$

Using Hamilton's canonical Eqs. (7.25), the above expression reduces to

$$\frac{dH}{dt} = \frac{\partial H}{\partial t}.$$

Now, if H is not an explicit function of time t, then,

$$\frac{\partial H}{\partial t} = 0,$$

implying

$$\frac{dH}{dt} = 0. \tag{7.34}$$

In other words, H is a constant.

7.2.1 Hamilton's Principle for a Conservative System

In conservative systems for which $\partial H/\partial t = 0$, Hamilton's variational principle can be stated as

$$\delta \int p \, dq = 0$$

for variations from a natural motion, provided that the end-configurations are fixed and H has, in the varied motion, the same constant value which it has in the natural motion.

Proof Consider

$$\delta \int_A^B p \, dq = \int_A^B (\delta p \, dq + p \delta \, dq)$$

$$= \int_A^B [\delta p\, dq + p\, d(\delta q)] \quad \text{(interchanging } d \text{ and } \delta\text{)}$$

$$= \int_A^B \delta p\, dq + [p\delta q]_A^B - \int_A^B \delta q\, dp$$

$$= [p\delta q]_A^B + \int_A^B (\delta p\, dq - \delta q\, dp). \tag{7.35}$$

Since C is a natural motion, Hamilton's canonical equations hold true and, therefore,

$$\dot{q} = \frac{\partial H}{\partial p}, \qquad \dot{p} = -\frac{\partial H}{\partial q}, \qquad \frac{\partial H}{\partial t} = 0.$$

That is,

$$dq = \frac{\partial H}{\partial p} dt, \qquad dp = -\frac{\partial H}{\partial q} dt. \tag{7.36}$$

Thus,

$$\delta \int p\, dq = [p\delta q]_A^B + \int_A^B \left[\delta p \left(\frac{\partial H}{\partial p} \right) + \delta q \left(\frac{\partial H}{\partial q} \right) \right] dt. \tag{7.37}$$

Now, using the chain rule of partial differentiation and also noting that $\partial H/\partial t = 0$, for a conservative system, Eq. (7.37) can be rewritten as

$$\delta \int p\, dq = [p\delta q]_A^B + \int_A^B \delta H\, dt. \tag{7.38}$$

Since $\delta q = 0$ at A and B, the end-events, the first term on the right-hand side of Eq. (7.38) vanishes. Also, since H is constant along C for any natural motion, $\delta H = 0$, thereby, the second term also vanishes and thus, Eq. (7.38) finally becomes

$$\delta \int_A^B p\, dq = 0, \tag{7.39}$$

which is the desired result.

7.2.2 Principle of Least Action

For a simple dynamical system,

$$\delta \int T\, dt$$

has a stationary value for the natural motion compared with the adjacent motions having the same end-events, provided H has in the varied motion the same constant value which it has in the natural motion.

Proof For a simple dynamical system, the Hamiltonian is defined as
$$H = T + V \tag{7.40}$$
and the generalized momenta
$$p = \frac{\partial H}{\partial \dot{q}} = \frac{\partial T}{\partial \dot{q}}$$
since V is not a function of $\dot{q}$. Then,
$$\int p \, dq = \int \frac{\partial T}{\partial \dot{q}} dq = \int \frac{\partial T}{\partial \dot{q}} \frac{dq}{dt} dt = \int \dot{q} \frac{\partial T}{\partial \dot{q}} dt. \tag{7.41}$$
Since T is a homogeneous quadratic function in $\dot{q}_\rho$'s, we have, using Euler's theorem,
$$\dot{q} \frac{\partial T}{\partial \dot{q}} = 2T. \tag{7.42}$$
Therefore,
$$\int p \, dq = \int 2T \, dt = 2 \int T \, dt. \tag{7.43}$$
But, the Hamilton's variational principle states that
$$\delta \int p \, dq = 0.$$
In other words,
$$\delta \int T \, dt = 0. \tag{7.44}$$
Hence the principle of least action.

7.3 CHARACTERISTIC FUNCTION AND HAMILTON–JACOBI EQUATION

Following Hamilton's principle for natural motion, let us consider two events A and B and actions for various motions which give us stationary value. Thus, we have
$$\delta \int_A^B (p \, dq - H \, dt) = 0. \tag{7.45}$$
On the other hand, the Hamilton's canonical equations for the natural motion are known to be
$$\dot{q} = \frac{\partial H}{\partial p}, \qquad \dot{p} = -\frac{\partial H}{\partial q}. \tag{7.46}$$
These canonical equations suggest that we can construct the natural motion

from initial data, called *Cauchy data* (q, p, t). The solution of the differential equation is called *Cauchy problem*.

Hamilton's characteristic function

Let C be the curve in E_{n+1}, representing a natural motion and also let A and B be two events on it. Then, the action on C from A to B is a function of these events (see Fig. 7.5). Thus,

$$S = S(q^*, t^*, q, t), \tag{7.47}$$

where (q^*, t^*) refer to the initial event A and (q, t) to the final event B. Here, q^*, q stand for n generalized coordinates.

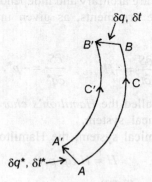

Figure 7.5 Variation from one natural motion to another.

Now, consider an adjacent natural motion represented by the curve C' with end-events A' and B'. Let the displacement from A to A' be $(\delta q^*, \delta t^*)$ and that from B to B' be $(\delta q, \delta t)$, as depicted in Fig. 7.5. Recalling the integral of action as

$$S = \int_A^B (p\,dq - H\,dt).$$

its variation is obtained as

$$\delta S = \int_A^B \delta p\,dq + p\delta(dq) - \delta H\,dt - H\delta(dt).$$

Interchanging d and δ and carrying out integration by parts, we have

$$\delta S = [p\delta q - H\delta t]_A^B + \int_A^B (\delta p\,dq - \delta q\,dp - \delta H\,dt + \delta t\,dH). \tag{7.48}$$

But,

$$\delta H = \frac{\partial H}{\partial q}\delta q + \frac{\partial H}{\partial p}\delta p + \frac{\partial H}{\partial t}\delta t.$$

Inserting this value of δH into Eq. (7.48), we get

$$\delta S = [p\delta q - H\delta t]_A^B + \int_A^B \left[\delta p\left(\dot{q} - \frac{\partial H}{\partial p}\right) - \delta q\left(\dot{p} + \frac{\partial H}{\partial q}\right) + \delta t\left(\dot{H} - \frac{\partial H}{\partial t}\right)\right] dt.$$

Since C represents natural motion, the integral vanishes and we are left with

$$\delta S = [p\delta q - H\delta t]_A^B$$

or

$$\delta S = p\delta q - H\delta t - p^*\delta q^* + H^*\delta t^*. \tag{7.49}$$

Here H^* represents the value of H at A etc. Noting that the $(2n+2)$ infinitesimals $(\delta q, \delta t, \delta q^*, \delta t^*)$ are arbitrary and independent, the partial derivatives of S with respect to the arguments, as given in Eq. (7.47), give us the following results:

$$\frac{\partial S}{\partial q} = p, \qquad \frac{\partial S}{\partial t} = -H, \qquad \frac{\partial S}{\partial q^*} = -p^*, \qquad \frac{\partial S}{\partial t^*} = H^*. \tag{7.50}$$

Here, the function S is called the *Hamilton's characteristic function*, which characterizes the dynamical system.

Suppose for a dynamical system, the Hamiltonian is given by

$$H = H(q, p, t).$$

Then, the second of Eqs. (7.50) gives

$$\frac{\partial S}{\partial t} + H(q, p, t) = 0.$$

Replacing p by $\partial S/\partial q$, that is, using the first of Eqs. (7.50), the above equation can be written as

$$\frac{\partial S}{\partial t} + H\left(q, \frac{\partial S}{\partial q}, t\right) = 0 \tag{7.51}$$

More explicitly,

$$\frac{\partial S}{\partial t} + H\left(q_1, q_2, \ldots, q_n, \frac{\partial S}{\partial q_1}, \frac{\partial S}{\partial q_2}, \ldots, \frac{\partial S}{\partial q_n}, t\right) = 0 \tag{7.52}$$

This is called the *Hamilton–Jacobi equation*, which is a partial differential equation, satisfied by the characteristic function S.

Complete integrals of the Hamilton–Jacobi equation

Let the Hamilton–Jacobi equation for a typical dynamical system be

$$\frac{\partial S}{\partial t} + H\left(q, \frac{\partial S}{\partial q}, t\right) = 0. \tag{7.53}$$

For a moment, let us forget how this equation has been derived and simply treat it as a partial differential equation in one dependent variable S and $(n+1)$ independent variables q and t. The derivatives appearing in Eq. (7.53) are only of first order. However, the derivatives occur in second or higher degrees (see Examples 7.1 and 7.2). Therefore, Eq. (7.53) is in general a first-order nonlinear partial differential equation. In view of the above observations, its complete solution must involve $(n+1)$ independent constants.

Let us assume that the Hamilton characteristic function of the form

$$S = S(q, t, a) + c \qquad (7.54)$$

is a complete integral of Eq. (7.53), where a stands for a set of n constants $(a_1, a_2, \ldots, a_n)$ and c for a single constant. Introducing

$$\frac{\partial S}{\partial q} = p \quad \text{and} \quad \frac{\partial S}{\partial a} = -b, \qquad (7.55)$$

where b stands for a new set of n constants, we shall prove the following Jacobi's theorem which will help us determine the natural motion for a given dynamical system.

Theorem 7.2 (*Jacobi's theorem*) Let $S(q, p, t)$ be any complete integral of the Hamilton–Jacobi equation

$$\frac{\partial S}{\partial t} + H\left(q, \frac{\partial S}{\partial q}, t\right) = 0,$$

then each motion with associated momenta given by

$$\frac{\partial S}{\partial q} = p \quad \text{and} \quad \frac{\partial S}{\partial a} = -b$$

is a natural motion satisfying the canonical equations.

Proof The Hamilton–Jacobi equation for a dynamical system is

$$\frac{\partial S}{\partial t} + H\left(q, p = \frac{\partial S}{\partial q}, t\right) = 0. \qquad (7.56)$$

Let

$$S = S(q, t, a) + c \qquad (7.57)$$

be the complete integral of Eq. (7.56), which contains $(2n+2)$ independent quantities

$$q_1, q_2, \ldots, q_n, t, a_1, a_2, \ldots, a_n, c.$$

Differentiating Eq. (7.56) partially with respect to a_ρ and q_ρ, respectively, we obtain

$$\frac{\partial^2 S}{\partial a_\rho \partial t} + \sum_{r=1}^{n} \frac{\partial H}{\partial p_r} \frac{\partial^2 S}{\partial a_\rho \partial q_r} = 0. \qquad (7.58)$$

and

$$\frac{\partial^2 S}{\partial q_\rho \partial t} + \frac{\partial H}{\partial q_\rho} + \sum_{r=1}^{n} \frac{\partial H}{\partial p_r} \frac{\partial^2 S}{\partial q_\rho \partial q_r} = 0. \quad (7.59)$$

Further, we are given

$$\frac{\partial S}{\partial a} = -b, \quad (7.60)$$

which represents n equations.

Now, differentiating Eq. (7.60) with respect to t, we get

$$\frac{\partial^2 S}{\partial t \partial a_\rho} + \sum_{r=1}^{n} \frac{\partial^2 S}{\partial q_r \partial a_\rho} \dot{q}_r = 0. \quad (7.61)$$

This equation represents n linear equations and can be solved to get $\dot{q}_1, \dot{q}_2, \ldots, \dot{q}_n$.

It may be noted that Eqs. (7.58) and (7.61) are precisely of the same form and therefore Eq. (7.58) can be solved to get

$$\frac{\partial H}{\partial p_1}, \frac{\partial H}{\partial p_2}, \ldots, \frac{\partial H}{\partial p_n}.$$

Thus, we have the first Hamilton's canonical equation

$$\dot{q}_\rho = \frac{\partial H}{\partial p_\rho}. \quad (7.62)$$

Also, we note that

$$\frac{\partial S}{\partial q} = p. \quad (7.63)$$

Differentiating this equation partially with respect to t, we obtain

$$\frac{\partial^2 S}{\partial t \partial q_\rho} + \sum_{r=1}^{n} \frac{\partial^2 S}{\partial q_r \partial q_\rho} \dot{q}_r = \dot{p}_\rho. \quad (7.64)$$

Comparing Eqs. (7.59) and (7.64), we have at once, the second canonical equation, that is,

$$\dot{p}_\rho = -\frac{\partial H}{\partial q_\rho}. \quad (7.65)$$

Thus, we find that Eqs. (7.62) and (7.65) constitute Hamilton's canonical equations and hence each motion described by the characteristic function S is a natural motion.

For illustration of Hamilton–Jacobi method, we consider the following couple of examples:

Example 7.1 (*Kepler's problem*) Suppose that a particle of mass m is attracted by an inverse-square gravitational force to a fixed point O, and the particle is moving in a plane whose position in polar coordinates is given by (r, θ). Discuss the motion of the particle.

Solution The kinetic energy of the system is found to be

$$T = \frac{1}{2}m(\dot{r}^2 + r^2\dot{\theta}^2), \tag{1}$$

while the potential energy is given by

$$V = -\int F\, dr.$$

From the given data, $F = -m\mu/r^2$. Therefore,

$$V = \int \frac{m\mu}{r^2} dr = -\frac{m\mu}{r}. \tag{2}$$

Taking r, θ as generalized coordinates, the Lagrangian for the given system is

$$L = T - V = \frac{1}{2}m(\dot{r}^2 + r^2\dot{\theta}^2) + \frac{m\mu}{r}. \tag{3}$$

Then, the generalized momenta is given as

$$p_r = \frac{\partial L}{\partial \dot{r}} = m\dot{r}, \qquad p_\theta = \frac{\partial L}{\partial \dot{\theta}} = mr^2\dot{\theta}. \tag{4}$$

However, the Hamiltonian is obtained as

$$H = T + V = H(q, p, t).$$

Thus, from Eqs. (1), (2) and (4), we have

$$H = \frac{1}{2}m\left(\frac{p_r^2}{m^2} + \frac{r^2 p_\theta^2}{m^2 r^4}\right) - \frac{m\mu}{r}$$

or

$$H = \frac{1}{2m}\left(p_r^2 + \frac{p_\theta^2}{r^2}\right) - \frac{m\mu}{r}. \tag{5}$$

But H should be a function of r, θ, p_r, p_θ, and t. However, we notice that θ is absent in H and therefore it is an ignorable coordinate.

To write down the Hamilton–Jacobi equation for the given problem, we have to substitute $\partial S/\partial r$ and $\partial S/\partial \theta$ for p_r and p_θ in H and thus obtain

$$\frac{\partial S}{\partial t} + \frac{1}{2m}\left[\left(\frac{\partial S}{\partial r}\right)^2 + \frac{1}{r^2}\left(\frac{\partial S}{\partial \theta}\right)^2\right] - \frac{m\mu}{r} = 0. \tag{6}$$

This being a nonlinear first-order partial differential equation, we use separation of variables method to find the complete integral of Eq. (6). Thus, we assume tentatively

$$S = -a_1 t + a_2 \theta + f(r), \tag{7}$$

where a_1 and a_2 are arbitrary constants and $f(r)$ is a function of r alone. Substituting Eq. (7) into Eq. (6), we get at once

$$-a_1 + \frac{1}{2m}\left(\frac{df}{dr}\right)^2 + \frac{a_2^2}{2mr^2} - \frac{m\mu}{r} = 0.$$

Solving for $f(r)$, we get

$$\left(\frac{df}{dr}\right)^2 = 2ma_1 - \frac{a_2^2}{r^2} + \frac{2m^2\mu}{r}.$$

On integration with respect to r, we find that

$$f(r) = \int \sqrt{2ma_1 + \frac{2m^2\mu}{r} - \frac{a_2^2}{r^2}}\, dr + c, \tag{8}$$

where c is an arbitrary constant. Now, substituting Eq. (8) into Eq. (7), we find the complete integral as

$$S = -a_1 t + a_2 \theta + \int \sqrt{2ma_1 + \frac{2m^2\mu}{r} - \frac{a_2^2}{r^2}}\, dr + c, \tag{9}$$

which contains three arbitrary constants.

In order to determine the natural motion of the given system, we may recall Eq. (7.60), which states

$$\frac{\partial S}{\partial a} = -b.$$

In this example, we have a_1 and a_2 as constants and therefore obtain after differentiating under integral sign

$$-b_1 = \frac{\partial S}{\partial a_1} = -t + m\int \frac{dr}{\sqrt{2ma_1 + 2m^2\mu/r - a_2^2/r^2}} \tag{10}$$

and

$$-b_2 = \frac{\partial S}{\partial a_2} = \theta - a_2 \int \frac{dr}{r^2\sqrt{2ma_1 + 2m^2\mu/r - a_2^2/r^2}}. \tag{11}$$

Here, Eq. (10) is a relation connecting r and t, which when solved for r gives us r in terms of t. Similarly, Eq. (11) gives us θ in terms of t.

The corresponding associated momenta is found to be

$$p_r = \frac{\partial S}{\partial r} = \sqrt{2ma_1 + \frac{2m^2\mu}{r} - \frac{a_2^2}{r^2}}, \qquad (12)$$

and

$$p_\theta = \frac{\partial S}{\partial \theta} = a_2. \qquad (13)$$

Noting that

$$H = -\frac{\partial S}{\partial t} = a_1. \qquad (14)$$

From Eq. (13), we conclude that the constant a_2 corresponds to that of constant angular momentum $mr^2\dot\theta$ of the system, while Eq. (14) indicates that the constant a_1 corresponds to that of the total energy of the system.

Example 7.2 Discuss the problem of Harmonic oscillator, that is, simple mass-spring system using the Hamilton–Jacobi method.

Solution For description, see Fig. 1.10. This being a natural system, its kinetic energy is given by

$$T = \frac{1}{2}m\dot x^2, \qquad (1)$$

while the potential energy V of this system is

$$V = \frac{1}{2}kx^2. \qquad (2)$$

Taking x as the generalized coordinate q, the Lagrangian for the given system is

$$L = T - V = \frac{1}{2}m\dot x^2 - \frac{1}{2}kx^2.$$

Then the generalized momenta is

$$p_x = \frac{\partial L}{\partial \dot q} = \frac{\partial L}{\partial \dot x} = m\dot x. \qquad (3)$$

The Hamiltonian of the system from Eqs. (1)–(3) is

$$H = T + V = \frac{1}{2}m\dot x^2 + \frac{1}{2}kx^2 = \frac{1}{2}\frac{p_x^2}{m} + \frac{1}{2}kx^2$$

or

$$H(q, p, t) = H(x, p, t) = \frac{p_x^2}{2m} + \frac{1}{2}kx^2. \tag{4}$$

Now, the Hamilton–Jacobi equation for the system is

$$\frac{\partial S}{\partial t} + H(q, p, t) = \frac{\partial S}{\partial t} + \frac{p_x^2}{2m} + \frac{1}{2}kx^2 = 0.$$

Substituting $\partial S/\partial x$ for p_x in H, we obtain

$$\frac{\partial S}{\partial t} + \frac{1}{2m}\left(\frac{\partial S}{\partial x}\right)^2 + \frac{1}{2}kx^2 = 0. \tag{5}$$

This is a nonlinear partial differential equation of first order. To get its complete integral using the variables separable method, we assume tentatively

$$S = -a_1 t + f(x), \tag{6}$$

where a_1 is an arbitrary constant and $f(x)$ is a function of x only. Substituting Eq. (6) into Eq. (5), we get

$$-a_1 + \frac{1}{2m}\left(\frac{df}{dx}\right)^2 + \frac{1}{2}kx^2 = 0$$

or

$$\frac{df}{dx} = \sqrt{2ma_1 - mkx^2}.$$

Therefore, on integration, we obtain

$$f(x) = \int \sqrt{2ma_1 - mkx^2}\, dx + c = m\int \sqrt{\frac{2a_1}{m} - \frac{k}{m}x^2}\, dx + c.$$

Let $k/m = \omega^2$, then we can rewrite the above equation as

$$f(x) = m\int \sqrt{\frac{2a_1\omega^2}{k} - \omega^2 x^2}\, dx + c$$

or

$$f(x) = m\omega \int \sqrt{\frac{2a_1}{k} - x^2}\, dx + c$$

which can be recast as

$$f(x) = m\omega \int \sqrt{\alpha^2 - x^2}\, dx + c \tag{7}$$

where

$$\alpha^2 = \frac{2a_1}{k} = \frac{2a_1}{m\omega^2}. \tag{8}$$

Substituting Eq. (7) into Eq. (6), we get

$$S = -a_1 t + m\omega \int \sqrt{\alpha^2 - x^2}\, dx + c. \tag{9}$$

To determine the natural motion of the system, we may note that, there is only one constant a, in this problem. Now, recalling the relation

$$-b_1 = \frac{\partial S}{\partial a_1} = -t + \frac{1}{\omega} \int_{x_0}^{x_1} \frac{dx}{\sqrt{\alpha^2 - x^2}}$$

or

$$t - b_1 = \frac{1}{\omega}\left[\cos^{-1}\left(\frac{x}{\alpha}\right)\right]_{x_0}^{x_1}. \tag{10}$$

The corresponding associated momenta is obtained from Eq. (9) as

$$p_x = \frac{\partial S}{\partial x} = m\omega \sqrt{\alpha^2 - x^2}. \tag{11}$$

7.4 PHASE SPACE AND LIOUVILLE'S THEOREM

Consider a space of $2n$ dimensions in which the coordinates of a point are specified by n generalized coordinates q_i and n generalized momenta p_i. Such a space is known as *phase space* of $2n$ dimensions.

If we represent the state of a mechanical system at a given time t in phase space, as

$$r = (q_i, p_i) \tag{7.66}$$

then the time evolution of the state of the system may be described by a curve $r(t)$ in phase space, which is called *phase path*. Now, if we consider many mechanical systems with slightly different q's and p's, all of them can be represented by a set of distinct points in the phase space, occupying an infinitesimal volume in it. In course of time, the system points move along different paths in the phase space due to different initial conditions. Suppose that the initial points corresponding to all the systems considered at time t_1, are contained in a volume dV_1 of the phase space, and after sometime, say at $t = t_2$, the same points occupy the region dV_2. The curves dV_1 and dV_2 describing schematically the evolution of the infinitesimal volume are shown in Fig. 7.6, with a typical point $A[q_i(t_1), p_i(t_1)]$ to $B[q_i(t_2), p_i(t_2)]$.

We may also imagine the points in phase space as particles of an incompressible fluid which moves from dV_1 region to dV_2 region during the time interval $(t_2 - t_1)$. Then, the Liouville's theorem states that the $2n$ dimensional space of volumes dV_1 and dV_2 occupied by the ensemble of points does not change with time, though its shape may change. In other words, the density of points remains constant with time. That is,

$$\frac{d\rho}{dt} = 0 \tag{7.67}$$

or

$$\rho = \text{constant}. \tag{7.68}$$

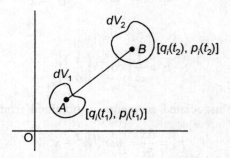

Figure 7.6 Motion of a volume in phase space.

Proof At first, we shall prove the theorem for a system with one degree of freedom. For one degree of freedom, q and p are not vectors but are single numbers and, therefore, we have two-dimensional phase space described by q and p coordinates. We may note that the volume element in this case can be seen as an area element $dq\,dp$ of phase space as depicted in Fig. 7.7.

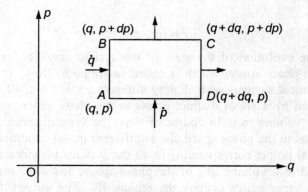

Figure 7.7 Area element in phase space.

Let the speed with which the representative points enter the element $ABCD$ through AB be $\dot{q}$. Then, the number of representative points entering the element through AB per unit time will be $\rho\dot{q}\,dp$. Similarly, the number of points leaving the element through CD will be

$$\rho\dot{q}\,dp + \frac{\partial}{\partial q}(\rho\dot{q})\,dq\,dp. \tag{7.69}$$

Phase Space and Liouville's Theorem (Sec. 7.4) | 269

Thus, the number of representative points that remain in the element $ABCD$ are found to be

$$\rho \dot{q} \, dp - [\rho \dot{q} \, dp + \frac{\partial}{\partial q}(\rho \dot{q}) \, dq \, dp] = -\frac{\partial}{\partial q}(\rho \dot{q}) \, dq \, dp. \qquad (7.70)$$

Similarly considering the number of representative points entering through AD and leaving through BC, we find that the number of points that remain in the element $ABCD$ are found to be

$$-\frac{\partial}{\partial p}(\rho \dot{p}) \, dq \, dp. \qquad (7.71)$$

Summing Eqs. (7.70) and (7.71), we obtain the increase in the representative points per unit time in the typical element $ABCD$ as

$$-\left[\frac{\partial}{\partial q}(\rho \dot{q}) + \frac{\partial}{\partial p}(\rho \dot{p})\right] dq \, dp. \qquad (7.72)$$

But the rate of increase of points per unit time in the above typical element is

$$\frac{\partial \rho}{\partial t} \, dq \, dp. \qquad (7.73)$$

Thus, equating Eqs. (7.72) and (7.73), we get

$$\frac{\partial \rho}{\partial t} = -\left[\frac{\partial}{\partial q}(\rho \dot{q}) + \frac{\partial}{\partial p}(\rho \dot{p})\right]$$

or

$$\frac{\partial \rho}{\partial t} + \frac{\partial}{\partial q}(\rho \dot{q}) + \frac{\partial}{\partial p}(\rho \dot{p}) = 0. \qquad (7.74)$$

Equivalently, we write

$$\frac{\partial \rho}{\partial t} + \rho \frac{\partial \dot{q}}{\partial q} + \frac{\partial \rho}{\partial q} \dot{q} + \rho \frac{\partial \dot{p}}{\partial p} + \frac{\partial \rho}{\partial p} \dot{p} = 0. \qquad (7.75)$$

Now, recalling Hamilton's canonical equations

$$\dot{q} = \frac{\partial H}{\partial p}, \qquad \dot{p} = -\frac{\partial H}{\partial q}$$

and their differentiation gives us

$$\frac{\partial \dot{q}}{\partial q} = \frac{\partial^2 H}{\partial q \partial p} \quad \text{and} \quad \frac{\partial \dot{p}}{\partial p} = -\frac{\partial^2 H}{\partial p \partial q},$$

which in turn yields

$$\frac{\partial \dot{q}}{\partial q} = -\frac{\partial \dot{p}}{\partial p}. \qquad (7.76)$$

Utilizing this result into Eq. (7.75), we get at once

$$\frac{\partial \rho}{\partial t} + \frac{\partial \rho}{\partial q}\dot{q} + \frac{\partial \rho}{\partial p}\dot{p} = 0. \tag{7.77}$$

But $\rho = \rho(q, p, t)$, and therefore, its total derivative is

$$\frac{d\rho}{dt} = \frac{\partial \rho}{\partial t} + \frac{\partial \rho}{\partial q}\frac{dq}{dt} + \frac{\partial \rho}{\partial p}\frac{dp}{dt}$$

or

$$\frac{d\rho}{dt} = \frac{\partial \rho}{\partial t} + \frac{\partial \rho}{\partial q}\dot{q} + \frac{\partial \rho}{\partial p}\dot{p}. \tag{7.78}$$

In view of Eq. (7.77), Eq. (7.78) simply becomes

$$\frac{d\rho}{dt} = 0, \tag{7.79}$$

implying, ρ = constant. Hence, the proof is complete.

If we extend these arguments to the case with n degrees of freedom, whose phase space is of $2n$ dimensions, the volume of a typical element is given by

$$dV = dq_1 dq_2 \ldots dq_n dp_1 dp_2 \ldots dp_n, \tag{7.80}$$

while the increase in the representative points in unit time, in the volume element dV is found to be

$$\frac{\partial \rho}{\partial t} dV = -\left[\frac{\partial}{\partial q_1}(\rho \dot{q}_1) + \frac{\partial}{\partial q_2}(\rho \dot{q}_2) + \cdots + \frac{\partial}{\partial q_n}(\rho \dot{q}_n)\right.$$

$$\left. + \frac{\partial}{\partial p_1}(\rho \dot{p}_1) + \frac{\partial}{\partial p_2}(\rho \dot{p}_2) + \cdots + \frac{\partial}{\partial p_n}(\rho \dot{p}_n)\right] dV$$

or

$$\frac{\partial \rho}{\partial t} + \sum_{i=1}^{n}\left[\frac{\partial (\rho \dot{q}_i)}{\partial q_i} + \frac{\partial (\rho \dot{p}_i)}{\partial p_i}\right] = 0.$$

Its expansion gives

$$\frac{\partial \rho}{\partial t} + \sum_{i=1}^{n}\left(\frac{\partial \rho}{\partial q_i}\dot{q}_i + \rho\frac{\partial \dot{q}_i}{\partial q_i} + \frac{\partial \rho}{\partial p_i}\dot{p}_i + \rho\frac{\partial \dot{p}_i}{\partial p_i}\right) = 0. \tag{7.81}$$

From Hamilton's canonical equations, we can easily see that

$$\sum_i \frac{\partial \dot{q}_i}{\partial q_i} + \sum_i \frac{\partial \dot{p}_i}{\partial p_i} = 0.$$

Using this result, Eq. (7.81) simplifies to

$$\frac{\partial \rho}{\partial t} + \sum_{i=1}^{n} \left(\frac{\partial \rho}{\partial q_i} \dot{q}_i + \frac{\partial \rho}{\partial p_i} \dot{p}_i \right) = 0,$$

that is,

$$\frac{d\rho}{dt} = 0 \quad \text{or} \quad \rho = \text{constant}. \tag{7.82}$$

Finally, it may be noted that our consideration of phase space has been more like Lagrangian picture, where the individual particles are followed in time.

7.5 SPECIAL TRANSFORMATIONS

Point transformation

If there exists a transformation from one set of generalized coordinates $q_1, q_2, \ldots, q_n$ to another set of generalized coordinates $Q_1, Q_2, \ldots, Q_n$, such a transformation is called *point* transformation. For instance, a point transformation is described by a set of equations

$$Q_i = Q_i(q_1, q_2, \ldots, q_n). \tag{7.83}$$

The inverse transformation is given by

$$q_i = q_i(Q_1, Q_2, \ldots, Q_n). \tag{7.84}$$

If we substitute the relations (7.84) into the Lagrangian function $L(\vec{q}, \dot{\vec{q}}, t)$, we obtain a Lagrange function $L(\vec{Q}, \dot{\vec{Q}}, t)$. It means that the Lagrangian is invariant under point transformation. Similarly, substituting Eq. (7.84) into Lagrange's equations

$$\frac{d}{dt}\left[\frac{\partial L(\vec{q}, \dot{\vec{q}}, t)}{\partial \dot{q}_\rho}\right] - \frac{\partial L(\vec{q}, \dot{\vec{q}}, t)}{\partial q_\rho} = 0, \tag{7.85}$$

we get

$$\frac{d}{dt}\left[\frac{\partial L(\vec{Q}, \dot{\vec{Q}}, t)}{\partial \dot{Q}_\rho}\right] - \frac{\partial L(\vec{Q}, \dot{\vec{Q}}, t)}{\partial Q_\rho} = 0, \tag{7.86}$$

which also indicates that the Lagrange's equations are invariant under point transformation.

This property of invariance is not preserved in the Hamiltonian formulation, which can be seen through the following theorem:

Theorem 7.3 Suppose the configuration of a dynamical system is defined by the generalized coordinates $q_1, q_2, \ldots, q_n$, whose behaviour under the action of a given force is described by the Hamiltonian $H(\vec{q}, \vec{p}, t)$ with the generalized momenta $\vec{p}$. If a new set of generalized coordinates $Q_1, Q_2, \ldots, Q_n$ are introduced which are related to $\vec{q}$ by the point transformation

$$Q_\rho = Q_\rho(\vec{q}, t) \qquad (7.87)$$

then the corresponding momenta $\vec{P}$ and the Hamiltonian K are given by

$$P_j = \sum_{\rho=1}^{n} p_\rho \frac{\partial q_\rho(\vec{Q}, t)}{\partial Q_j} \qquad (7.88)$$

and

$$K = H - \sum_{\rho=1}^{n} p_\rho \frac{\partial q_\rho(\vec{Q}, t)}{\partial t}. \qquad (7.89)$$

Proof For a dynamical system with n degrees of freedom, the Hamiltonian is defined as

$$H = \sum_{\rho=1}^{n} \dot{q}_\rho p_\rho - L \quad \text{or} \quad L = \sum \dot{q}_\rho p_\rho - H.$$

We have already observed that the Lagrangian is invariant under point transformation. Therefore, we can write

$$\sum p_\rho \dot{q}_\rho - H = \sum P_\rho \dot{Q}_\rho - K$$

or

$$\sum_{\rho=1}^{n} p_\rho \, dq_\rho - H dt = \sum_{\rho=1}^{n} P_\rho \, dQ_\rho - K dt \qquad (7.90)$$

But, from calculus, we have

$$dq_\rho = \sum_{j=1}^{n} \frac{\partial q_\rho(\vec{Q}, t)}{\partial Q_j} dQ_j + \frac{\partial q_\rho(\vec{Q}, t)}{\partial t} dt. \qquad (7.91)$$

Substituting Eq. (7.91) into Eq. (7.90), we obtain

$$\sum_{\rho=1}^{n} \sum_{j=1}^{n} p_\rho \frac{\partial q_\rho(\vec{Q}, t)}{\partial Q_j} dQ_j + \sum_{\rho=1}^{n} p_\rho \frac{\partial q_\rho(\vec{Q}, t)}{\partial t} dt - H dt = \sum_{j=1}^{n} P_j \, dQ_j - K dt.$$

$$(7.92)$$

Equating the coefficient of dQ_j and dt on both sides of Eq. (7.92), we find

$$P_j = \sum_{\rho=1}^{n} p_\rho \frac{\partial q_\rho(\vec{Q}, t)}{\partial Q_j}$$

and

$$K = H - \sum_{\rho=1}^{n} p_\rho \frac{\partial q_\rho(\vec{Q}, t)}{\partial t}.$$

Hence the proof is complete.

It may be noted that if the point transformation from $\vec{q}$ to $\vec{Q}$ is independent of time, then $K = H$, while if the transformation depends on time, then, $K \neq H$.

Example 7.3 The configuration of a known dynamical system is described by the generalized coordinate q and its behaviour by the Hamiltonian function $H(q, p, t) = ap^2 + bp(q+t)^2$, where a and b are constants. A point transformation is made to a new generalized coordinate $Q = q + t$. Find the corresponding Hamiltonian function $K(Q, P, t)$.

Solution Following Theorem 7.3 and Eqs. (7.88) and (7.89), we have from $q = Q - t$,

$$P = \frac{p \, \partial q(Q, t)}{\partial Q} = (p)(1) = p \tag{1}$$

and

$$K = H - \frac{p \, \partial q(Q, t)}{\partial t} = H - p(-1) = H + p$$

or

$$K = ap^2 + bp(q+t)^2 + p. \tag{2}$$

Now, using Eq. (1) into Eq. (2), we find

$$K = aP^2 + bPQ^2 + P,$$

which is the required result.

Canonical transformation

For solving many complex problems in mechanics, it is often found to be advantageous to transform from one set of generalized coordinates $q_1, q_2, ..., q_n$ and generalized momenta $p_1, p_2, ..., p_n$ to another set of variables $Q_1, Q_2, ..., Q_n$ and $P_1, P_2, ..., P_n$.

A transformation of the type

$$\vec{Q} = \vec{Q}(\vec{q}, \vec{p}), \qquad \vec{P} = \vec{P}(\vec{q}, \vec{p}) \tag{7.93}$$

is called a *canonical transformation*, if and only if there exists a function of the type $F(\vec{q}, \vec{p})$ satisfying the property

$$dF(\vec{q}, \vec{p}) = \sum_{\rho=1}^{n} p_\rho dq_\rho - \sum_{\rho=1}^{n} P_\rho dQ_\rho(\vec{q}, \vec{p}). \tag{7.94}$$

However, a time-dependent transformation

$$\vec{Q} = \vec{Q}(\vec{q}, \vec{p}, t), \qquad \vec{P} = \vec{P}(\vec{q}, \vec{p}, t) \tag{7.95}$$

is said to be a canonical transformation, if and only if any one of the following three equivalent properties are found to be true:

(i) The transformation is canonical in the sense of Eq. (7.94) for every value of time.

(ii) There exists a function $F(\vec{q}, \vec{p}, t)$ such that for every arbitrary fixed time, say $t = t_0$ satisfying

$$dF(\vec{q}, \vec{p}, t_0) = \sum_{\rho=1}^{n} p_\rho dq_\rho - \sum_{\rho=1}^{n} P_\rho(\vec{q}, \vec{p}, t_0) dQ_\rho(\vec{q}, \vec{p}, t_0), \quad (7.96)$$

where

$$dF(\vec{q}, \vec{p}, t_0) = \sum_{\rho=1}^{n} \frac{\partial F(\vec{q}, \vec{p}, t_0)}{\partial q_\rho} dq_\rho + \sum_{\rho=1}^{n} \frac{\partial F(\vec{q}, \vec{p}, t_0)}{\partial p_\rho} dp_\rho \quad (7.97)$$

and

$$dQ_\rho(\vec{q}, \vec{p}, t_0) = \sum_{j} \frac{\partial Q_\rho(\vec{q}, \vec{p}, t_0)}{\partial q_j} dq_j + \sum_{j} \frac{\partial Q_\rho(\vec{q}, \vec{p}, t_0)}{\partial p_j} dp_j. \quad (7.98)$$

(iii) There exists a function $F(\vec{q}, \vec{p}, t)$ such that

$$dF(\vec{q}, \vec{p}, t) = \sum_{\rho} p_\rho dq_\rho - \sum_{\rho} P_\rho(\vec{q}, \vec{p}, t) dQ_\rho(\vec{q}, \vec{p}, t) + R(\vec{q}, \vec{p}, t), \quad (7.99)$$

where

$$R \equiv \sum P_\rho(\vec{q}, \vec{p}, t) \frac{\partial Q_\rho(\vec{q}, \vec{p}, t)}{\partial t} + \frac{\partial F(\vec{q}, \vec{p}, t)}{\partial t}. \quad (7.100)$$

The equivalence of (i) and (ii) is obvious. The equivalence of (ii) and (iii) follows from Taylor's expansion

$$dF(\vec{q}, \vec{p}, t) = dF(\vec{q}, \vec{p}, t_0) + \left.\frac{\partial F(\vec{q}, \vec{p}, t)}{\partial t} dt\right|_{t=t_0} \quad (7.101)$$

and

$$dQ_\rho(\vec{q}, \vec{p}, t) = dQ_\rho(\vec{q}, \vec{p}, t_0) + \left.\frac{\partial Q_\rho(\vec{q}, \vec{p}, t)}{\partial t} dt\right|_{t=t_0}. \quad (7.102)$$

7.5.1 Lagrange Brackets

Let u, v be any two members of the set of variables $q_1, q_2, \ldots, q_n$ and $p_1, p_2, \ldots, p_n$; then the Lagrange bracket associated with the transformation

$$\vec{Q} = \vec{Q}(\vec{q}, \vec{p}, t), \quad \vec{P} = \vec{P}(\vec{q}, \vec{p}, t) \quad (7.103)$$

denoted by $[u, v]$ is defined as

$$[u, v] = \sum_i \left(\frac{\partial Q_i}{\partial u} \frac{\partial P_i}{\partial v} - \frac{\partial Q_i}{\partial v} \frac{\partial P_i}{\partial u} \right). \tag{7.104}$$

Theorem 7.4 The transformation

$$\vec{Q} = \vec{Q}(\vec{q}, \vec{p}, t), \qquad \vec{P} = \vec{P}(\vec{q}, \vec{p}, t) \tag{7.105}$$

is a canonical transformation, if and only if the Lagrange brackets $[q_i, q_j]$, $[q_i, p_j]$, $[p_i, p_j]$ satisfy

$$[q_i, q_j] = 0, \qquad [q_i, p_j] = \delta_{ij}, \qquad [p_i, p_j] = 0, \tag{7.106}$$

where δ_{ij} is a Kronecker delta with the property

$$\delta_{ij} = \begin{cases} 1, & i = j \\ 0, & i \neq j \end{cases} \tag{7.107}$$

Proof For simplicity, let us assume that t is a constant, then the transformation (7.105) is canonical if and only if, there exists a function F, such that

$$dF = \sum_\rho P_\rho \, dq_\rho - \sum_\rho P_\rho \, dQ_\rho$$

$$= \sum_\rho P_\rho \, dq_\rho - \sum_\rho P_\rho \left(\sum_j \frac{\partial Q_\rho}{\partial q_j} dq_j + \sum_j \frac{\partial Q_\rho}{\partial p_j} dp_j \right)$$

or

$$dF = \sum_j \left(p_j - \sum_k P_k \frac{\partial Q_k}{\partial q_j} \right) dq_j + \sum_j \left(-\sum_k P_k \frac{\partial Q_k}{\partial p_j} \right) dp_j. \tag{7.108}$$

Now, we define

$$A_j = p_j - \sum_k P_k \frac{\partial Q_k}{\partial q_j} \tag{7.109}$$

and

$$B_j = -\sum_k P_k \frac{\partial Q_k}{\partial p_j}, \tag{7.110}$$

so that Eq. (7.108) can be rewritten as

$$dF = \sum_j A_j \, dq_j + \sum_j B_j \, dp_j. \tag{7.111}$$

But, the necessary and sufficient conditions that there exists a function F such that Eq. (7.111) is a perfect differential are given by

$$\frac{\partial A_j}{\partial q_i} = \frac{\partial A_i}{\partial q_j}, \qquad (7.112)$$

$$\frac{\partial A_j}{\partial p_i} = \frac{\partial B_i}{\partial q_j}, \qquad (7.113)$$

$$\frac{\partial B_j}{\partial p_i} = \frac{\partial B_i}{\partial p_j}. \qquad (7.114)$$

Substituting Eq. (7.109) into Eq. (7.112), we obtain

$$\frac{\partial}{\partial q_i}\left(p_j - \sum_k P_k \frac{\partial Q_k}{\partial q_j}\right) = \frac{\partial}{\partial q_j}\left(p_i - \sum_k P_k \frac{\partial Q_k}{\partial q_i}\right).$$

On carrying out the differentiation, we get

$$-\sum_k \frac{\partial P_k}{\partial q_i}\frac{\partial Q_k}{\partial q_j} - \sum_k P_k \frac{\partial^2 Q_k}{\partial q_i \partial q_j} = -\sum_k \frac{\partial P_k}{\partial q_j}\frac{\partial Q_k}{\partial q_i} - \sum_k P_k \frac{\partial^2 Q_k}{\partial q_j \partial q_i}.$$

Further simplification gives us

$$\sum_k \frac{\partial Q_k}{\partial q_i}\frac{\partial P_k}{\partial q_j} - \sum_k \frac{\partial Q_k}{\partial q_j}\frac{\partial P_k}{\partial q_i} = 0, \qquad (7.115)$$

which is just the Lagrange bracket $[q_i, q_j]$ and thus we obtain

$$[q_i, q_j] = 0. \qquad (7.116)$$

Similarly, substituting Eqs. (7.109) and (7.110) into Eqs. (7.113) and (7.114), respectively, we get

$$[q_i, p_j] = \delta_{ij} \qquad (7.117)$$

and

$$[p_i, p_j] = 0. \qquad (7.118)$$

Thus, the proof is complete.

It may be observed that

$$[u, v] = \sum_i \left(\frac{\partial Q_i}{\partial u}\frac{\partial P_i}{\partial v} - \frac{\partial Q_i}{\partial v}\frac{\partial P_i}{\partial u}\right)$$

$$= -\sum_i \left(\frac{\partial Q_i}{\partial v}\frac{\partial P_i}{\partial u} - \frac{\partial Q_i}{\partial u}\frac{\partial P_i}{\partial v}\right)$$

$$= -[v, u]. \qquad (7.119)$$

Thus, the commutative law does not hold true in the case of Lagrange brackets.

Properties of canonical transformation

P1: If the transformation $\vec{Q} = \vec{Q}(\vec{q}, \vec{p}, t)$, $\vec{P} = \vec{P}(\vec{q}, \vec{p}, t)$ is a canonical transformation, the Jacobian of the transformation is equal to ± 1. That is,

$$J\left(\frac{\vec{Q}, \vec{P}}{\vec{q}, \vec{p}}\right) = \pm 1. \tag{7.120}$$

Verification: We note that

$$J\left(\frac{\vec{Q}, \vec{P}}{\vec{q}, \vec{p}}\right) = \begin{vmatrix} \frac{\partial Q_1}{\partial q_1} & \cdots & \frac{\partial Q_n}{\partial q_1} & \frac{\partial P_1}{\partial q_1} & \cdots & \frac{\partial P_n}{\partial q_1} \\ \vdots & & \vdots & \vdots & & \vdots \\ \frac{\partial Q_1}{\partial q_n} & \cdots & \frac{\partial Q_n}{\partial q_n} & \frac{\partial P_1}{\partial q_n} & \cdots & \frac{\partial P_n}{\partial q_n} \\ \frac{\partial Q_1}{\partial p_1} & \cdots & \frac{\partial Q_n}{\partial p_1} & \frac{\partial P_1}{\partial p_1} & \cdots & \frac{\partial P_n}{\partial p_1} \\ \vdots & & \vdots & \vdots & & \vdots \\ \frac{\partial Q_1}{\partial p_n} & \cdots & \frac{\partial Q_n}{\partial p_n} & \frac{\partial P_1}{\partial p_n} & \cdots & \frac{\partial P_n}{\partial p_n} \end{vmatrix}. \tag{7.121}$$

On performing the following sequence of operations on the determinant (7.121), that is,

(i) multiplying the bottom set of n rows by (-1),
(ii) multiplying the right-hand set of n columns by (-1),
(iii) interchanging the top set of n rows with the bottom set of n rows,

we obtain

$$J\left(\frac{\vec{Q}, \vec{P}}{\vec{q}, \vec{p}}\right) = \begin{vmatrix} -\frac{\partial Q_1}{\partial p_1} & \cdots & -\frac{\partial Q_n}{\partial p_1} & \frac{\partial P_1}{\partial p_1} & \cdots & \frac{\partial P_n}{\partial p_1} \\ \vdots & & \vdots & \vdots & & \vdots \\ -\frac{\partial Q_1}{\partial p_n} & \cdots & -\frac{\partial Q_n}{\partial p_n} & \frac{\partial P_1}{\partial p_n} & \cdots & \frac{\partial P_n}{\partial p_n} \\ \frac{\partial Q_1}{\partial q_1} & \cdots & \frac{\partial Q_n}{\partial q_1} & -\frac{\partial P_1}{\partial q_1} & \cdots & -\frac{\partial P_n}{\partial q_1} \\ \vdots & & \vdots & \vdots & & \vdots \\ \frac{\partial Q_1}{\partial q_n} & \cdots & \frac{\partial Q_n}{\partial q_n} & -\frac{\partial P_1}{\partial q_n} & \cdots & -\frac{\partial P_n}{\partial q_n} \end{vmatrix}.$$

278 | Hamiltonian Methods (Ch. 7)

Performing further the following series of operations in order, that is,
(i) interchanging the left hand n columns with the right-hand n columns,
(ii) taking the transpose of the resultant, the value of the determinant remains unchanged and the result is

$$J\left(\frac{\vec{Q}, \vec{P}}{\vec{q}, \vec{p}}\right) = \begin{vmatrix} \frac{\partial P_1}{\partial p_1} & \cdots & \frac{\partial P_1}{\partial p_n} & -\frac{\partial P_1}{\partial q_1} & \cdots & -\frac{\partial P_1}{\partial q_n} \\ \vdots & & \vdots & \vdots & & \vdots \\ \frac{\partial P_n}{\partial p_1} & \cdots & \frac{\partial P_n}{\partial p_n} & -\frac{\partial P_n}{\partial q_1} & \cdots & -\frac{\partial P_n}{\partial q_n} \\ -\frac{\partial Q_1}{\partial p_1} & \cdots & -\frac{\partial Q_1}{\partial p_n} & \frac{\partial Q_1}{\partial q_1} & \cdots & \frac{\partial Q_1}{\partial q_n} \\ \vdots & & \vdots & \vdots & & \vdots \\ -\frac{\partial Q_n}{\partial p_1} & \cdots & -\frac{\partial Q_n}{\partial p_n} & \frac{\partial Q_n}{\partial q_1} & \cdots & \frac{\partial Q_n}{\partial q_n} \end{vmatrix}. \quad (7.122)$$

Multiplying Eq. (7.121) by Eq. (7.122), we get in terms of Lagrange brackets,

$$\left[J\left(\frac{\vec{Q}, \vec{P}}{\vec{q}, \vec{p}}\right)\right]^2 = \begin{vmatrix} [q_1, p_1] & \cdots & [q_1, p_n] & [q_1, q_1] & \cdots & [q_n, q_1] \\ \vdots & & \vdots & \vdots & & \vdots \\ [q_n, p_1] & \cdots & [q_n, p_n] & [q_1, q_n] & \cdots & [q_n, q_n] \\ [p_1, p_1] & \cdots & [p_1, p_n] & [q_1, p_1] & \cdots & [q_n, p_1] \\ \vdots & & \vdots & \vdots & & \vdots \\ [p_n, p_1] & \cdots & [p_n, p_n] & [q_1, p_n] & \cdots & [q_n, p_n] \end{vmatrix}.$$

$$= \begin{vmatrix} 1 & 0 & \cdots & 0 & 0 \\ 0 & 1 & \cdots & 0 & 0 \\ \vdots & \vdots & & \vdots & \vdots \\ 0 & 0 & \cdots & 1 & 0 \\ 0 & 0 & \cdots & 0 & 1 \end{vmatrix} = 1.$$

Therefore,
$$J\left(\frac{\vec{Q}, \vec{P}}{\vec{q}, \vec{p}}\right) = \pm 1.$$

P2: If the transformation $\vec{Q} = \vec{Q}(\vec{q}, \vec{p}, t)$, $\vec{P} = \vec{P}(\vec{q}, \vec{p}, t)$ is a canonical transformation, then, the inverse transformation $\vec{q} = \vec{q}(\vec{Q}, \vec{P}, t)$, $\vec{p} = \vec{p}(\vec{Q}, \vec{P}, t)$ is also a canonical transformation.

Verification: The transformation $\vec{Q} = \vec{Q}(\vec{q}, \vec{p}, t)$, $\vec{P} = \vec{P}(\vec{q}, \vec{p}, t)$ possesses inverse if the Jacobian satisfies the condition

$$J\left(\frac{\vec{Q}, \vec{P}}{\vec{q}, \vec{p}}\right) \neq 0.$$

In fact, we have verified the truth of this condition under Pl. Therefore, the function $F(\vec{q}, \vec{p}, t)$ exists satisfying

$$dF(\vec{q}, \vec{p}, t_0) = \sum_\rho p_\rho dq_\rho - \sum_\rho P_\rho(\vec{q}, \vec{p}, t_0) \, dQ_\rho(\vec{q}, \vec{p}, t_0). \qquad (7.123)$$

If we substitute the inverse transformation into Eq. (7.123), we find that the function $-F(\vec{Q}, \vec{P}, t_0)$ satisfies the relation

$$d[-F(\vec{Q}, \vec{P}, t_0)] = \sum_\rho P_\rho dQ_\rho - \sum_\rho p_\rho(\vec{Q}, \vec{P}, t_0) \, dq_\rho(\vec{Q}, \vec{P}, t_0), \qquad (7.124)$$

which shows that the inverse transformation is also a canonical transformation.

P3: If the transformations $\vec{q}'' = \vec{q}''(\vec{q}', \vec{p}', t), \vec{p}'' = \vec{p}''(\vec{q}', \vec{p}', t)$ and $\vec{q}' = \vec{q}'(\vec{q}, \vec{p}, t)$, $\vec{p}' = \vec{p}'(\vec{q}, \vec{p}, t)$ are canonical transformations, then the product of these two transformations, that is, the transformation

$$\vec{q}'' = \vec{q}''(\vec{q}, \vec{p}, t), \qquad \vec{p}'' = \vec{p}''(\vec{q}, \vec{p}, t)$$

is also a canonical transformation.

P4: The set of canonical transformations obeys the associative law of multiplication.

P5: From the properties P2 to P4, we can conclude that the canonical transformations form a group.

Example 7.4 Show that the transformation

$$Q = \log\left(\frac{1}{q} \sin p\right), \qquad P = q \cot p$$

is a canonical transformation and hence find the function F.

Solution To show that the given transformation is canonical, the necessary and sufficient conditions are, that the Lagrange brackets satisfy the following relations:

$$[q, q] = [p, p] = 0 \quad \text{and} \quad [q, p] = 1. \tag{1}$$

We note from the given data

$$\frac{\partial Q}{\partial q} = \frac{q}{\sin p}\left(-\frac{1}{q^2}\sin p\right) = -\frac{1}{q}, \qquad \frac{\partial Q}{\partial p} = \frac{q}{\sin p}\left(\frac{1}{q}\cos p\right) = \cot p$$

and

$$\frac{\partial P}{\partial q} = \cot p, \qquad \frac{\partial P}{\partial p} = -q\operatorname{cosec}^2 p \tag{2}$$

From (1) and (2), we find

$$[q, q] = \frac{\partial Q}{\partial q}\frac{\partial P}{\partial q} - \frac{\partial Q}{\partial q}\frac{\partial P}{\partial q} = 0$$

$$[p, p] = \frac{\partial Q}{\partial p}\frac{\partial P}{\partial p} - \frac{\partial Q}{\partial p}\frac{\partial P}{\partial p} = 0$$

$$[q, p] = \frac{\partial Q}{\partial q}\frac{\partial P}{\partial p} - \frac{\partial Q}{\partial p}\frac{\partial P}{\partial q} = \operatorname{cosec}^2 p - \cot^2 p = 1.$$

Therefore, the given transformation is a canonical transformation.

The function F is given by

$$dF = p\, dq - P\, dQ$$

$$= p\, dq - q\cot p\, d\left[\log\left(\frac{1}{q}\sin p\right)\right]$$

$$= p\, dq - q\cot p\, \frac{q}{\sin p}\left(\frac{1}{q}\cos p\, dp - \frac{1}{q^2}\sin p\, dq\right)$$

$$= p\, dq - q\cot^2 p\, dp + \cot p\, dq$$

$$= (p + \cot p)\, dq - q\cot^2 p\, dp$$

$$= (p + \cot p)\, dq + q\, dp - q\operatorname{cosec}^2 p\, dp$$

or

$$dF = p\, dq + q\, dp + \cot p\, dq - q\operatorname{cosec}^2 p\, dp$$

$$= d(pq) + d(q\cot p).$$

Hence,

$$F = pq + q\cot p.$$

Example 7.5 Show that the transformation $Q = p^2 + q$, $P = p + t$ is canonical and hence find the function $F(q, p, t)$ such that $dF(q, p, t_0) = pdq - P(q, p, t_0)\, dQ(q, p, t_0)$

Solution To show that the given transformation is canonical, the necessary and sufficient condition is that the Lagrange brackets satisfy:

$$[q, q] = [p, p] = 0 \quad \text{and} \quad [q, p] = 1.$$

From the given data, we have

$$[q, q] = \frac{\partial Q}{\partial q}\frac{\partial P}{\partial q} - \frac{\partial Q}{\partial q}\frac{\partial P}{\partial q} = 0 \quad \text{and} \quad [p, p] = 0,$$

where

$$\frac{\partial Q}{\partial q} = 1, \qquad \frac{\partial Q}{\partial p} = 2p, \qquad \frac{\partial P}{\partial q} = 0, \qquad \frac{\partial P}{\partial p} = 1.$$

Therefore,

$$[q, p] = \frac{\partial Q}{\partial q}\frac{\partial P}{\partial p} - \frac{\partial Q}{\partial p}\frac{\partial P}{\partial q} = (1)(1) - (2p)(0) = 1.$$

Hence, the given transformation is canonical and there exists a function $F(q, p, t)$ such that

$$dF(q, p, t_0) = pdq - P(q, p, t_0)\, dQ(q, p, t_0)$$

Using the given data, we have

$$dF(q, p, t_0) = pdq - (p + t_0)(2pdp + dq)$$

or

$$dF(q, p, t_0) = -t_0\, dq - 2p(p + t_0)dp.$$

Integrating the above result at a time $t = t_0$ along any path from some reference point (q_0, p_0) to (q, p), we get

$$F(q, p, t_0) - F(q_0, p_0, t_0) = -t_0 q - \frac{2}{3}p^3 - t_0 p^2$$

or

$$F(q, p, t_0) = -qt_0 - \frac{2}{3}p^3 - p^2 t_0 + F(q_0, p_0, t_0),$$

which is the required result.

7.5.2 Poisson Brackets

Suppose we are given two functions f and g of the dynamical variables $\vec{q}$, $\vec{p}$ and time t, then the Poisson bracket expression for the two functions is defined as

$$(f, g) = \sum_{i=1}^{n}\left(\frac{\partial f}{\partial q_i}\frac{\partial g}{\partial p_i} - \frac{\partial f}{\partial p_i}\frac{\partial g}{\partial q_i}\right). \tag{7.125}$$

An important property of Poisson bracket is Jacobi's identity, namely

$$(f,(g,h)) + (g,(h,f)) + (h,(f,g)) = 0,$$

where f, g and h are functions of $\vec{q}$ and $\vec{p}$.

To verify this identity, let us recall the Hamilton canonical equations for a dynamical system, that is,

$$\dot{q}_i = \frac{\partial H}{\partial p_i}, \qquad \dot{p}_i = -\frac{\partial H}{\partial q_i}. \tag{7.126}$$

Using the Poisson bracket notation and (7.126), we write

$$(q_i, H) = \sum_{k=1}^{n} \left(\frac{\partial q_i}{\partial q_k} \frac{\partial H}{\partial p_k} - \frac{\partial q_i}{\partial p_k} \frac{\partial H}{\partial q_k} \right) = \frac{\partial H}{\partial p_i} \tag{7.127}$$

and

$$(p_i, H) = \sum_{k=1}^{n} \left(\frac{\partial p_i}{\partial q_k} \frac{\partial H}{\partial p_k} - \frac{\partial p_i}{\partial p_k} \frac{\partial H}{\partial q_k} \right) = -\frac{\partial H}{\partial q_i}. \tag{7.128}$$

Hence, the canonical equations can be written as

$$\dot{q}_i = (q_i, H), \qquad \dot{p}_i = (p_i, H). \tag{7.129}$$

Since f and g are functions of $\vec{q}$, $\vec{p}$ and t, we have from calculus

$$\frac{df}{dt} = \sum_i \left(\frac{\partial f}{\partial q_i} \dot{q}_i + \frac{\partial f}{\partial p_i} \dot{p}_i \right) + \frac{\partial f}{\partial t}.$$

Using Eq. (7.126), the above expression can be recast as

$$\frac{df}{dt} = \sum_i \left(\frac{\partial f}{\partial q_i} \frac{\partial H}{\partial p_i} - \frac{\partial f}{\partial p_i} \frac{\partial H}{\partial q_i} \right) + \frac{\partial f}{\partial t}$$

or

$$\frac{df}{dt} = (f, H) + \frac{\partial f}{\partial t}. \tag{7.130}$$

Similarly, we can show that

$$\frac{dg}{dt} = (g, H) + \frac{\partial g}{\partial t}. \tag{7.131}$$

If f and g are not explicit functions of time, then, we at once write

$$(f, H) + \frac{\partial f}{\partial t} = 0. \tag{7.132}$$

$$(g, H) + \frac{\partial g}{\partial t} = 0. \tag{7.133}$$

Utilizing the result given by Eq. (7.130), we can also write

$$\frac{d}{dt}(f, g) = ((f, g), H) + \frac{\partial}{\partial t}(f, g) = 0$$

or

$$\frac{d}{dt}(f, g) = ((f, g), H) + \left(\frac{\partial f}{\partial t}, g\right) + \left(f, \frac{\partial g}{\partial t}\right) = 0. \quad (7.134)$$

Using Eqs. (7.132) and (7.133), the above expression can be rewritten as

$$((f, g), H) + (-(f, H), g) + (f, -(g, H)) = 0.$$

Writing in reverse order, we have

$$-(f, (g, H)) + (g, (f, H)) - (H, (f, g)) = 0.$$

That is,

$$-(f, (g, H)) - (g, (H, f)) - (H, (f, g)) = 0$$

or

$$(f, (g, H)) + (g, (H, f)) + (H, (f, g)) = 0. \quad (7.135)$$

Noting that H is a function of q's and p's, the Jacobi's identity follows.

It can be shown that the Lagrange and Poisson brackets are reciprocal quantities. The interested student can take it as an exercise.

Example 7.6 If q_1 and q_2 are generalized coordinates and p_1, p_2 are the corresponding generalized momenta, then find the Poisson bracket (X, Y) when $X = q_1^2 + q_2^2$ and $Y = 2p_1 + p_2$.

Solution Here $X = q_1^2 + q_2^2$ and $Y = 2p_1 + p_2$. Then

$$(X, Y) = \left(\frac{\partial X}{\partial q_1}\frac{\partial Y}{\partial p_1} - \frac{\partial X}{\partial p_1}\frac{\partial Y}{\partial q_1}\right) + \left(\frac{\partial X}{\partial q_2}\frac{\partial Y}{\partial p_2} - \frac{\partial X}{\partial p_2}\frac{\partial Y}{\partial q_2}\right)$$

$$= (2q_1 \times 2 - 0) + (2q_2 \times 1 - 0)$$

$$= 4q_1 + 2q_2.$$

7.6 CALCULUS OF VARIATIONS

In this section, we shall discuss the fundamentals of calculus of variations, where the basic problem is to find the curve for which a given line integral has extremum or stationary value. We shall also demonstrate that the Hamilton's principle is a special case of the general formulation.

In physics, we often encounter problems, where one has to find the maxima or minima of special quantities called functionals. What is a functional?

Definition 7.1 Let $y_1, y_2, ..., y_n$ be functions of a single variable x defined in an interval (a, b). Let S be the set of these functions. Then, *functionals* are defined as variable quantities, whose value depends on the choice of one or more functions in S.

For example, in mechanics of deformable bodies the potential energy is a functional in the sense, that it is a function of displacement components u, v, w which themselves are functions of coordinates x, y and z. Alternatively, a functional is defined as a function of several functions.

Fundamental lemma of variational calculus

If $\phi(x)$ is a continuous function in $[x_0, x_1]$ and if

$$\int_{x_0}^{x_1} \phi(x)\eta(x)\,dx = 0 \tag{7.136}$$

for all continuous functions $\eta(x)$, which is also continuously differentiable, then $\phi(x) = 0$ on $[x_0, x_1]$.

Now, consider a function $F(x, y, y')$ defined on the curve given by

$$y = y(x) \tag{7.137}$$

between the points $A(x_0, y_0)$, $B(x_1, y_1)$, where $y' = dy/dx$. The fundamental problem is to find a particular curve $y(x)$ for which the functional

$$v[y(x)] = \int_{x_0}^{x_1} F(x, y, y')\,dx \tag{7.138}$$

has an extremum value subject to the conditions

$$y(x_0) = y_0, \qquad y(x_1) = y_1. \tag{7.139}$$

Let $y = y(x)$ be a curve for which $v[y(x)]$ has an extremum, subject to the conditions (7.139). Let $F(x, y, y')$ have continuous second derivatives with respect to all its arguments and let $y''(x)$ be continuous in $[x_0, x_1]$. We choose any continuously differentiable function $\eta(x)$ which vanishes at the end points, that is, $\eta(x_0) = \eta(x_1) = 0$ and construct a neighbouring curve with the same end points and hence a function (see Fig. 7.8):

$$y = y(x) + \varepsilon\eta(x), \tag{7.140}$$

where ε is a parameter independent of x and y.

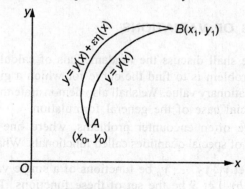

Figure 7.8 δ-variation.

Using Eq. (7.140) into Eq. (7.138), we get the integral on the varied path as

$$v = v(\varepsilon) = \int_{x_0}^{x_1} F(x, y + \varepsilon\eta, y' + \varepsilon\eta')\, dx. \tag{7.141}$$

This is treated as a function of ε, once y and η are assigned. It takes on its stationary value when $\varepsilon = 0$ and the necessary condition is

$$\left.\frac{dv}{d\varepsilon}\right|_{\varepsilon=0} = 0. \tag{7.142}$$

But,

$$\frac{dv}{d\varepsilon} = \int_{x_0}^{x_1} \left[\frac{\partial F}{\partial(y+\varepsilon\eta)} \frac{\partial(y+\varepsilon\eta)}{\partial \varepsilon} + \frac{\partial F}{\partial(y'+\varepsilon\eta)} \frac{\partial(y'+\varepsilon\eta')}{\partial \varepsilon} \right] dx$$

$$= \int_{x_0}^{x_1} \left[\frac{\partial F}{\partial(y+\varepsilon\eta)} \eta + \frac{\partial F}{\partial(y'+\varepsilon\eta')} \eta' \right] dx.$$

The condition (7.142) implies

$$\int_{x_0}^{x_1} \left(\frac{\partial F}{\partial y} \eta + \frac{\partial F}{\partial y'} \eta' \right) dx = 0. \tag{7.143}$$

Noting that,

$$\int_{x_0}^{x_1} \frac{\partial F}{\partial y'} \eta'\, dx = \int_{x_0}^{x_1} \frac{\partial F}{\partial y'} \frac{d\eta}{dx}\, dx = \left[\frac{\partial F}{\partial y'} \eta \right]_{x_0}^{x_1} - \int_{x_0}^{x_1} \eta \frac{d}{dx}\left(\frac{\partial F}{\partial y'} \right) dx.$$

And since $\eta(x_0) = \eta(x_1) = 0$, the above equation becomes

$$\int_{x_0}^{x_1} \frac{\partial F}{\partial y'} \eta'\, dx = -\int_{x_0}^{x_1} \frac{d}{dx}\left(\frac{\partial F}{\partial y'} \right) \eta(x)\, dx.$$

Therefore, Eq. (7.143) can be rewritten as

$$\int_{x_0}^{x_1} \left[\frac{\partial F}{\partial y} - \frac{d}{dx}\left(\frac{\partial F}{\partial y'} \right) \right] \eta(x)\, dx = 0. \tag{7.144}$$

Since $\eta(x)$ is an arbitrary and continuously differentiable function, using the fundamental lemma of variational calculus, we at once get

$$\frac{\partial F}{\partial y} - \frac{d}{dx}\left(\frac{\partial F}{\partial y'} \right) = 0. \tag{7.145}$$

This equation is known as Euler–Lagrange equation. Since $\partial F/\partial y'$ is, in general, a function of x explicitly and also implicitly through y and y', the second term in Eq. (7.145) can be written in the expanded form as

$$\frac{d}{dx}\left(\frac{\partial F}{\partial y'}\right) = \frac{\partial}{\partial x}\left(\frac{\partial F}{\partial y'}\right) + \frac{\partial}{\partial y}\left(\frac{\partial F}{\partial y'}\right)\frac{dy}{dx} + \frac{\partial}{\partial y'}\left(\frac{\partial F}{\partial y'}\right)\frac{dy'}{dx}.$$

Using subscript notation for partial derivatives, Eq. (7.145) is equivalent to

$$F_{y'y'}\frac{d^2 y}{dx^2} + F_{y'y}\frac{dy}{dx} + (F_{y'x} - F_y) = 0. \tag{7.146}$$

This equation is of second order in y unless $F_{y'y'} = 0$. Thus, its general solution contains two arbitrary constants.

This result can be generalized to the case of a function, F, involving several independent variables y_i, y'_i, which are of course functions of x. Thus, the necessary condition for the extremum of the functional of the form

$$v[y_1, y_2, \ldots, y_n] = \int_{x_0}^{x_1} F(x, y_1, y_2, \ldots, y_n, y'_1, y'_2, \ldots, y'_n)\, dx \tag{7.147}$$

subject to the corresponding boundary conditions is

$$\frac{\partial F}{\partial y_i} - \frac{d}{dx}\left(\frac{\partial F}{\partial y'_i}\right) = 0, \quad i = 1, 2, \ldots, n. \tag{7.148}$$

Here, we shall vary only one of the functions, say $y_i(x)$, and treat v as a functional of a single function y_i and a similar argument is applicable to all functions y_i ($i = 1, 2, \ldots, n$). Thus, we get a system of second-order differential equations and they determine a set of $2n$ parameter family of integral curves in the space of $(x, y_1, y_2, \ldots, y_n)$. This constitutes a family of extremals of the given general variational problem.

Hamilton's principle and Lagrange's equations

If we identify the Lagrangian as

$$L = L(q_1, q_2, \ldots, q_n, \dot{q}_1, \dot{q}_2, \ldots, \dot{q}_n, t) \tag{7.149}$$

to the functional

$$F = F(y_1, y_2, \ldots, y_n, \dot{y}_1, \dot{y}_2, \ldots, \dot{y}_n, x) \tag{7.150}$$

with the transformation $x \to t$, $y_i \to q_i$, $\dot{y}_i \to \dot{q}_i$, then the functional $v \to S$ defines the action integral

$$S = \int_{t_0}^{t_1} L\, dt. \tag{7.151}$$

This function is known as *Hamilton's characteristic function*. Its variation

$$\delta S = \delta \int_{t_0}^{t_1} L\, dt = 0, \tag{7.152}$$

which corresponds to Eq. (7.29), that is, the motion of the system from time t_0 to t_1, so that the integral (7.151) has an extremum value for the correct path of the motion. This is what is already referred to as Hamilton's principle. The necessary condition for the Hamilton's principle, as given in Eq. (7.33), is then

$$\frac{\partial L}{\partial q_i} - \frac{d}{dt}\left(\frac{\partial L}{\partial \dot{q}_i}\right) = 0, \quad i = 1, 2, \ldots, n. \tag{7.153}$$

where q_i are the generalized coordinates.

Example 7.7 Prove that the arc length of the curve

$$l[y(x)] = \int_{x_0}^{x_1} \sqrt{1 + y'^2}\, dx.$$

Joining two points in a plane gives straight lines of the form $y = ax + b$, as its extremal, given $y(x_0) = y_0$, $y(x_1) = y_1$.

Solution In this problem, $F = \sqrt{(1 + y')^2}$ and therefore we observe that F depends on y' only. Thus, the Euler–Lagrange equation to this problem becomes

$$\frac{d}{dx}\left(\frac{\partial F}{\partial y'}\right) = 0.$$

That is,

$$\frac{d}{dx}\left[\frac{y'}{\sqrt{1 + y'^2}}\right] = 0.$$

On integration with respect to x, we get

$$\frac{y'}{\sqrt{1 + y'^2}} = \text{constant}$$

implying

$$y' = \text{constant} = a \text{ (say)}$$

Integrating once again, we get

$$y = ax + b. \tag{1}$$

This is the possible extremal, provided (x_0, y_0) and (x_1, y_1) lie on it. Thus, we have

$$y_0 = ax_0 + b, \quad y_1 = ax_1 + b.$$

On solving for a and b, we get

$$a = \frac{y_1 - y_0}{x_1 - x_0}, \quad b = \frac{y_1 x_0 - y_0 x_1}{x_0 - x_1}.$$

With these values of a and b, the extremals are the straight lines, $y = ax + b$.

Example 7.8 Find the extremal of the following functional

$$\int_{x_0}^{x_1} \frac{\sqrt{1+y'^2}}{x}\, dx.$$

Solution In this problem, F is independent of y and depends on x and y' only. Therefore, the Euler–Lagrange equation becomes

$$\frac{d}{dx}\left(\frac{\partial F}{\partial y'}\right) = 0.$$

On integration, we find

$$\frac{\partial F}{\partial y'} = \frac{\partial}{\partial y'}\left(\frac{\sqrt{1+y'^2}}{x}\right) = \text{constant}.$$

That is,

$$\frac{y'}{x\sqrt{1+y'^2}} = \text{constant} = c_1 \text{ (say)}$$

Let $y' = \tan\theta$, then

$$x = \frac{\sin\theta}{c_1} = A\sin\theta \tag{1}$$

and

$$dy = \tan\theta\, dx = \tan\theta \frac{dx}{d\theta}\, d\theta = A\sin\theta\, d\theta.$$

On integration, we get

$$y = -A\cos\theta + B, \tag{2}$$

where A and B are constants of integration. Eliminating θ between (1) and (2), we find

$$\frac{x^2}{A^2} + \frac{(y-B)^2}{A^2} = 1 \quad \text{or} \quad x^2 + (y-B)^2 = A^2. \tag{3}$$

Hence the required extremal is a circle with centre on y-axis.

Example 7.9 Find a curve with given boundary points, whose rotation about the x-axis generates a surface of minimum area.

Solution We know that the area of surface of revolution is given by (see Fig. 7.9)

$$S[y(x)] = 2\pi \int_{x_0}^{x_1} y\, ds = 2\pi \int_{x_0}^{x_1} y\sqrt{1+y'^2}\, dx \tag{1}$$

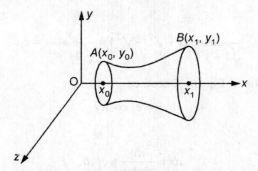

Figure 7.9 Description of Example 7.9.

with $y(x_0) = y_0$, $y(x_1) = y_1$. In this example

$$F = y\sqrt{1+y'^2}, \tag{2}$$

which is a function of y and y' only. Therefore,

$$\frac{dF}{dx} = \frac{\partial F}{\partial y}\frac{dy}{dx} + \frac{\partial F}{\partial y'}\frac{dy'}{dx} \tag{3}$$

or

$$\frac{dF}{dx} = \frac{\partial F}{\partial y}y' + \frac{\partial F}{\partial y'}y''. \tag{4}$$

Also,

$$\frac{d}{dx}\left(y'\frac{\partial F}{\partial y'}\right) = y'\frac{d}{dx}\left(\frac{\partial F}{\partial y'}\right) + \frac{\partial F}{\partial y'}y''. \tag{5}$$

Subtracting Eq. (5) from Eq. (4), we get

$$\frac{dF}{dx} - \frac{d}{dx}\left(y'\frac{\partial F}{\partial y'}\right) = y'\left[\frac{\partial F}{\partial y} - \frac{d}{dx}\left(\frac{\partial F}{\partial y'}\right)\right].$$

Using the Euler–Lagrange equation, the right-hand side of the above equation vanishes. Thus, we have

$$\frac{d}{dx}\left(F - y'\frac{\partial F}{\partial y'}\right) = 0.$$

That is,

$$F - y'\frac{\partial F}{\partial y'} = \text{constant} = c_1 \text{ (say)}$$

In other words,

$$y\sqrt{1+y'^2} - \frac{y\,y'^2}{\sqrt{1+y'^2}} = c_1$$

or

$$\frac{y}{\sqrt{1+y'^2}}(1+y'^2 - y'^2) = \frac{y}{\sqrt{1+y'^2}} = c_1. \tag{6}$$

Let $y' = \sinh t$, then, Eq. (6) becomes

$$y = c_1 \cosh t \tag{7}$$

Also

$$dx = \frac{dy}{\sinh t} = c_1 \, dt,$$

from which it follows that

$$\frac{dx}{dt} = c_1 \quad \text{implying} \quad x = c_1 t + c_2. \tag{8}$$

Eliminating t between Eqs. (7) and (8), we find

$$y = c_1 \cosh\left(\frac{x - c_2}{c_1}\right). \tag{9}$$

This is a catenary with vertex at (c_2, c_1). Equation (9) is a family of catenaries, the revolution of which forms surfaces called *catenoids*.

Example 7.10 Show that the path followed by a particle in sliding from one point A to another point B, in the absence of friction and in the shortest time, is a cycloid.

Solution Let the origin coincide with A. Since the particle moves only under gravity, the motion is confined to a vertical plane (see Fig. 7.10)

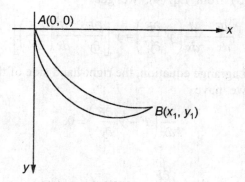

Figure 7.10 Description of Example 7.10.

In the vertical plane containing A and B, the element length is given as

$$ds = \sqrt{1+y'^2} \, dx.$$

The speed of the particle is easily seen to be

$$\left(\frac{ds}{dt}\right)^2 = 2gy \quad \text{or} \quad \frac{ds}{dt} = \sqrt{2gy}.$$

Therefore,

$$dt = \sqrt{\frac{1+y'^2}{2gy}}\, dx.$$

Thus, the time spent in moving from $A(0,0)$ to $B(x_1, y_1)$ is

$$t[y(x)] = \frac{1}{\sqrt{2g}} \int_0^{x_1} \frac{\sqrt{1+y'^2}}{\sqrt{y}}\, dx.$$

In this example,

$$F = \frac{\sqrt{1+y'^2}}{\sqrt{y}}$$

and it does not contain x explicitly but is a function of y and y'. Therefore, the Euler–Lagrange equation gives us (refer to Example 7.9)

$$F - y' F_{y'} = \text{constant} = c \text{ (say)}. \tag{1}$$

That is,

$$\frac{\sqrt{1+y'^2}}{\sqrt{y}} - \frac{y'^2}{\sqrt{y(1+y'^2)}} = c.$$

On simplification, we get

$$y(1+y'^2) = c_1. \tag{2}$$

Let $y' = \cot\theta$, then, Eq. (2) gives

$$y(1+\cot^2\theta) = y\,\text{cosec}^2\theta = c_1$$

or

$$y = c_1 \sin^2\theta = \frac{c_1}{2}(1-\cos 2\theta). \tag{3}$$

We also have

$$dx = \frac{dy}{y'} = \frac{2c_1 \sin\theta \cos\theta\, d\theta}{\cot\theta}$$

or

$$\frac{dx}{d\theta} = 2c_1 \sin^2\theta = c_1(1-\cos 2\theta).$$

Therefore,
$$x = c_1\left(\theta - \frac{\sin 2\theta}{2}\right) + c_2. \qquad (4)$$

Suppose, we assume $2\theta = \phi$, then, Eqs. (3) and (4) can be written as
$$\left.\begin{array}{l} x - c_2 = a(\phi - \sin\phi) \\ y = a(1 - \cos\phi) \end{array}\right\}, \qquad (5)$$

where $c_1/2 = a$. Since the curve passes through $A(0, 0)$, we find $0 = a(1 - \cos\phi)$ or $\cos\phi = 1$. That is,
$$\phi = 0.$$

Also $0 - c_2 = a(0 - 0)$. Therefore,
$$c_2 = 0.$$

Hence, the equation of the required curve is
$$\left.\begin{array}{l} x = a(\phi - \sin\phi) \\ y = a(1 - \cos\phi), \end{array}\right\} \qquad (6)$$

which is a cycloid.

It is John Bernoulli in the year 1696, who first talked about the problem of quickest descent called *Brachistochrone*.

Example 7.11 A bead of mass m slides freely on a frictionless circular wire of radius r that rotates in a horizontal plane about a point on the circular wire with a constant angular velocity ω. Derive the Lagrange's equation of motion for the bead. Also show that the bead oscillates as a simple pendulum with length $l = g/\omega^2$.

Solution Suppose that the circular wire is rotating in the xy-plane about a point O, with constant angular velocity ω, as shown in Fig. 7.11, and the rotation is in the anticlockwise direction.

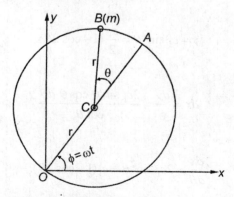

Figure 7.11 Description of Example 7.11.

Let C be the centre of the circular wire, and B the bead. As the wire rotates in the anticlockwise direction with a constant angular velocity ω, at a given instant let OA make an angle ϕ with the x-axis and let $\angle ACB = \theta$ so that $\phi = \omega t$. Now, the x, y coordinates of the bead at any time t are

$$\left. \begin{array}{l} x = r\cos \omega t + r \cos(\theta + \omega t) \\ y = r\sin \omega t + r \sin(\theta + \omega t). \end{array} \right\} \quad (1)$$

Thus, this problem involves only a single degree of freedom and we take the generalized coordinate as θ. The potential energy V of the bead can be taken to be zero. However, its kinetic energy T is given by

$$T = \frac{1}{2}m(\dot{x}^2 + \dot{y}^2)$$

$$= \frac{mr^2}{2}[\omega^2 + (\dot{\theta} + \omega)^2 + 2\omega(\dot{\theta} + \omega)\cos\theta]. \quad (2)$$

Recalling Eq. (7.153), the Lagrange's equation of motion for the bead is

$$\frac{\partial T}{\partial \theta} - \frac{d}{dt}\left(\frac{\partial T}{\partial \dot{\theta}}\right) = 0. \quad \text{(Since } L = T\text{).} \quad (3)$$

Now, using the result given by Eq. (2), Eq. (3) becomes

$$mr^2[(\ddot{\theta} - \omega\dot{\theta}\sin\theta) + \omega(\dot{\theta} + \omega)\sin\theta] = 0$$

or

$$\ddot{\theta} + \omega^2 \sin\theta = 0, \quad (4)$$

which shows tha the bead oscillates about the line OA, like a pendulum of length

$$l = \frac{g}{\omega^2}.$$

EXERCISES

7.1 Define canonical transformation. Show that the canonical transformations form a group.

7.2 Prove that the transformation defined by

$$P = \frac{1}{2}(q^2 + p^2), \qquad \tan Q = \frac{q}{p}$$

is canonical and hence find the function $F(q, p)$ such that $dF = pdq - PdQ$.

7.3 Given the following transformation,

$$Q = \log p, \qquad P = -(1 + q + \log p)p$$

show that it is canonical by applying the Lagrange bracket test. Also, Find the function $F(q, p)$ such that

$$dF(q, p) = p\,dq - P(q, p)\,dQ(q, p).$$

7.4 Show that the transformation

$$Q = q + te^p, \qquad P = p$$

is a canonical transformation and hence find a function $F(q, p, t)$ such that

$$dF(q, p, t_0) = p\,dq - P(q, p, t_0)\,dQ(q, p, t_0).$$

7.5 Prove that the transformation

$$Q = \log\left(\frac{\sin p}{q}\right), \qquad P = q \cot p$$

is a canonical transformation by using the Lagrange bracket test.

7.6 Show that the transformation

$$Q = \sqrt{e^{-2q} - p^2}, \qquad P = \cos^{-1}(pe^q)$$

is a canonical transformation.

7.7 State and prove the Hamilton's principle for a conservative system.

7.8 Explain the principle of least action for a simple dynamical system.

7.9 Define the Hamilton's characteristic function S and derive Hamilton–Jacobi equation for a dynamical system.

7.10 Given that f, g, h are three functions of generalized coordinates and momenta, prove the following Jacobi's identity.

$$((f, g), h) + ((g, h), f) + ((h, f), g) = 0,$$

where $(.\,,.)$ denotes Poisson bracket.

7.11 What is phase space? State and prove Liouville's theorem.

7.12 If q_k and p_k ($k = 1, 2, 3$) represent generalized coordinates and the corresponding momenta of a particle, what is the dimension of phase space and the configuration space?

7.13 Derive the Euler–Lagrange equations of motion using calculus of variation.

7.14 State and prove Brachistochrone problem.

8

Special Theory of Relativity

I have used the shoulders of giants as a spring board.
—Stephen Hawking

8.1 SOME FUNDAMENTAL CONCEPTS

The most difficult task for a human being is to accept a change. Our minds have learned to follow a particular pattern in our daily life which can be changed only by a conscious effort, but it is certainly not easy. We have become slaves of our own habits. It is better to have a habit of changing habits.

Newton proposed his famous laws in 1686 in his book, *Principia*. Those were the laws of motion with respect to an inertial frame of reference. They went unchallenged for over two centuries. In the beginning of the 20th century, Einstein proposed an entirely different concept of space and time. It was difficult for many to understand and accept this new concept. The fact is, that certain ideas were hammered into the heads of many for two centuries. How can they change their thinking overnight?

The obvious question anyone whould ask in this situation would be, "Why do you want us to change our thinking?" Here, we have a yardstick to measure distance, a clock to measure time, an inertial frame to position an object and Newton's laws to predict the position of an object as a function of time. This system has been working well for a couple of centuries. In fact, it predicted the position of planets, their eclipses, seasons on Earth, and so on. Then, why should we change our thinking?

Although Newton's laws have served us quite well, there are certain things it cannot explain, as discussed below:

(i) Newton formulated his laws with respect to an inertial frame—a system fixed in space. But, where does such a system exist? on Earth? Sun? stars? No, it cannot be located on these objects, because they too are moving. In fact, there is no object that is truly stationary. If that is so, what is the use of formulating laws with respect to an inertial system that is non-existing? In practice, since all our observations are from the moving object—the Earth— the Newton's laws cannot be applied to these observations, but we can introduce corrections.

(ii) It has been recognized that Newton's laws of motion are invariant under Galilean transformation. In other words, Newton's laws are valid not only with respect to the coordinate system fixed in inertial frame, but also with respect to the system, moving with a uniform speed along a straight line. Then, what is Galilean transformation? The Galilean transformation relates the positions of a point as measured in two coordinate systems which are in relative translational motion.

Suppose a coordinate frame S' with origin O' moves with a constant velocity $\vec{v}$ relative to the origin O of the fixed coordinate frame S (inertial frame), as shown in Fig. 8.1. If $\vec{r}'$ is the position vector of a point P in space as measured in S' frame and $\vec{r}$ is the position vector of the same point P as measured in S frame, then, the *Galilean transformation* is

$$\vec{r} = \vec{r}' + \vec{v}t. \tag{8.1}$$

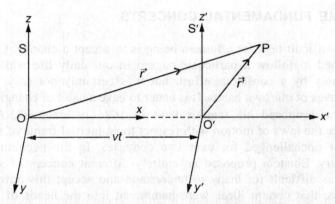

Figure 8.1 Galilean transformation between two coordinate frames.

Differentiating Eq. (8.1) twice with respect to t, we get velocity and acceleration as

$$\vec{V} = \vec{V}' + \vec{v} \tag{8.2}$$

and

$$\frac{d^2\vec{r}}{dt^2} = \frac{d^2\vec{r}'}{dt^2} \quad \text{(since } \vec{v} \text{ is constant).} \tag{8.3}$$

We asssume that the mass of the particle m is constant and therefore, remains as such under Galilean transformation. Thus, we get from Eq. (8.3)

$$m\frac{d^2\vec{r}}{dt^2} = m\frac{d^2\vec{r}'}{dt^2},$$

which indicates that Newton's second law is invariant under Galilean transformation. In other words, the force $F'(\vec{r}', \vec{v}', t)$ on the particle in S' frame is the same as the force $F(\vec{r}, \vec{v}, t)$ in the S frame. If the force is independent of velocity, that is, if

$$F'(\vec{r}', t) = F(\vec{r}, t) \tag{8.4}$$

then, the equation of motion and all consequences of Newton's laws such as conservation of energy and momentum are invariant under Galilean transformation. Thus, if mass m remains constant, then, the Newton's laws hold true in both the systems. But, this has created further problems. It has been observed that an object moving with a velocity $\vec{V}$, comparable to that of light, does behave differently in both the systems. Consider the following examples:

(a) Let there be two identical clocks of finest make, reading the same time, and continue to run in perfect unison as long as they are kept side by side. Let one of them be on Earth and the other be put in a space shuttle on a long flight. When the shuttle comes back to Earth, will the two clocks read the same?

The readings will be different. The clock which is on flight will be slower in comparison with the clock that is kept at home/Earth. Of course, it is assumed that a great care is taken of the clock in flight and all accidental causes are ruled out. But according to Newton's theory, both the clocks should show the same reading.

(b) A medium is necessary for propagation of a physical phenomenon. Man requires ground to run. Waves require water for their propagation. Similarly, sound waves require air for their propagation. These examples leads us to think that light waves too require a medium for their travel or propagation, which is ether. It is also a common observation that waves travel faster in the direction of motion and slower in the direction opposite to the direction of motion of the medium.

If v is the velocity of the medium and c is the velocity of the wave in stationary medium, then the velocity of the wave in the direction of flow is $c + v$ and the velocity of the wave in the direction opposite to the flow is $c - v$.

In the year 1887, Michelson and Morley conducted a very carefully planned experiment, which proved that the velocity of light is same in all directions irrespective of measurement being carried out from a moving or stationary object. Why is this so? Does Newtonian mechanics have an answer for such an extraordinary behaviour of light? No! But, Einstein recognized the fact that the motion of light did not fit into the Newtonian mechanics, and a new mechanics was needed to bring light into its fold, which gave birth to the special theory of relativity. Thus, there was no doubt that the theory of relativity gives us a mathematical model closer to nature than the Newtonian model.

The theory of relativity is divided into two parts:

(i) *Special theory of relativity*—deals with the phenomena independent of gravitational attraction.

(ii) *General theory of relativity*—may be called Einstein's theory of gravitation.

In this chapter we shall restrict ourselves to the study of special theory of relativity. It is based on the following two postulates enunciated by Einstein, in the year 1905:
1. The laws of physics should be expressed in equations having the same form in all frames of reference moving with uniform velocity with respect to one another.
2. The speed of light in free space has the same value for all observers irrespective of their state of motion.

Equivalence of space and time

Although we do not know yet complete details of the theory of relativity, certainly, we have a feeling that it has something to do with space and time; a four-dimensional coordinate system where one of the coordinates represents, 'time', and the other three 'length'.

Obviously, a question arises—how can time be equivalent to distance (length)? In a three-dimensional space, we can talk of a cube, say a unit cube which extends through 1 metre in each direction. How would a unit cube in four-dimensional space look like? How much is 1 metre equal to in time scale? One second? Ten seconds? One month? Is one hour longer than or shorter than one metre?

These questions seems ridiculous, but they are not. We often say that downtown is only 15 minutes away by car or Japan is only 8 hours away by jet plane, although what we are trying to express is our distance from downtown or from Japan. We are able to talk in this manner, because we imply velocity. A jet plane flying at 500 km per hour (kph) would take 8 hours to reach Japan which is nearly 4000 km from here. Thus, if we can agree on some standard velocity, we should be able to express length in terms of time. Because of some extraordinary property of light, we have decided to take 'velocity of light' as the standard velocity. It has proved quite effective. Instead of telling that a distant star is so many billions of kilometre away, we are able to specify its distance in terms of light years:

1 Light year = Distance travelled by light in one year.

So, now we know what we mean by equivalence of space and time. The special theory of relativity binds space and time into space–time.

8.2 THE LORENTZ TRANSFORMATION

Let us consider two inertial frames of reference S and S' in uniform translatory motion with respect to each other. We also assume that the motion of S' frame with respect to S frame is along the x-axis only. For convenience, let us assume that the origins O and O' and S and S' frames coincide at $t = t' = 0$. Let P be a certain physical point which can be specified in terms of either the position vector $\vec{r}$ and t or $\vec{r}'$ and t'. For Newton's laws to hold true in both the inertial frames, the Galilean transformation equation (see Fig. 8.1) is

$$\vec{r}' = \vec{r} - \vec{V}t. \tag{8.5}$$

As there is no relative motion in y and z directions, we can express the above equation in component form as

$$x' = x - vt, \qquad y' = y, \qquad z' = z, \qquad t' = t. \tag{8.6}$$

If the velocity is measured in S and S' frames, then, they are related as

$$v'_x = v_x - v, \qquad v'_y = v_y, \qquad v'_z = v_z. \tag{8.7}$$

That is, if the velocity of light in S frame is c and the velocity of light in S' frame is c', then, we have

$$c' = c - v, \tag{8.8}$$

which violates the fundamental postulate of special theory of relativity.

For example, consider that a source of light is situated at the origin O, in S frame, from which a wavefront of light is emitted, when $t = t' = 0$, that is, when the origins of S and S' frames coincide. When the light reaches the point P, let the positions and times measured by observers in frames S and S' be (x, y, z, t) and (x', y', z', t'), respectively. Then the time measured in S frame is

$$t = \frac{OP}{c} = \frac{(x^2 + y^2 + z^2)^{1/2}}{c}$$

or

$$x^2 + y^2 + z^2 = c^2 t^2. \tag{8.9}$$

Thus, an observer in S sees that the light propagates as a spherical wave of radius ct, where c is the speed of light. According to the postulates of the special theory of relativity, the velocity of light should be only c, even in the S' frame. Thus, an observer in S' frame also sees the light pulse expanding with velocity c and the time taken by the light pulse in travelling the distance O'P is

$$t' = \frac{O'P}{c} = \frac{(x'^2 + y'^2 + z'^2)^{1/2}}{c}$$

or

$$x'^2 + y'^2 + z'^2 = c^2 t'^2. \tag{8.10}$$

Now, we shall find the coordinate transformation that relates Eqs. (8.9) and (8.10). For, let us recall the Galilean transformation described through Eqs. (8.6) which connects the measurements in S and S' frames and by substituting these values into Eq. (8.10), we get

$$(x - vt)^2 + y^2 + z^2 = c^2 t^2, \tag{8.11}$$

which is not in agreement with Eq. (8.9). Thus, the Galilean transformation fails. It is, therefore, necessary that we look for different transformation equations, consistent with the postulates of special theory of relativity.

Let us assume that the possible transformation relating S and S' is of the type

$$x' = k(x - vt), \qquad y' = y, \qquad z' = z, \qquad t' = k'(t + \alpha x). \qquad (8.12)$$

Here, k, k' and α are constants. Substituting this transformation into Eq. (8.10), we obtain

$$k^2(x - vt)^2 + y^2 + z^2 = c^2 k'^2 (t + \alpha x)^2$$

or

$$k^2(x^2 - 2xvt + v^2 t^2) + y^2 + z^2 = c^2 k'^2 (t^2 + 2t\alpha x + \alpha^2 x^2)$$

or

$$x^2(k^2 - c^2 k'^2 \alpha^2) - 2xt(k^2 v + \alpha c^2 k'^2) + y^2 + z^2 = \left(k'^2 - \frac{v^2 k^2}{c^2}\right) c^2 t^2. \qquad (8.13)$$

This resulting equation will be identical with Eq. (8.9), provided the following conditions are satisfied. That is,

$$k^2 - c^2 k'^2 \alpha^2 = 1, \qquad (8.14)$$

$$k^2 v + \alpha c^2 k'^2 = 0, \qquad (8.15)$$

$$k'^2 - \frac{v^2 k^2}{c^2} = 1. \qquad (8.16)$$

Now, Eq. (8.15) gives

$$\alpha = \frac{-k^2 v}{c^2 k'^2}.$$

Substituting this value of α into Eq. (8.14), we get

$$k^2 - \frac{c^2 k'^2 k^4 v^2}{c^4 k'^4} = 1 \quad \text{or} \quad 1 - \frac{k^2 v^2}{c^2 k'^2} = \frac{1}{k^2}.$$

Utilizing the value of k'^2 from Eq. (8.16), the above equation reduces to

$$\frac{1 - \dfrac{k^2 v^2}{c^2}}{1 + \dfrac{v^2 k^2}{c^2}} = \frac{1}{k^2}$$

or

$$\frac{1}{1 + \dfrac{v^2 k^2}{c^2}} = \frac{1}{k^2}$$

or
$$k^2 = 1 + \frac{v^2 k^2}{c^2}.$$
that is,
$$k^2 = \frac{1}{1 - \frac{v^2}{c^2}}.$$
Therefore,
$$k = \frac{1}{\sqrt{1 - \frac{v^2}{c^2}}} \qquad (8.17)$$

Using this value of k, Eq. (8.16) gives
$$k'^2 = 1 + \frac{v^2 k^2}{c^2} = 1 + \frac{v^2}{c^2 \left(1 - \frac{v^2}{c^2}\right)}$$
or
$$k' = \frac{1}{\sqrt{1 - \frac{v^2}{c^2}}} \qquad (8.18)$$

Thus, we find
$$k = k' = \frac{1}{\sqrt{1 - \frac{v^2}{c^2}}}.$$

Using these values of k and k', Eq. (8.15) gives at once
$$\alpha = \frac{-v}{c^2}. \qquad (8.19)$$

Hence, we found that the new transformation equations which are in agreement with the constancy of the speed of light in all frames of reference are

$$\left. \begin{array}{l} x' = \dfrac{(x - vt)}{\sqrt{1 - \dfrac{v^2}{c^2}}} \\ y' = y \\ z' = z \\ t' = \dfrac{\left(t - \dfrac{vx}{c^2}\right)}{\sqrt{1 - \dfrac{v^2}{c^2}}}. \end{array} \right\} \qquad (8.20)$$

These equations are known as *Lorentz transformation*. It may be observed that, when $v \ll c$, then $v/c \to 0$ and we get the Galilean transformation from the Lorentz transformation. Its inverse transformation is found to be

$$\left.\begin{array}{l} x = \dfrac{(x' + vt')}{\sqrt{1 - \dfrac{v^2}{c^2}}} \\[4pt] y = y' \\[4pt] z = z' \\[4pt] t = \dfrac{\left(t' + \dfrac{vx'}{c^2}\right)}{\sqrt{1 - \dfrac{v^2}{c^2}}} \end{array}\right\} \qquad (8.21)$$

We may observe that from Eq. (8.20) or Eq. (8.21) that $v < c$, corresponds to physically meaningful values; otherwise if $v > c$, quantities become imaginary. Hence, the relative velocity between S and S' should be less than c. In other words, no particle in the universe can move with velocity $\geq c$ with respect to any other particle.

8.3 IMMEDIATE CONSEQUENCES OF LORENTZ TRANSFORMATIONS

For people habituated to think in Newtonian way, the concepts that follow, such as contraction of length, time dilation and velocity addition are fascinating, which are the immediate consequences of Lorentz transformations.

Contraction of length

Let us consider a measuring rod AB placed along the x'-axis in S' frame which is moving with velocity v, relative to S frame. Suppose, an observer in S' frame is at rest with respect to S' frame and hence with respect to the rod. In other words, the rod is at rest with respect to this observer. Then the length L' of the rod measured by the observer in S' is

$$L' = x'_B - x'_A.$$

But, for the observer in S, the length appears to be

$$L = x_B - x_A$$

at that instant. From Lorentz transformation Eqs. (8.20), we have

$$x'_B = \frac{x_B - vt}{\sqrt{1 - \dfrac{v^2}{c^2}}}, \qquad x'_A = \frac{x_A - vt}{\sqrt{1 - \dfrac{v^2}{c^2}}}.$$

Therefore,

$$L' = x'_B - x'_A = \frac{x_B - x_A}{\sqrt{1 - \dfrac{v^2}{c^2}}} = \frac{L}{\sqrt{1 - \dfrac{v^2}{c^2}}}.$$

Immediate Consequences of Lorentz Transformations (Sec. 8.3)

Equivalently

$$L = L'\sqrt{1 - \frac{v^2}{c^2}} < L'. \tag{8.22}$$

That is, the length of the rod AB fixed in S' frame as seen from S appears to be less than that as seen from S'.

Similarly, if S' views the rod fixed in S, then, we find that

$$L' = L\sqrt{1 - \frac{v^2}{c^2}} < L. \tag{8.23}$$

That is, for an observer in S', the length of the rod appears to be less than that as seen from S. This phenomenon is known as *Fitzgerald contraction*.

Time dilation

Consider a frame of reference S, in which at a point x, a clock is located. Imagine an observer fixed in the moving frame S', where a second identical clock is fixed. Assume that two events are recorded by the clock in S at times t_1 and t_2. At time t_1, on the clock, the observer in S' notes the time as

$$t'_1 = \frac{t_1 - \frac{vx}{c^2}}{\sqrt{1 - \frac{v^2}{c^2}}} \tag{8.24}$$

and also at time t_2 on the clock in S the observer in S' frame notes the time as

$$t'_2 = \frac{t_2 - \frac{vx}{c^2}}{\sqrt{1 - \frac{v^2}{c^2}}}. \tag{8.25}$$

Thus, the time interval is

$$t'_2 - t'_1 = \frac{t_2 - t_1}{\sqrt{1 - \frac{v^2}{c^2}}}. \tag{8.26}$$

If we denote the time intervals recorded by the two clocks as τ and τ', then,

$$\tau' = \frac{\tau}{\sqrt{1 - \frac{v^2}{c^2}}}$$

or

$$\tau = \tau'\sqrt{1 - \frac{v^2}{c^2}} < \tau'. \tag{8.27}$$

Therefore, the time interval between the events as recorded by S' is greater than the time interval, as recorded by S. Hence for an observer in S' frame the clock in S will appear to run slow. Thus, each observer considers the clock of the other to be running slow. In other words, the proper time is less than the relativisitic time. This concept is called *time dilation* or *time dilatation*.

Composition of velocities (velocity addition)

Suppose P and Q are two particles travelling with uniform velocities u_1 and u_2 along OX axis of the Galilean frame S. Then, in Newtonian mechanics, the relative velocity of the two particles is $(u_1 - u_2)$. However, in relativity the velocity of Q relative to P is the velocity of Q as estimated by a Galilean observer in S' frame travelling with P in which P is fixed.

Recalling the Lorentz transformations, as described in Eqs. (8.20) between the frames S and S', we have

$$\left. \begin{array}{c} x' = \dfrac{x - u_1 t}{\sqrt{1 - \dfrac{u_1^2}{c^2}}}, \\ \\ t' = \dfrac{t - \dfrac{u_1 x}{c^2}}{\sqrt{1 - \dfrac{u_1^2}{c^2}}}. \end{array} \right\} \qquad (8.28)$$

Now, consider the motion of Q, its velocity in S frame is

$$\frac{dx}{dt} = u_2. \qquad (8.29)$$

But, the velocity of Q as estimated by an observer in S' frame is say u'. Then, using Eq. (8.28), we get

$$u' = \frac{dx'}{dt'} = \frac{dx - u_1 dt}{dt - u_1 \dfrac{dx}{c^2}} = \frac{\dfrac{dx}{dt} - u_1}{1 - \left(\dfrac{u_1}{c^2}\right)\left(\dfrac{dx}{dt}\right)}$$

or

$$u' = \frac{u_2 - u_1}{1 - \dfrac{u_1 u_2}{c^2}}. \qquad (8.30)$$

However, Newton's law gives

$$u' = u_2 - u_1. \qquad (8.31)$$

We can verify that, if u_1 and u_2 are very small when compared with c, the velocity of light, that is, if $u_1 \ll c, u_2 \ll c$, that is, $u_1/c \ll 1, u_2/c \ll 1$, implying $u_1 u_2/c^2 \ll 1$, then, $(1 - u_1 u_2/c^2) \simeq 1$. Hence, the difference between Eq. (8.30) and Eq. (8.31) is negligibly small. Solving Eq. (8.30) for u_2, we find

$$u_2 - u_1 = u' - \frac{u_1 u_2 u'}{c^2}$$

or

$$u_2 \left(1 + \frac{u_1 u'}{c^2}\right) = u' + u_1.$$

Therefore,

$$u_2 = \frac{u_1 + u'}{1 + \frac{u_1 u'}{c^2}}. \qquad (8.32)$$

This is the relativistic law of composition of velocities or the law of addition of velocities in special theory of relativity.

8.4 THE MASS OF A MOVING PARTICLE

In relativistic mechanics, the mass of a moving particle depends on its velocity. To demonstrate this fact, we should note that the conservation of mass and momentum holds true in any coordinate system. For illustration, consider a simple head-on collision between two similar and perfectly elastic particles.

First, let us consider two coordinate systems S and S′ in which S′ is moving with a velocity v relative to S. Let the two particles be moving with velocities u' and $-u'$, before collision along the X-axis (see Fig. 8.2), such that a head-on collision occurs. Since the particles are perfectly similar and elastic, on collision, they come to rest, and then rebound under the action of elastic forces, and move back with reversed velocities $-u'$ and u', of course with the same magnitude.

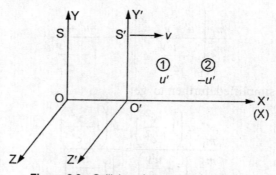

Figure 8.2 Collision of two elastic particles.

Let us now consider the second system S, moving relative to S' in the X-direction with velocity $-v$. Assume that m_1 and m_2 are the masses of the two particles and u_1, u_2 be their velocities before collision. Let M be the sum of the two masses at the instant, in the course of collision, when they have to come to relative rest and then move with velocity v, relative to S. For the conservation laws to hold in this new system of coordinates, the total mass and the total momentum of the two particles must be same before impact and at the instant of relative rest. Thus, we write

$$m_1 + m_2 = M, \tag{8.33}$$

$$m_1 u_1 + m_2 u_2 = Mv. \tag{8.34}$$

Using the addition law of velocities, we find

$$u_1 = \frac{u' + v}{1 + \frac{u'v}{c^2}}, \quad u_2 = \frac{-u' + v}{1 - \frac{u'v}{c^2}}. \tag{8.35}$$

Further, Eq. (8.34) can be rewritten as

$$m_1 u_1 + m_2 u_2 = (m_1 + m_2) v.$$

Substituting the values of u_1 and u_2 from Eq. (8.35), the above equation gives

$$m_1 \frac{u' + v}{1 + \frac{u'v}{c^2}} + m_2 \frac{-u' + v}{1 - \frac{u'v}{c^2}} = (m_1 + m_2) v$$

or

$$\frac{m_1}{m_2} \frac{u' + v}{1 + \frac{u'v}{c^2}} + \frac{-u' + v}{1 - \frac{u'v}{c^2}} = \left(1 + \frac{m_1}{m_2}\right) v$$

or

$$\frac{m_1}{m_2} \left(\frac{u' + v}{1 + \frac{u'v}{c^2}} - v \right) = v - \frac{-u' + v}{1 - \frac{u'v}{c^2}},$$

which can be simplified further to get

$$\frac{m_1}{m_2} \frac{u'\left(1 - \frac{v^2}{c^2}\right)}{\left(1 + \frac{u'v}{c^2}\right)} = \frac{u'\left(1 - \frac{v^2}{c^2}\right)}{\left(1 - \frac{u'v}{c^2}\right)}$$

that is,

$$\frac{m_1}{m_2} = \frac{1+\dfrac{u'v}{c^2}}{1-\dfrac{u'v}{c^2}}. \tag{8.36}$$

Now, using Eqs. (8.35), let us consider

$$\left(1-\frac{u_1^2}{c^2}\right) = 1 - \frac{1}{c^2}\left(\frac{u'+v}{1+\dfrac{u'v}{c^2}}\right)^2$$

$$= \frac{c^2\left(1+\dfrac{u'v}{c^2}\right)^2 - (u'+v)^2}{c^2\left(1+\dfrac{u'v}{c^2}\right)^2}$$

$$= \frac{c^2\left(1-\dfrac{v^2}{c^2}\right) - u'^2\left(1-\dfrac{v^2}{c^2}\right)}{c^2\left(1+\dfrac{u'v}{c^2}\right)^2}$$

$$= \frac{(c^2 - u'^2)\left(1-\dfrac{v^2}{c^2}\right)}{c^2\left(1+\dfrac{u'v}{c^2}\right)^2}$$

$$= \frac{\left(1-\dfrac{v^2}{c^2}\right)\left(1-\dfrac{u'^2}{c^2}\right)}{\left(1+\dfrac{u'v}{c^2}\right)^2}.$$

Therefore,

$$1 + \frac{u'v}{c^2} = \frac{\sqrt{1-\dfrac{v^2}{c^2}}\sqrt{1-\dfrac{u'^2}{c^2}}}{\sqrt{1-\dfrac{u_1^2}{c^2}}}. \tag{8.37}$$

Similarly, it can be shown that

$$1 - \frac{u'v}{c^2} = \frac{\sqrt{1 - \frac{v^2}{c^2}}\sqrt{1 - \frac{u'^2}{c^2}}}{\sqrt{1 - \frac{u_2^2}{c^2}}}. \qquad (8.38)$$

Utilizing Eqs. (8.37) and (8.38) into Eq. (8.36), we arrive at

$$\frac{m_1}{m_2} = \frac{\sqrt{1 - \frac{u_2^2}{c^2}}}{\sqrt{1 - \frac{u_1^2}{c^2}}} \qquad (8.39)$$

or

$$m_1\sqrt{1 - \frac{u_1^2}{c^2}} = m_2\sqrt{1 - \frac{u_2^2}{c^2}} = m_0 \text{ (say),}$$

an absolute constant. In other words, we may write

$$m_1 = \frac{m_0}{\sqrt{1 - \frac{u_1^2}{c^2}}} \quad \text{and} \quad m_2 = \frac{m_0}{\sqrt{1 - \frac{u_2^2}{c^2}}}. \qquad (8.40)$$

This result proves that if a particle of mass m is moving with velocity u relative to a system, then we find

$$m = \frac{m_0}{\sqrt{1 - \frac{u^2}{c^2}}}. \qquad (8.41)$$

Here, m_0 is called the *rest mass* (mass of the body at rest). If $u = 0$, $m = m_0$ is obvious.

8.5 EQUIVALENCE OF MASS AND ENERGY

Consider two systems S and S', where S' is moving with velocity v relative to S along X-axis. In the system S, let a particle of mass m be moving with velocity v along X-axis (see Fig. 8.3).
Then, we have established in the previous section that

$$m = \frac{m_0}{\sqrt{1 - \frac{v^2}{c^2}}}. \qquad (8.42)$$

Equivalence of Mass and Energy (Sec. 8.5)

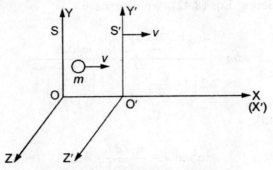

Figure 8.3 Motion of a particle.

If a force $\vec{F}$ is applied on this particle, there will be a change in speed which in effect produces a change in kinetic energy, denoted by dT, and the work is done on the particle. Therefore, we have

$$dT = F\, ds = F\frac{ds}{dt}\, dt = Fv\, dt. \tag{8.43}$$

Now, the relativistic form of Newton's second law states that

$$F = \frac{d}{dt}(mv) = \frac{d}{dt}\left(\frac{m_0 v}{\sqrt{1-\frac{v^2}{c^2}}}\right). \tag{8.44}$$

Then, Eq. (8.43) can be rewritten as

$$dT = Fv\, dt = m_0 v\, d\left(\frac{v}{\sqrt{1-\frac{v^2}{c^2}}}\right).$$

On carrying out differentiation using the quotient rule, we have

$$dT = m_0 v \frac{\sqrt{1-\frac{v^2}{c^2}}\, dv - \left(\frac{v}{2}\right)\left(\frac{-2v}{c^2}\right)\frac{dv}{\sqrt{1-\frac{v^2}{c^2}}}}{1-\frac{v^2}{c^2}}$$

or

$$dT = \frac{m_0 v\, dv}{\left(1-\frac{v^2}{c^2}\right)^{3/2}}. \tag{8.45}$$

Now, differentiating Eq. (8.42), we also have

$$dm = \frac{-m_0 \left(\dfrac{-2v\,dv}{c^2}\right)}{2\left(1 - \dfrac{v^2}{c^2}\right)^{3/2}} = \frac{m_0 v\, dv}{c^2 \left(1 - \dfrac{v^2}{c^2}\right)^{3/2}}.$$

Using Eq. (8.45),

$$c^2 dm = \frac{m_0 v\, dv}{\left(1 - \dfrac{v^2}{c^2}\right)^{3/2}} = dT. \tag{8.46}$$

Suppose that the particle is initially at rest and at that time, its mass is m_0. After the application of the force, let its mass be

$$m = m_0 + dm,$$

and the total kinetic energy T acquired by the particle is given by

$$T = \int dT = \int_{m_0}^{m} c^2\, dm = c^2(m - m_0)$$

or

$$T + m_0 c^2 = mc^2. \tag{8.47}$$

Let E be the total energy, that is, the sum of the kinetic energy of the moving particle and rest energy, then

$$E = T + m_0 c^2 = mc^2. \tag{8.48}$$

This is the most famous and significant result derived by Einstein in special theory of relativity, demonstrating that mass and energy are identical. This result has many practical implications.

The relativistic formula for kinetic energy, as given in Eq. (8.47), can be rewritten as

$$T = mc^2 - m_0 c^2 = \frac{m_0 c^2}{\sqrt{1 - \dfrac{v^2}{c^2}}} - m_0 c^2.$$

If $v^2/c^2 \ll 1$, then using binomial expansion, we obtain

$$\frac{1}{\sqrt{1 - \dfrac{v^2}{c^2}}} = 1 + \frac{1}{2}\frac{v^2}{c^2}, \qquad v \ll c.$$

Thus, we have the result

$$T \simeq \left(1 + \frac{1}{2}\frac{v^2}{c^2}\right)m_0c^2 - m_0c^2 \simeq \frac{1}{2}m_0v^2,$$

which means that at low speeds, the relativistic expression for the kinetic energy of a moving particle does indeed reduce to the classical one.

For illustration of various concepts of special theory of relativity, we shall consider below several examples:

Example 8.1 If $\vec{p}$ stands for linear momentum, prove that $E^2 - c^2p^2 = m_0^2c^4$.

Solution Consider

$$E^2 - c^2p^2 = (mc^2)^2 - c^2(p_x^2 + p_y^2 + p_z^2).$$

Here, p_x, p_y, p_z are the components of linear momentum ($\vec{p}$) of an object whose total energy is E. Therefore,

$$E^2 - c^2p^2 = m^2c^4 - c^2[(mu_x)^2 + (mu_y)^2 + (mu_z)^2]$$

$$= m^2c^2[c^2 - (u_x^2 + u_y^2 + u_z^2)]$$

$$= m^2c^2(c^2 - u^2)$$

$$= m^2c^4\left(1 - \frac{u^2}{c^2}\right)$$

$$= \left(\frac{m_0}{\sqrt{1 - \frac{u^2}{c^2}}}\right)^2 c^4\left(1 - \frac{u^2}{c^2}\right)$$

or

$$E^2 - c^2p^2 = m_0^2c^4.$$

Hence, the result follows.

Example 8.2 A body of mass m disintegrates into two parts m_1 and m_2, while at rest. Show that the energies E_1 and E_2 of the two parts satisfy the relation

$$\frac{E_1}{E_2} = \frac{m^2 + m_1^2 - m_2^2}{m^2 + m_2^2 - m_1^2}.$$

Solution Given

$$E = E_1 + E_2 = \text{Total energy} = mc^2. \quad (1)$$

After disintegration, the masses m_1 and m_2 move in opposite directions with equal momenta p. Thus,

$$E_1^2 = c^2 p^2 + m_1^2 c^4, \qquad E_2^2 = c^2 p^2 + m_2^2 c^4.$$

Therefore,
$$E_1^2 - E_2^2 = c^4 (m_1^2 - m_2^2)$$

or
$$(E_1 + E_2)(E_1 - E_2) = c^4 (m_1^2 - m_2^2). \tag{2}$$

Now, using Eq. (1) into Eq. (2), we get
$$E(E_1 - E_2) = c^4 (m_1^2 - m_2^2)$$

or
$$mc^2 (E_1 - E_2) = c^4 (m_1^2 - m_2^2).$$

That is,
$$E_1 - E_2 = \frac{c^2}{m}(m_1^2 - m_2^2). \tag{3}$$

Adding Eqs. (1) and (3), we obtain
$$2E_1 = mc^2 + \frac{c^2}{m}(m_1^2 - m_2^2) = \frac{c^2}{m}(m^2 + m_1^2 - m_2^2). \tag{4}$$

Similarly, subtracting Eq. (3) from Eq. (1), we find
$$2E_2 = mc^2 - \frac{c^2}{m}(m_1^2 - m_2^2) = \frac{c^2}{m}(m^2 + m_2^2 - m_1^2). \tag{5}$$

From Eqs. (4) and (5), we get at once
$$\frac{E_1}{E_2} = \frac{m^2 + m_1^2 - m_2^2}{m^2 + m_2^2 - m_1^2}.$$

Hence the result.

Example 8.3 The S' frame has a speed $v = 0.6c$ relative to S frame. Clocks are adjusted so that $t = t' = 0$ at $x = x' = 0$. Two events occur. Event-1 occurs at $x_1 = 10$ m, $t_1 = 2 \times 10^{-7}$ s. ($y_1 = 0, z_1 = 0$). Event-2 occurs at $x_2 = 50$ m, $t_2 = 3 \times 10^{-7}$ s. ($y_2 = 0, z_2 = 0$). What is the distance between the events as measured in S' frame? What is the time difference as measured in S'?

Solution We are given $v = 0.6c$. Therefore,
$$\frac{v^2}{c^2} = \left(\frac{6}{10}\right)^2 = \left(\frac{3}{5}\right)^2 = \frac{9}{25}.$$

Thus,

$$\sqrt{1-\frac{v^2}{c^2}} = \sqrt{1-\frac{9}{25}} = \frac{4}{5}. \tag{1}$$

Now, using Lorentz transformation, the distance between the events as measured in S' is

$$x'_2 - x'_1 = \frac{(x_2 - x_1) - v(t_2 - t_1)}{\sqrt{1-\frac{v^2}{c^2}}}.$$

From the given data and also taking the value of $c = 3 \times 10^8$ m/s, we find

$$x'_2 - x'_1 = \frac{5}{4}[(50-10) - (0.6)(3 \times 10^8)(3-2)10^{-7}] = 27.5 \text{ m}. \tag{2}$$

To compute the time difference as measured in S' frame we shall use the formula of time dilation and get

$$t'_2 - t'_1 = \frac{\left[(t_2 - t_1) - \left(\frac{v}{c}\right)\frac{(x_2 - x_1)}{c}\right]}{\sqrt{1-\frac{v^2}{c^2}}}$$

$$= \frac{5}{4}\left[(3-2)10^{-7} - (0.6)\left(\frac{50-10}{3 \times 10^8}\right)\right]$$

$$= 2.5 \times 10^{-8}.$$

Example 8.4 The length of the space vehicle is 100 m on the ground. When it is in flight, its length as observed on the ground is 99 m. Calculate its speed.

Solution Using Fitzgerald contraction formula, we have

$$L' = L\sqrt{1-\frac{v^2}{c^2}}.$$

Substituting the given data, we find

$$99 = 100\left(1-\frac{v^2}{c^2}\right)^{1/2},$$

where c is the velocity of light. Taking $c = 3 \times 10^8$ ms, we have

$$\left(\frac{99}{100}\right)^2 = 1 - \frac{v^2}{c^2} = 1 - \frac{v^2}{(3 \times 10^8)^2}.$$

That is,

$$\frac{v^2}{(3\times10^8)^2} = 1 - \left(\frac{99}{100}\right)^2 = \left(1+\frac{99}{100}\right)\left(1-\frac{99}{100}\right) = \frac{199}{(100)^2}.$$

Therefore,

$$v = \frac{\sqrt{199}}{100} \times 3 \times 10^8 = 42.32 \times 10^6 \text{ m/s}.$$

Example 8.5 A rocket is chasing an enemy's spaceship. An observer on the Earth observes the speed of the rocket as 2.5×10^8 m/s and that of the spaceship as 2×10^8 m/s. Calculate the velocity of the enemy's spaceship (ESS) seen by the rocket.

Solution Let us consider the rocket to be in S' frame and Earth in S frame, as shown in Fig. 8.4.

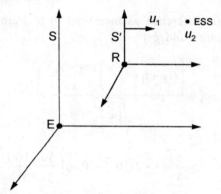

Figure 8.4 Description of Example 8.5.

Suppose u' be the velocity of the enemy's spaceship (ESS) as seen by the rocket; u_1 the velocity of the rocket relative to Earth; u_2 the velocity of the spaceship (ESS) relative to Earth.

We have from the addition law of velocities

$$u_2 = \frac{u_1 + u'}{1 + \frac{u_1 u'}{c^2}}. \qquad (1)$$

From the given data, we have

$$u_1 = 2.5 \times 10^8 \text{ m/s},$$
$$u_2 = 2.0 \times 10^8 \text{ m/s},$$
$$c = 3 \times 10^8 \text{ m/s}$$
$$u' = ?.$$

Substituting these values into Eq. (1), we get on simplification

$$u_2 + \frac{u_1 u_2 u'}{c^2} = u_1 + u'$$

or
$$u'\left(1 - \frac{u_1 u_2}{c^2}\right) = u_2 - u_1$$

that is,
$$u'\left[1 - \frac{2.5 \times 10^8 \times 2 \times 10^8}{(3 \times 10^8)^2}\right] = (2.0 - 2.5)10^8$$

or
$$u'\left(1 - \frac{5}{9}\right) = -0.5 \times 10^8.$$

Therefore,
$$u' = -\frac{9 \times 0.5 \times 10^8}{4} = -1.12 \times 10^8 \text{ m/s}.$$

The negative sign indicates that the enemy's spaceship is approaching the rocket with velocity 1.12×10^8 m/s.

Example 8.6 Two electrons move towards each other. The speed of each being $0.9c$ in a Galilean frame of reference. What is their speed relative to each other?

Solution The situation in the example is described in Fig. 8.5.

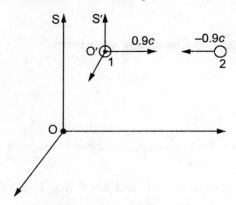

Figure 8.5 Description of Example 8.6.

From the given data, we have $u_1 = 0.9c$, which is the speed of electron 1 relative to S; $u_2 = -0.9c$, which is the speed of electron 2 relative to S.

Let u' be the speed of the second electron with respect to the first electron. Then, from the addition law of velocities, we have

$$u_2 = \frac{u_1 + u'}{1 + \frac{u_1 u'}{c^2}} \quad \text{or} \quad u'\left(1 - \frac{u_1 u_2}{c^2}\right) = u_2 - u_1.$$

Substituting the given data, we find

$$u'\left[1 - \frac{(0.9)(-0.9)c^2}{c^2}\right] = -1.8c$$

or

$$u'(1 + 0.81) = -1.8c$$

or

$$u' = -\frac{1.8c}{1.81} = -0.995c.$$

Example 8.7 Show that the Lorentz transformation may be regarded as a rotation of axes through an imaginary angle.

Solution Let us consider a four-dimensional space described by x_1, x_2, x_3, x_4, where $x_4 = ict$ and $i = \sqrt{-1}$. Suppose x_1 and x_4 axes are rotated through an angle θ, so that new axes are x_1' and x_4', while x_2 and x_3 axes remain unchanged. Then, we have

$$x_1' = x_1 \cos\theta + x_4 \sin\theta \tag{1}$$

$$x_4' = -x_1 \sin\theta + x_4 \cos\theta \tag{2}$$

$$x_2' = x_2 \tag{3}$$

$$x_3' = x_3 \tag{4}$$

Let us take angle θ to be imaginary such that

$$\tan\theta = \frac{iv}{c} \tag{5}$$

then,

$$\sin\theta = \frac{iv}{\sqrt{c^2 - v^2}}, \qquad \cos\theta = \frac{c}{\sqrt{c^2 - v^2}}. \tag{6}$$

Substituting these values of $\sin\theta$ and $\cos\theta$ into Eqs. (1) and (2), we get

$$x_1' = \frac{x_1 c}{\sqrt{c^2 - v^2}} + \frac{x_4(iv)}{\sqrt{c^2 - v^2}}.$$

Since $x_4 = ict$,

$$x_1' = \frac{x_1}{\sqrt{1 - \frac{v^2}{c^2}}} + \frac{(ict)(iv)}{c\sqrt{1 - \frac{v^2}{c^2}}} = \frac{x_1 - vt}{\sqrt{1 - \frac{v^2}{c^2}}} \tag{7}$$

and

$$x'_4 = \frac{-x_1(iv)}{\sqrt{c^2-v^2}} + \frac{(ict)c}{\sqrt{c^2-v^2}}$$

or

$$ict' = -\frac{ix_1 v}{c\sqrt{1-\frac{v^2}{c^2}}} + \frac{ic^2 t}{c\sqrt{1-\frac{v^2}{c^2}}}.$$

Cancelling ic on both sides, we get

$$t' = \frac{t - \frac{vx_1}{c^2}}{\sqrt{1-\frac{v^2}{c^2}}}. \tag{8}$$

Thus, we have

$$\left.\begin{array}{l} x'_1 = \dfrac{x_1 - vt}{\sqrt{1-\dfrac{v^2}{c^2}}} \\[2mm] x'_2 = x_2 \\[2mm] x'_3 = x_3 \\[2mm] t' = \dfrac{t - \dfrac{vx_1}{c^2}}{\sqrt{1-\dfrac{v^2}{c^2}}}, \end{array}\right\} \tag{9}$$

which constitute Lorentz transformation.

EXERCISES

8.1 Prove the relativistic formula

$$m = \frac{m_0}{\sqrt{1-\frac{v^2}{c^2}}}$$

and then obtain the relation $E = mc^2$.

8.2 Find the relativistic expression for the kinetic energy of a particle, whose rest mass is m_0, moving with velocity v. Obtain the relation $E^2 = p^2c^2 + m_0^2c^4$.

8.3 Show that the rest mass of a particle of momentum p and kinetic energy T is given by

$$m_0 = \frac{p^2c^2 - T^2}{2Tc^2}.$$

Hint: Recall $E = T + m_0c^2$ and $E^2 = p^2c^2 + m_0^2c^4$. Squarring the first equation and equating to the second, we get the result.

8.4 The rest mass of a proton is 1.6725×10^{-27} kg. Find its mass and momentum, when it is moving with a velocity of 2.7×10^8 m/s. If it collides with a stationary neucleus of mass 2.5×10^{-26} kg and coalesces, find the velocity of the combined particle.

8.5 At what velocity, the mass of the particle will be double of its rest mass.

8.6 Show that the equation

$$dx^2 + dy^2 + dz^2 - c^2dt^2$$

is invariant under Lorentz transformation.

8.7 Show that the relativistic form of Newton's second law, when $\vec{F}$ is parallel to $\vec{V}$ is

$$F = m_0 \frac{dv}{dt}\left(1 - \frac{v^2}{c^2}\right)^{-3/2}$$

9

Rocket Dynamics

Learning gives creativity; creativity leads to thinking; thinking provides knowledge; knowledge makes you great.

—A.P.J. ABDUL KALAM

9.1 INTRODUCTION

A rocket is a variable mass vehicle, which acquires thrust by the ejection of high-speed particles. In order to place a satellite into a required orbit, we should have a rocket to carry the satellite and to inject it with the required velocity, at a specified location in space. For all Earth-bound satellites, the minimum injection velocity required is about 7.6 km/s whereas, for lunar missions or interplanetary travel, the minimum injection velocity required is about 11.2 km/s. If we look into the historical development of rocketry, it was first used by Chinese during the 13th century as a weapon of warfare. Firework rockets have been commonly seen in India for centuries. The development of modern rocketry was due to Robert H. Goddard who used them to conduct high-altitude experiments in the USA. During the Second World War, Germans, Russians and British developed several types of rockets of which German V-2 rocket was well known. This had a range of about 320 km and attained an altitude of 100 km. Then, came the modern spaceships such as Cosmos, Soyuz series developed by the USSR and the space shuttle developed by the USA for exploration of outerspace and interplanetary travel. The rockets employed in space flight, also called *space vehicles*, generally have 2 to 4 stages. It was von Braun of the USA, the chief architect for the design of modern Saturn-5 rocket, who was responsible for launching and landing man on the Moon and subsequent return to Earth in the year 1969. Most recent developments include ESRO–Ariane vehicle, space shuttles of the USA and the USSR for placing multiple payloads simultaneously in different orbits. India, too, has developed several launch vehicles, such as SLV-3E2(1980), ASLV-D3(1992), PSLV-D2(1994), and more recently, the geo-synchronous satellite launch vehicle GSLV (2001, 2003). GSLV-D2 flight, carried an experimental communication satellite GSAT-2 with a payload of 1800 kg. It may be noted that GSLV is a three-stage vehicle and 49 m tall. The first stage comprises a solid propellant motor and four liquid propellant strap-on motors. The second stage is powered by a single liquid propellant engine, while the third stage is a cryogenic one. Its lift-off weight is 414 ton.

Advanced and developed countries are already planning to bring back to Earth, ship loads of Helium-3, a valuable fuel for thermonuclear reactors, which is found to be abundant on the Moon, but practically absent on the Earth. A couple of shuttle spacecrafts can bring to Earth enough liquid Helium-3 to meet all global energy needs for a year.

9.2 EQUATION OF MOTION FOR VARIABLE MASS

A *rocket* is defined as a variable-mass vehicle, which is propelled by the ejection of burnt-out fuel, thereby reducing the mass of the rocket as it accelerates. Another illustration for the motion of variable mass is a raindrop falling through a damp atmosphere and increasing its mass during its onward motion, as it coalesces with smaller droplets. Therefore, the term 'variable mass' should be understood only in the sense of being continuously added to or removed from the body.

Let us imagine a body of variable mass $m(t)$ travelling with velocity $\vec{V}$, which coalesces with a small mass δm moving with velocity $\vec{u}$ after an interval of time δt. Then, at time $(t + \delta t)$ the body has a mass $(m + \delta m)$ moving with velocity $(\vec{V} + \delta \vec{V})$. Further, we also assume that the body is only translating and no rotation is imparted to the body due to collision (see Fig. 9.1). Suppose the body is subject to an external force $\vec{F}$, which may be due to gravitational and atmospheric resistance.

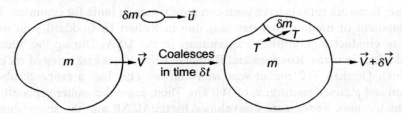

Figure 9.1 Illustration of variable-mass motion.

Following the Newton's second law, the applied force will balance the rate of change of linear momentum of the system. But, this force should also include a term due to continuous impulsive action by each body on the other, in view of the Newton's third law. Thus, an equal and oppositely directed force $\pm T$ will be exerted by each body on the other. Therefore, the equation of motion of the small mass δm is given by

T = Rate of change of linear momentum.

That is,

$$T = \lim_{\delta t \to 0} \left[\frac{\delta m (\vec{V} + \delta \vec{V}) - \delta m (\vec{u})}{\delta t} \right] = (\vec{V} - \vec{u}) \frac{dm}{dt}. \qquad (9.1)$$

During the same interval δt, the change in linear momentum of the main body can be written as

$$\vec{F} - \vec{T} = \lim_{\delta t \to 0} \left[\frac{m(\vec{V} + \delta \vec{V}) - m\vec{V}}{\delta t} \right] = m \frac{d\vec{V}}{dt}. \qquad (9.2)$$

Substituting Eq. (9.1) into Eq. (9.2), we find

$$\vec{F} = (\vec{V} - \vec{u}) \frac{dm}{dt} + m \frac{d\vec{V}}{dt}, \qquad (9.3)$$

which is the equation of motion for a variable mass.

9.3 PERFORMANCE OF A SINGLE-STAGE ROCKET

Certain parameters of primary importance can be identified by the preliminary study of the behaviour of a rocket in its vertical flight, neglecting the external forces like aerodynamic as well as gravity forces. For this purpose, let us imagine a simple model for a single-stage rocket, as made up of a cylinder closed at the upper end by a nose cone to house and to protect the payload, say a satellite, during its flight from aerodynamic and heating effects as it passes through the dense atmospheric layers. As the fuel (propellant) is burnt in the rocket motor, the exhaust gases come out of the nozzle, fitted at the other end of the rocket. This simple model is shown in Fig. 9.2.

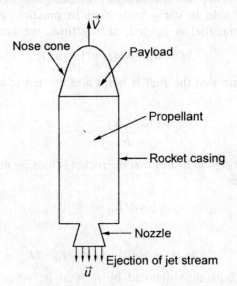

Figure 9.2 Simple model of a single-stage rocket.

Let the exhaust velocity, that is, the velocity of the burnt fuel or propellant relative to the rocket casing be $\vec{v}_e$, then,

$$\vec{V} - \vec{u} = \vec{v}_e, \qquad (9.4)$$

where $\vec{V}$ is the velocity of the rocket and $\vec{u}$ is the velocity of the fuel. Substituting Eq. (9.4) into Eq. (9.3) and considering the motion in vertical direction, the equation of motion of the rocket of mass $M(t)$ moving with velocity $\vec{V}$, subject to an external force $\vec{F}$, is given as

$$\vec{F} = M\frac{d\vec{V}}{dt} + \vec{v}_e\frac{dM}{dt}. \qquad (9.5)$$

Here, M is the total mass of the rocket including the payload. To understand some of the important characteristics of a rocket, we shall, for the time being, neglect all external forces and thus, the equation of motion of the rocket becomes

$$M(t)\frac{d\vec{V}}{dt} = -\vec{v}_e\frac{dM}{dt}, \qquad (9.6)$$

where

$$M = M_p + M_f(t) + M_s. \qquad (9.7)$$

Here, M_p denotes the mass of the payload, M_f denotes the mass of the fuel and M_s the mass of the structure or casing of the rocket. Let $M_f + M_s = M_0$. Initially, let $M_f(0) = \varepsilon M_0$, so that $M_s = (1-\varepsilon)M_0$. Here, $(1-\varepsilon)$ is called the structural factor and the parameter ε is the ratio of initial fuel mass to the initial rocket mass. Obviously $0 < \varepsilon < 1$ and it is preferable to have ε, nearly equal to 1, in the sense that the structure should be as light as possible such that we shall be able to carry more fuel. In practice, ε lies in the range $(0.7, 0.9)$. Once the fuel is ignited, at any time, we have

$$M = M_p + M_f(t) + (1-\varepsilon)M_0. \qquad (9.8)$$

Now, let us assume that the fuel is burnt and ejected at a constant rate, say k, then

$$\frac{dM_f(t)}{dt} = -k. \qquad (9.9)$$

Here, the negative sign indicates that the rocket is loosing mass. On integration, Eq. (9.9) becomes

$$[M_f(t)]_0^t = -kt$$

or

$$M_f(t) = M_f(0) - kt = \varepsilon M_0 - kt. \qquad (9.10)$$

Suppose at $t = t$ (burn-out) denoted by t_b, that is, when the entire fuel is burnt, in which case $M_f(t_b) = 0$, Eq. (9.10) gives

$$t_b = \frac{\varepsilon M_0}{k}. \qquad (9.11)$$

Utilizing Eq. (9.10) into Eq. (9.8), we get the rocket mass at any time t in the interval $(0, t_b)$ as

$$M(t) = M_p + M_0 - kt. \qquad (9.12)$$

Substituting this expression for $M(t)$ into Eq. (9.6), we have

$$(M_p + M_0 - kt)\frac{d\vec{V}}{dt} = \vec{v}_e k \qquad (9.13)$$

or

$$[M_p + M_f(0) + M_s - kt]\frac{d\vec{V}}{dt} = \vec{v}_e k, \qquad (9.14)$$

which at once gives that

$$\frac{d\vec{V}}{dt} = \frac{\vec{v}_e k}{M_p + M_f(0) + M_s - kt}.$$

Integrating with respect to t between the limits 0 to t_b, we find

$$[\vec{V}]_0^{t_b} = -\vec{v}_e \{\log[M_p + M_f(0) + M_s - kt]\}_0^{t_b}$$

$$= -\vec{v}_e \log\left[\frac{M_p + M_f(0) + M_s - kt_b}{M_p + M_f(0) + M_s}\right].$$

Using Eq. (9.11), we have

$$\Delta \vec{V} = -\vec{v}_e \log\left[\frac{M_p + M_f(0) + M_s - \varepsilon M_0}{M_p + M_f(0) + M_s}\right]$$

$$= -\vec{v}_e \log\left[1 - \frac{\varepsilon M_0}{M_p + M_f(0) + M_s}\right].$$

Therefore,

$$\Delta \vec{V} = -\vec{v}_e \log\left[1 - \frac{\varepsilon M_0}{M_p + \varepsilon M_0 + (1-\varepsilon)M_0}\right].$$

Thus, the velocity-increment caused by the expenditure of the entire propellant is given by

$$\Delta \vec{V} = -\vec{v}_e \log\left(1 - \frac{\varepsilon M_0}{M_p + M_0}\right). \qquad (9.15)$$

In case of no payload

$$\Delta \vec{V} = -\vec{v}_e \log(1 - \varepsilon). \qquad (9.16)$$

From Eq. (9.15), it is clear that the performance of the rocket depends on
(i) $\vec{v}_e$, the exhaust velocity
(ii) $(1-\varepsilon)$, the structural factor
(iii) $M_p/M_0 = M_p/[M_f(0) + M_s] =$ mass ratio.

9.3.1 Exhaust Speed Parameters

Thrust The thrust T of a rocket motor (engine) is defined as the rate of change of momentum of propellant or fuel on burning (assuming that it is ejected at a constant rate, k). That is,

$$T = (\vec{V} - \vec{u})\frac{dM}{dt} = \vec{v}_e k. \tag{9.17}$$

Specific impulse On firing a rocket motor at sea level (on Earth), one can measure the thrust of a rocket motor with time until its propellant is exhausted, that is, until its burn-out time.

Thus, a rocket motor as it burns its propellant exerts a thrust $T(t)$ during the burn-out time, t_b (see Fig. 9.3) and delivers a total impulse I, which is defined as

$$I = \int_0^{t_b} T(t)\, dt. \tag{9.18}$$

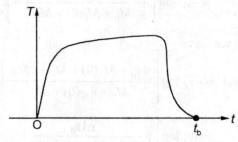

Figure 9.3 A typical thrust-time curve.

Let the propellant mass of a typical rocket engine be M_f. Then, the impulse developed by the rocket engine per unit weight of the propellant, that is per 1 kg of propellant (working fluid) at sea level is called *specific thrust* and is also written as

$$I = \int_0^{t_b} \frac{T(t)\, dt}{M_f g} = \frac{T}{M_f g}, \tag{9.19}$$

where g is the standard surface gravity of the Earth. But,

$$M_f = t_b k. \tag{9.20}$$

Therefore,

$$I = \frac{T}{t_b kg}. \tag{9.21}$$

Performance of a Single-stage Rocket (Sec. 9.3)

Now, the *specific impulse*, I_{sp}, is defined as the thrust produced per unit weight of fuel consumed per second. That is,

$$I_{sp} = \frac{T}{kg}. \tag{9.22}$$

Eliminating k between Eqs. (9.17) and (9.22), we at once get

$$I_{sp} = \frac{\vec{v}_e}{g}. \tag{9.23}$$

It may be noted that specific impulse is the most important characteristic of a rocket engine. There is a difference between specific impulse, I_{sp}, in vacuum and in atmosphere (e.g. on Earth).

At this point, let us look at the dimensions of I_{sp}.

$$\dim(T) = \frac{ML}{T^2},$$

$$\dim(k) = \dim\left(\frac{dM}{dt}\right) = \frac{M}{T},$$

$$\dim(g) = \frac{L}{T^2}.$$

Therefore, from Eq. (9.22), we have

$$\dim(I_{sp}) = \frac{ML}{T^2} \frac{T}{M} \frac{T^2}{L} = T(s) \tag{9.24}$$

which indicates that the units of I_{sp} in the SI system is seconds. The I_{sp}s for of few known solid and liquid propellants are listed in the Table 9.1.

Table 9.1 List of a Few Known Propellants

Rocket motor	Propellant	$I_{sp}(s)$	Remarks
Solid	Polyurethane and ammonium perchlorate	250	
Liquid	Hydrogen peroxide and kerosene	265	The storage tanks should be perfectly clean. Some fire and explosive hazards are reported.
Liquid	Liquid oxygen and liquid hydrogen	385	Fire and explosive risks and severe cold burns can occur.
Liquid	Liquid fluorine and liquid hydrogen	425	Liquid fluorine reacts vigorously with most substances.

For example, if a propellant has an I_{sp} of 385 s, the burning of 1 kg of propellant will produce a thrust of 385-kg force or kgf. In India, we have used solid propellant in our SLV programme, and both liquid and solid propellants in our GSLV programme. Solid fuels are relatively cheaper, easy to handle and simple in making solid propellant rocket motors. Solid engines have the disadvantage of low-specific impulse and their thrust is not easily controllable, while, in the case of liquid engines, we can achieve high I_{sp}, though the cost is high and its design is complex, with a particular advantage that its thrust is controllable. In the best solid engines of today, I_{sp} in vacuum reaches 250–300 s, while for liquid-propellant engines, it reaches 350–450 s. Still higher values of specific impulse can be achieved by designing nuclear or electric rocket engines.

9.3.2 Effect of Gravity

If we assume that the only external force acting on the rocket is the gravitational force, which can be taken as constant (approximately), the equation of motion as described in Eqs. (9.5) and (9.13) for a single-stage rocket ascending vertically can be written as

$$(M_p + M_0 - kt)\frac{d\vec{V}}{dt} = \vec{v}_e k - (M_p + M_0 - kt)g. \qquad (9.25)$$

Integrating with respect to t from 0 to t_b, we obtain

$$[\vec{V}]_0^{t_b} = \vec{v}_e \int_0^{t_b} \frac{k\,dt}{M_p + M_0 - kt} - \int_0^{t_b} g\,dt$$

or

$$\Delta \vec{V} = -\vec{v}_e \log [M_p + M_0 - kt]_0^{t_b} - gt_b$$

that is,

$$\Delta \vec{V} = -\vec{v}_e \log\left(1 - \frac{kt_b}{M_p + M_0}\right) - gt_b. \qquad (9.26)$$

Eliminating t_b with the help of Eq. (9.11), the above equation can be recast into

$$\Delta \vec{V} = -\vec{v}_e \log\left(1 - \frac{\varepsilon M_0}{M_p + M_0}\right) - \frac{g\varepsilon M_0}{k}. \qquad (9.27)$$

This equation gives the velocity increment caused by the expenditure of the entire propellant.

In order to demonstrate the effect of the external gravitational force on the attainable velocity of a single-stage rocket, we shall consider the following example:

Suppose $\varepsilon = 0.8$, $M_p/M_0 = 1/100$, $I_{sp} = 100$ s then, recalling $I_{sp} = \bar{v}_e/g$, the exhaust velocity $\bar{v}_e$ can be computed. Thus,
$$v_e = I_{sp}g = 300 \times 9.80 = 2940 \text{ m/s}.$$
With this data, in the absence of gravity, Eq. (9.27) gives

$$\Delta \vec{V} = -v_e \log\left(1 - \frac{\varepsilon}{\frac{M_p}{M_0} + 1}\right)$$

$$= -2940 \log\left(1 - \frac{0.8}{1.01}\right)$$

$$= (-2940)(-1.570598)$$

$$= 4617.6 \text{ m/s}.$$

Roughly, the velocity achieved by the single-stage rocket is 4.6 km/s.

Now, considering $M_0 = 10^5$ kg and $k = 5 \times 10^3$ kg/s, the second term in Eq. (9.27) due to gravity is

$$\frac{g \varepsilon M_0}{k} = \frac{9.8 \times 0.8 \times 10^5}{5 \times 10^3} = 156.8 \text{ m/s}$$

or roughly 0.16 km/s. This clearly demonstrates that the effect of external gravitational force on the attainable terminal velocity is very small. Thus, the predictions made by neglecting gravity are found to be reasonably accurate.

To compute the vertical height h attained by the rocket, at burn out, we go back to Eq. (9.26) and write

$$\Delta \vec{V} = \frac{dh}{dt} = -\bar{v}_e \log\left(1 - \frac{kt}{M_p + M_0}\right) - gt. \tag{9.28}$$

Integrating by parts rule from 0 to t, we have

$$h = -\frac{1}{2}gt^2 - v_e\left[\log\left(1 - \frac{kt}{M_p + M_0}\right)t + \int_0^t \frac{tk/(M_p + M_0)}{1 - kt/(M_p + M_0)} dt\right]$$

or

$$h = -\frac{1}{2}gt^2 - v_e\left[t\log\left(1 - \frac{kt}{M_p + M_0}\right) + \frac{k}{M_p + M_0}\int_0^t \frac{t\, dt}{1 - kt/(M_p + M_0)}\right]. \tag{9.29}$$

Let, $kt/(M_p + M_0) = U$, then,

$$\frac{k}{M_p+M_0}\int\frac{t\,dt}{1-kt/(M_p+M_0)}=\int\frac{U}{1-U}\frac{M_p+M_0}{k}\,dU$$

$$=\frac{M_p+M_0}{k}\int\frac{U\,dU}{1-U}$$

$$=\frac{M_p+M_0}{k}\left(\int -dU+\int\frac{dU}{1-U}\right)$$

$$=\frac{M_p+M_0}{k}[-U-\log(1-U)]$$

$$=-\frac{M_p+M_0}{k}\left[\frac{kt}{M_p+M_0}+\log\left(1-\frac{kt}{M_p+M_0}\right)\right].$$

Thus, Eq. (9.29) becomes

$$h=-\frac{1}{2}gt^2-v_e\left[t\log\left(1-\frac{kt}{M_p+M_0}\right)-t-\frac{M_p+M_0}{k}\log\left(1-\frac{kt}{M_p+M_0}\right)\right].$$

That is,

$$h=-\frac{1}{2}gt^2+v_e t-v_e\left(t-\frac{M_p+M_0}{k}\right)\log\left(1-\frac{kt}{M_p+M_0}\right). \qquad (9.30)$$

But, at burn out

$$t_b=\frac{\varepsilon M_0}{k}.$$

Hence, at burn out, the height attainable by a single-stage rocket is found to be

$$h=-\frac{g\varepsilon^2 M_0^2}{2k^2}+\frac{v_e\varepsilon M_0}{k}+\frac{v_e}{k}[M_0(1-\varepsilon)+M_p]\log\left(1-\frac{\varepsilon M_0}{M_p+M_0}\right). \qquad (9.31)$$

Using the typical data given earlier, this expression gives that the height attainable by a single-stage rocket is roughly 30 km, which is very small when compared with the dimensions of the Earth.

Example 9.1 A rocket engine of mass 3×10^5 kg can eject exhaust gases with velocity 3000 m/s and at a rate of 10^4 kg/s. The rocket is fired vertically from the surface of the Earth. Show that such a rocket is not capable of escaping from the Earth's gravitational field, assuming that the gravitational force acting on the rocket during the flight is constant. Also, assume that the rocket is not carrying any payload.

Performance of a Single-stage Rocket (Sec. 9.3)

Solution Given the following data: $F = -mg$, $v_e = 3000$ m/s, $M = 3 \times 10^5$ kg and $dm/dt = k = 10^4$ kg/s, the equation of motion of the rocket in the presence of external gravitational force, as given in Eq. (9.25), can be written as

$$(M_p + M_0 - kt)\frac{dV}{dt} = v_e k - (M_p + M_0 - kt)g. \tag{1}$$

In the absence of payload, the equation of motion simplifies to

$$(M_0 - kt)\frac{dV}{dt} = v_e k - (M_0 - kt)g \tag{2}$$

or

$$\frac{dV}{dt} = \frac{v_e k}{M_0 - kt} - g. \tag{3}$$

Integration of this equation yields

$$[V]_0^V = -v_e [\log(M_0 - kt)]_0^t - g[t]_0^t + \text{constant}.$$

Using the condition $V = 0$ when $t = 0$, we get

$$V = -v_e \log(M_0 - kt) + v_e \log M_0 - gt. \tag{4}$$

Now, noting that $V = dh/dt$, and integrating by parts rule, we get

$$h = -v_e \int_0^t \log(M_0 - kt)\, dt + v_e \int_0^t \log M_0 \, dt - \frac{1}{2}gt^2.$$

Noting that M_0 depends on time only implicitly but not explicitly, we have

$$h = -v_e \left[t \log(M_0 - kt) - \int_0^t \frac{t(-k)\, dt}{M_0 - kt} \right] + v_e [t \log M_0] - \frac{gt^2}{2}$$

$$= -v_e t \log\left(\frac{M_0 - kt}{M_0}\right) + v_e \int_0^t \left(1 - \frac{M_0}{M_0 - kt}\right) dt - \frac{gt^2}{2}$$

$$= -v_e t \log\left(\frac{M_0 - kt}{M_0}\right) + v_e t + \left[\frac{v_e M_0}{k} \log(M_0 - kt)\right]_0^t - \frac{gt^2}{2}.$$

That is,

$$h = -v_e t \log\left(\frac{M_0 - kt}{M_0}\right) + v_e t + \frac{v_e M_0}{k} \log\left(\frac{M_0 - kt}{M_0}\right) - \frac{gt^2}{2}. \tag{5}$$

Assuming that the maximum ratio of fuel mass to total mass of the rocket is about 0.9, which means that the maximum fuel capacity is $3 \times 10^5 \times 0.9 = 2.7 \times 10^5$ kg. Further, let the fuel be exhausted in t seconds, then,

$$t = \frac{\text{Total fuel}}{\text{Mass flow rate of the fuel}} = \frac{2.7 \times 10^5}{10^4} = 27 \text{ s}. \tag{6}$$

During this period, the height attained by the rocket can be computed from Eq. (5) as

$$h = -3000 \times 27 \log\left(\frac{3 \times 10^5 - 10^4 \times 27}{3 \times 10^5}\right) + (3000 \times 27)$$

$$+ \frac{3000 \times 3 \times 10^5}{10^4} \log\left(\frac{3 \times 10^5 - 10^4 \times 27}{3 \times 10^5}\right) - \frac{(9.8)(27)^2}{2}$$

or

$$h = 98128 \text{ m} \simeq 98 \text{ km}. \tag{7}$$

At this height, the rocket will be moving with a speed given by Eq. (9.16). In this example, it is found to be

$$V = -v_e \log(1-\varepsilon) = -3000 \log\left(1 - \frac{9}{10}\right)$$

or

$$V = 6900 \text{ m/s} = 6.9 \text{ km/s}. \tag{8}$$

But, the escape velocity from the surface of the Earth is about 11.2 km/s. Hence, the single-stage rocket with the given capacity is not capable of escaping from the gravitational field of the Earth.

Even this result, that is, $V = 6.9$ km/s is over-optimistic, because, we have ignored aerodynamic forces in our computations. Moreover, this vehicle is not carrying any payload. In fact, any practical orbital mission should necessarily carry some payload. From these arguments, it is clear that a single-stage rocket has a limited practical application in orbital studies. The reason being that we are accelerating the entire dead weight (structure weight) up to the final burn-out velocity. Therefore, what is needed is a vehicle or a rocket, which sheds useless weight progressively during the burn-out phase which is achieved by means of multi-staging.

Example 9.2 A rocket of total mass M contains a proportion, εM ($0 < \varepsilon < 1$), as fuel. If the exhaust speed, v_e, is constant, show that the final speed of the rocket is independent of the rate at which the fuel is burnt (neglect the payload mass).

Solution Neglecting the payload mass, the equation of motion of the rocket is

$$(M_0 - kt)\frac{dV}{dt} = v_e k - (M_0 - kt)g, \tag{1}$$

where k is the rate of fuel burnt. Thus,

$$\frac{dV}{dt} = \frac{v_e k}{M_0 - kt} - g.$$

Integration of this equation with respect to t, yields

$$V = -v_e \left[\log (M_0 - kt)\right]_0^t - gt$$

that is,

$$V = -v_e \log\left(\frac{M_0 - kt}{M_0}\right) - gt.$$

But $M_0 = M_f + M_s$. Suppose at $t = \tau$, the whole fuel is burnt, then, we find

$$V = -v_e \log\left(\frac{M_s}{M_0}\right) - g\tau, \qquad (2)$$

which is independent of k, the rate at which the fuel is burnt.

9.4 PERFORMANCE OF A TWO-STAGE ROCKET

A simple model for a typical two-stage rocket is depicted in Fig. 9.4. The first stage is also called *booster*. When all the fuel in the booster motor is exhausted, the booster along with its instruments will be detached from the upper second-stage and soon the second stage is ignited.

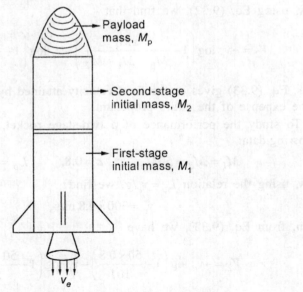

Figure 9.4 A simple two-stage rocket model.

Suppose that the mass of the first stage is M_1, with fuel mass εM_1 and the second stage has a mass M_2 with fuel mass εM_2. Utilizing Eq. (9.15), which we have derived for a single-stage rocket, in the absence of any external force, we can find the increment in velocity attained at the burnout of the entire propellant of the first stage, in time t_1 (say) by the expression

332 Rocket Dynamics (Ch. 9)

$$\Delta V_1 = -v_e \log\left(1 - \frac{\varepsilon M_1}{M_1 + M_2 + M_p}\right), \qquad (9.32)$$

where M_p is the mass of the payload. Assuming that the exhaust speed of the ejected gas for each stage is same, and equal to v_e, the time required for the burn-out of the second stage fuel is

$$t_2 = \frac{\varepsilon M_2}{k},$$

where k is the rate of fuel burnt. Now, integrating the equation of motion, that is, Eq. (9.25), ignoring all the external forces, between the limits 0 to t_2 (here 0 corresponds to the time of second-stage ignition), we at once get

$$\int_{V_1}^{V_2} dV = v_e \int_0^{t_2} \frac{k\,dt}{M_p + M_2 - kt}$$

that is,

$$V_2 - V_1 = -v_e \log[M_p + M_2 - kt]_0^{t_2} = -v_e \log\left(1 - \frac{\varepsilon M_2}{M_p + M_2}\right).$$

Now, using Eq. (9.32), we find that

$$V_2 = -v_e \log\left(1 - \frac{\varepsilon M_1}{M_p + M_1 + M_2}\right) - v_e \log\left(1 - \frac{\varepsilon M_2}{M_p + M_2}\right). \qquad (9.33)$$

Thus, Eq. (9.33) gives us the final velocity attained by a two-stage rocket, at the expense of the entire propellant.

To study the performance of a two-stage rocket, let us consider the following data:

$$M_1 = M_2 = 50 M_p, \qquad \varepsilon = 0.8, \qquad I_{sp} = 300 \text{ s}.$$

Now, using the relation $I_{sp} = v_e/g$, we find

$$v_e = 300 \times 9.8 \text{ m/s}.$$

Then, from Eq. (9.33), we have

$$V_2 = -v_e \log\left(1 - \frac{50 \times 0.8}{101}\right) - v_e \log\left(1 - \frac{50 \times 0.8}{51}\right)$$

$$= -300 \times 9.8 \left(\log \frac{61}{101} + \log \frac{11}{51}\right)$$

$$= -300 \times 9.8 \, (-2.038)$$

$$= 5991.72 \text{ m/s}$$

or
$$V_2 = 6.0 \text{ km/s}.$$
This is certainly a good improvement over a single-stage rocket, but still, this velocity is not sufficient to place a payload (satellite) into an Earth-bound orbit (we require at least 7.6 km/s).

In the above example, of course, we have chosen $M_1 : M_2 = 1 : 1$. Instead, if we choose some appropriate ratio, it might produce a considerable increase in the final velocity. This idea leads to an optimization problem as discussed below: Let $M_1 + M_2 = M$, then, Eq. (9.33) can be recast into

$$V_2 = -v_e \log\left(1 - \varepsilon \frac{M - M_2}{M + M_p}\right) - v_e \log\left(1 - \frac{\varepsilon M_2}{M_2 + M_p}\right). \quad (9.34)$$

Suppose, we assume that the total mass M is fixed, then we can see that V_2 is effectively a function of M_2 only. For maxima or minima of V_2, the necessary condition is

$$\frac{dV_2}{dM_2} = 0.$$

Therefore,

$$\frac{dV_2}{dM_2} = -v_e \left\{ \frac{\varepsilon/(M+M_p)}{1-\varepsilon(M-M_2)/(M+M_p)} \right\} + v_e \left\{ \frac{[(M_2+M_p)\varepsilon - \varepsilon M_2]/(M_2+M)^2}{1-\varepsilon M_2/(M_2+M_p)} \right\} = 0$$

That is,

$$\frac{\varepsilon}{M + M_p - \varepsilon M + \varepsilon M_2} = \frac{\varepsilon M_p}{(M_2 + M_p)(M_2 - \varepsilon M_2 + M_p)}.$$

Further simplification gives

$$(1-\varepsilon)(M_2^2 + 2M_p M_2 - M M_p) = 0.$$

Since $\varepsilon \neq 1$, obviously we should have

$$M_2^2 + 2M_p M_2 - M_p M = 0, \quad (9.35)$$

which is a quadratic in M_2 and its solution gives

$$M_2 = \frac{-2M_p \pm \sqrt{4M_p^2 + 4M_p M}}{2} = -M_p \pm \sqrt{M_p^2 + M_p M}. \quad (9.36)$$

Dividing both sides by M and considering the positive sign, we obtain

$$\frac{M_2}{M} = -\frac{M_p}{M} + \left(\frac{M_p^2}{M^2} + \frac{M_p}{M}\right)^{1/2}.$$

Assuming that the ratio (M_p/M) is small, which is in fact true always, the above expression can be rewritten and expanded, using binomial theorem, as

$$\frac{M_2}{M} = -\frac{M_p}{M} + \left(\frac{M_p}{M}\right)^{1/2}\left(1 + \frac{1}{2}\frac{M_p}{M} + \cdots\right)$$

or

$$\frac{M_2}{M} = \left(\frac{M_p}{M}\right)^{1/2} - \frac{M_p}{M} + \frac{1}{2}\left(\frac{M_p}{M}\right)^{3/2} + \cdots \quad (9.37)$$

As a first approximation, we take

$$\frac{M_2}{M} = \left(\frac{M_p}{M}\right)^{1/2}. \quad (9.38)$$

Now, suppose we assume that $(M_p/M) = (1/100)$, then Eq. (9.38) gives

$$\frac{M_2}{M} = \frac{1}{10} \quad (9.39)$$

and

$$\frac{M_1}{M} = 1 - \frac{1}{10} = \frac{9}{10}. \quad (9.40)$$

Evidently, Eqs. (9.39) and (9.40) suggest that

$$M_1 : M_2 = 9 : 1. \quad (9.41)$$

This optimal choice of the ratio clearly dictates that the mass of the boosting first stage must be very large compared to that of the second stage.

To compute the optimal velocity attained at the burn-out of the second stage with the above data, we proceed as follows: Let

$$\frac{M_p}{M} = \beta, \quad (9.42)$$

which is of course small, then Eq. (9.37) becomes

$$\frac{M_2}{M} = \beta^{1/2} - \beta + O(\beta)^{3/2}. \quad (9.43)$$

Now, we shall demonstrate below that the optimal choice of β will lead to a substantital increase in the terminal velocity at the burn-out of the second stage.

Using Eqs. (9.42) and (9.43) into Eq. (9.34), we get

$$V_2 = -v_e \log\left[1 - \frac{\varepsilon(1 - \beta^{1/2} + \beta + O(\beta)^{3/2}}{1 + \beta}\right]$$
$$- v_e \log\left\{1 - \frac{\varepsilon[\beta^{1/2} - \beta + O(\beta)^{3/2}]}{\beta^{1/2} + O(\beta)^{3/2}}\right\}.$$

or

$$V_2 \simeq -v_e \log\{1-\varepsilon[1-\beta^{1/2}+O(\beta)]\} - v_e \log\left\{1-\frac{\varepsilon[1-\beta^{1/2}+O(\beta)]}{1+O(\beta)}\right\}$$

that is,

$$V_2 \simeq -2v_e \log[1-\varepsilon(1-\beta^{1/2})]. \tag{9.44}$$

Substituting $M_p/M = 1/100$, $\varepsilon = 0.8$, $I_{sp} = 300$ s into Eq. (9.44), we find the optimal value of V_2 as 7.6 km/s, which is a considerable improvement in the previously computed result. It is therefore suggested that adding another stage would further increase the terminal velocity to take care of any planned orbital mission around the Earth.

9.5 OPTIMIZATION OF A MULTI-STAGE ROCKET

For designing a multi-stage rocket, the most crucial problem is to plan in advance the rocket thrust programme of various stages so as to give the payload the necessary injection velocity, say V_τ, at a specified position, in order to realize or achieve the objective of a space mission. To study this problem, let the masses of the individual stages of a multi-stage rocket (see Fig. 9.5), say, that of the ith stage be denoted by M_i. Let the total mass of the multi-stage rocket be M, excluding the payload.

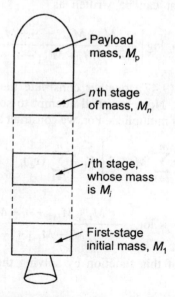

Figure 9.5 A simple model of a multi-stage rocket.

Let the total initial mass of the ith stage be M_i, containing the fuel $\varepsilon_i M_i$ ($1 \le i \le n$). Let $(v_e)_i$ stand for the exhaust speed of the ith-stage motor. The optimization problem in this case can be stated as:

How to choose the individual masses M_i so as to minimize

$$M = \sum_{i=1}^{n} M_i \qquad (9.45)$$

subject to the constraint that the final velocity attained by the rocket, at the burn-out of the nth stage is V_τ.

Following the discussion of Section 9.3, we can write that the velocity increment at the burn-out of the ith stage, as

$$\Delta V_i = -(v_e)_i \log\left(1 - \frac{\varepsilon_i M_i}{M_i + M_{i+1} + \cdots + M_n + M_p}\right),$$

where M_p is the mass of the payload. That is,

$$\Delta V_i = -(v_e)_i \log\left[\frac{(1-\varepsilon_i)M_i + M_{i+1} + \cdots + M_n + M_p}{M_i + M_{i+1} + \cdots + M_n + M_p}\right]$$

or

$$\Delta V_i = -(v_e)_i \log\left[\frac{M_i + M_{i+1} + \cdots + M_n + M_p}{(1-\varepsilon_i)M_i + M_{i+1} + \cdots + M_n + M_p}\right]. \qquad (9.46)$$

Therefore, the constraint can be written as

$$V_\tau = \sum_{i=1}^{n} (v_e)_i \log\left[\frac{M_i + M_{i+1} + \cdots + M_n + M_p}{(1-\varepsilon_i)M_i + M_{i+1} + \cdots + M_n + M_p}\right]. \qquad (9.47)$$

Thus, Eqs. (9.45) and (9.47) together constitute the optimization problem with equality constraint. Now, we shall attempt to solve this problem using the Lagrange's method of multipliers. For, we construct a Lagrangian function as

$$L(M_1, M_2, \ldots, M_n, \lambda) = \sum_{i=1}^{n} M_i + \lambda\left\{V_\tau - \sum_{i=1}^{n} (v_e)_i \right.$$
$$\left. \times \log\left[\frac{M_i + M_{i+1} + \cdots + M_n + M_p}{(1-\varepsilon_i)M_i + M_{i+1} + \cdots + M_n + M_p}\right]\right\} \qquad (9.48)$$

and seek the minima of this function by solving the necessary conditions. That is, by solving

$$\frac{\partial L}{\partial M_i} = 0, \qquad (i = 1, 2, \ldots, n). \qquad (9.49)$$

Before, we do that, for simplicity, let us introduce the notation

$$\mu_i = \frac{M_i + M_{i+1} + \cdots + M_n + M_p}{(1-\varepsilon_i)M_i + M_{i+1} + \cdots + M_n + M_p} \qquad (9.50)$$

so that the constraint, that is, Eq. (9.47) reduces to

$$V_\tau - \sum_{i=1}^{n} (v_e)_i \log \mu_i = 0. \qquad (9.51)$$

Also, let us express M, in terms of μ_i, by noting that

$$\frac{M_i + M_{i+1} + \cdots + M_n + M_p}{M_{i+1} + M_{i+2} + \cdots + M_n + M_p}$$

$$= \frac{\varepsilon_i(M_i + M_{i+1} + \cdots + M_n + M_p)}{(1-\varepsilon_i)M_i + M_{i+1} + \cdots + M_n + M_p - (1-\varepsilon_i)(M_i + M_{i+1} + \cdots + M_n + M_p)}$$

$$= \frac{\varepsilon_i \mu_i}{1 - (1-\varepsilon_i)\mu_i}. \qquad (9.52)$$

Further, we rewrite $(M + M_p)/M_p$ as

$$\frac{M + M_p}{M_p} = \frac{M_1 + M_2 + \cdots + M_n + M_p}{M_2 + M_3 + \cdots + M_n + M_p} \cdot \frac{M_2 + M_3 + \cdots + M_n + M_p}{M_3 + M_4 + \cdots + M_n + M_p}$$

$$\times \frac{M_3 + M_4 + \cdots + M_n + M_p}{M_4 + M_5 + \cdots + M_n + M_p} \cdots \frac{M_{n-1} + M_n + M_p}{M_n + M_p} \cdot \frac{M_n + M_p}{M_p}$$

that is, using compact notation, we write

$$\frac{M + M_p}{M_p} = \prod_{i=1}^{n} \left(\frac{M_i + M_{i+1} + \cdots + M_n + M_p}{M_{i+1} + M_{i+2} + \cdots + M_n + M_p} \right).$$

Now, introducing Eq. (9.52), the above equation becomes

$$\frac{M + M_p}{M_p} = \prod_{i=1}^{n} \frac{\varepsilon_i \mu_i}{1 - (1-\varepsilon_i)\mu_i}. \qquad (9.53)$$

As M_p is constant, the problem of minimizing of M is equivalent to minimizing $(M + M_p)/M_p$. That is also equivalent to minimizing $\log[(M + M_p)/M_p]$. But,

$$\log\left(\frac{M + M_p}{M_p}\right) = \sum_{i=1}^{n} [\log \varepsilon_i + \log \mu_i - \log[1 - (1-\varepsilon_i)\mu_i]]. \qquad (9.54)$$

Adding λ times the constraint, that is, λ times of Eq. (9.51) to Eq. (9.54), we can recast the Lagrange function as

$$L(\mu_1, \mu_2, \ldots, \mu_n, \lambda) = \sum_{i=1}^{n} \{\log \varepsilon_i + \log \mu_i - \log[1 - \mu_i(1 - \varepsilon_i)]\}$$

$$+ \lambda \left[\sum_{i=1}^{n} (v_e)_i \log \mu_i\right] - \lambda V_\tau$$

$$= \sum_{i=1}^{n} \{[1 + \lambda (v_e)_i] \log \mu_i + \log \varepsilon_i$$

$$- \log[1 - \mu_i(1 - \varepsilon_i)]\} - \lambda V_\tau, \tag{9.55}$$

where λ is a Lagrange multiplier. Differentiating Eq. (9.55) with respect to μ_i and equating to zero, we can determine the values of μ_i, which minimizes M, of course in terms of λ. Thus, we have

$$\frac{\partial L}{\partial \mu_i} = \frac{1 + \lambda (v_e)_i}{\mu_i} + \frac{1 - \varepsilon_i}{1 - \mu_i(1 - \varepsilon_i)} = 0.$$

On solving for μ_i, we get

$$[1 + \lambda (v_e)_i][1 - \mu_i(1 - \varepsilon_i)] = \mu_i \varepsilon_i - \mu_i,$$

which gives

$$\mu_i = \frac{1 + \lambda (v_e)_i}{\lambda (v_e)_i (1 - \varepsilon_i)}. \tag{9.56}$$

In order to find λ, we substitute this value of μ_i into Eq. (9.51) and get

$$V_\tau = \sum_{i=1}^{n} (v_e)_i \log\left[\frac{1 + \lambda (v_e)_i}{\lambda (v_e)_i (1 - \varepsilon_i)}\right]. \tag{9.57}$$

In this equation, everything is known except λ, which can be computed numerically. Having known λ, we can determine μ_i from Eq. (9.56). For simplicity, we may take $(v_e)_1 = (v_e)_2 = \cdots = (v_e)_n = v_e$ (the exhaust speeds) and also equal-structural factors, that is,

$$\varepsilon_1 = \varepsilon_2 = \cdots = \varepsilon_n = \varepsilon \text{ (say)}.$$

Then, Eq. (9.56) becomes

$$\mu_1 = \mu_2 = \cdots = \mu_n = \frac{1 + \lambda v_e}{\lambda v_e(1 - \varepsilon)} = \mu \text{ (say)}. \tag{9.58}$$

Under these conditions, Eq. (9.51) simplifies to

$$V_\tau = n v_e \log \mu$$

or

$$\mu = \exp\left(\frac{V_\tau}{nv_e}\right). \tag{9.59}$$

Using Eq. (9.58) into Eq. (9.59), we have

$$\frac{1+\lambda v_e}{\lambda v_e(1-\varepsilon)} = \exp\left(\frac{V_\tau}{nv_e}\right)$$

or

$$\lambda\left[v_e(1-\varepsilon)\exp\left(\frac{V_\tau}{nv_e}\right) - v_e\right] = 1$$

which gives

$$\lambda = \frac{1}{\{(1-\varepsilon)\exp[V_\tau/(nv_e)] - 1\}v_e}. \tag{9.60}$$

Using Eq. (9.59) into Eq. (9.53), we find that

$$\frac{M+M_p}{M_p} = \left(\frac{\varepsilon\exp[V_\tau/nv_e]}{\{1-(1-\varepsilon)\exp[V_\tau/(nv_e)]\}}\right)^n$$

or

$$M = M_p\left(\frac{\varepsilon^n\exp(V_\tau/v_e)}{\{1-(1-\varepsilon)\exp[V_\tau/(nv_e)]\}^n} - 1\right). \tag{9.61}$$

This equation gives us the optimal mass of a typical multi-stage vehicle needed to give to the specified payload M_p, the required injection velocity V_τ.

Launch-site selection The Earth rotates from west to east. The man-made satellites also go round the Earth in the same direction. When satellites are launched towards east, the satellite gains incremental velocity equivalent to that of the Earth's velocity as it leaves the Earth's atmosphere. To achieve this, the launch site should be ideally located on the east coast, preferably near the Equator for launching communication satellites in particular.

Also, the launch site must be situated on a sea coast, as during the course of rocket flight, in particular during stage separations of a multi-stage rocket, a lot of spent hardware of lower stages will be shed which will fall into the sea, from the safety point of view.

Example 9.3 Using the typical values for the parameters such as $V_\tau = 7.8$ km/s, $I_{sp} = 300$ s, $\varepsilon = 0.8$, find the optimal mass M required for 2-, 3-, 4-; and 5-stage vehicles in terms of payload mass, M_p. Give comments on the best choices of these vehicles. Also, find the ratio of masses in the case of 2- and 3-stage vehicle (rocket).

Solution We are given the following data:
$$V_\tau = 7.8 \text{ km/s}, \qquad I_{sp} = 300 \text{ s}, \qquad \varepsilon = 0.8 \tag{1}$$

Noting that $I_{sp} = v_e/g$, the exhaust speed v_e is
$$v_e = I_{sp} g = 300 \times 9.8 = 2940 \text{ m/s}.$$

Suppose, we are looking for a four-stage vehicle, that is, given $n = 4$. Using the formula given by Eq. (9.61), we have

$$M = M_p \left(\frac{\varepsilon^n \exp(V_\tau/v_e)}{\{1-(1-\varepsilon)\exp[V_\tau/(nv_e)]\}^n} - 1 \right). \tag{2}$$

Substituting the given data into Eq. (2) for $n = 4$, we get

$$M = M_p \left(\frac{(0.8)^4 \exp(7.8 \times 1000/2940)}{1 - \{1 - 0.8 \exp[(7.8 \times 1000)/(4 \times 2940)]\}^4} - 1 \right)$$

or

$$M = M_p \left[\frac{0.4096 \times 14.197}{(1 - 0.2 \times 1.941)^4} - 1 \right] = M_p (41.506 - 1) \approx 40 M_p. \tag{3}$$

Similarly, we can find the total mass M required for 2-, 3- and 5-stage vehicles, and these are tabulated below:

$$\begin{aligned} n &= 2, & M &\approx 148 M_p \\ n &= 3, & M &\approx 52 M_p \\ n &= 4, & M &\approx 40 M_p \\ n &= 5, & M &\approx 36 M_p \end{aligned} \tag{4}$$

The results shown in Eq. (4) reveal that the decrease in total mass from two-stage to three-stage is quite considerable, while from 3-stage to 4-stage, the reduction in total mass is relatively small. However, it may be noted that the cost of vehicle goes up substantially as the number of stages increases. Hence, it is preferable to have a three-stage rocket to achieve the required terminal velocity V_τ.

Now, to find the ratio of masses in this simple model, we may recall Eq. (9.50), that is

$$\mu_i = \frac{M_i + M_{i+1} + \cdots + M_n + M_p}{(1-\varepsilon_i) M_i + M_{i+1} + \cdots + M_n + M_p}$$

and assume that
$$\mu_1 = \mu_2 = \cdots = \mu_n = \mu$$
and
$$\varepsilon_1 = \varepsilon_2 = \cdots = \varepsilon_n = \varepsilon,$$
so that
$$\mu = \frac{M_i + M_{i+1} + \cdots + M_n + M_p}{(1-\varepsilon)M_i + M_{i+1} + \cdots + M_n + M_p}. \tag{5}$$

Further, we define
$$\beta = \frac{\mu - 1}{[1 - \mu(1-\varepsilon)]}. \tag{6}$$

Substituting Eq. (5) into Eq. (6) we find on simplification that
$$\beta = \frac{M_i}{M_{i+1} + M_{i+2} + \cdots + M_n + M_p}, \tag{7}$$

which is true for all $i = 1, 2, \ldots, n$. Thus, for

$$i = n, \qquad \beta = \frac{M_n}{M_p} \quad \text{or} \quad M_n = \beta M_p$$

$$i = n-1, \qquad \beta = \frac{M_{n-1}}{M_n + M_p} = \frac{M_{n-1}}{(\beta + 1)M_p} \tag{8}$$

or
$$M_{n-1} = \beta(\beta + 1)M_p. \tag{9}$$

For $i = n - 2$,
$$\beta = \frac{M_{n-2}}{M_{n-1} + M_n + M_p} = \frac{M_{n-2}}{[\beta(\beta+1) + \beta + 1]M_p}$$

or
$$M_{n-2} = \beta(\beta+1)^2 M_p \tag{10}$$

and so on. Lastly,
$$M_1 = \beta(\beta+1)^n M_p. \tag{11}$$

To find the mass ratios, we recall Eq. (9.60) to get
$$\lambda = \frac{1}{v_e \{(1-\varepsilon)\exp[V_\tau/(nv_e)] - 1\}}$$

or

$$\lambda = \frac{1}{2940\,[0.2\exp(2.65306/n) - 1]}.\tag{12}$$

Using the given data, we also find

$$\mu = \frac{1 + \lambda v_e}{\lambda v_e(1-\varepsilon)} = \frac{1 + \lambda(2940)}{(2940\lambda)(0.2)}.\tag{13}$$

Also from Eq. (6), we obtain

$$1 + \beta = \frac{\varepsilon\mu}{1 - \mu(1-\varepsilon)} = \frac{0.8\mu}{1 - \mu(0.2)}.\tag{14}$$

Thus, for a two-stage rocket, that is, when $n = 2$, Eqs. (12)–(14) give us

$$\lambda = -0.00138, \qquad \mu = 3.767196, \qquad 1 + \beta = 12.22318.\tag{15}$$

Hence,

$$M_1 : M_2 = (1 + \beta) : 1 = 12.22 : 1.\tag{16}$$

Similarly, when $n = 3$, we find

$$\lambda = -0.0006595, \qquad \mu = 2.42142, \qquad 1 + \beta = 3.75621$$

Hence,

$$M_1 : M_2 : M_3 = (\beta + 1)^2 : (\beta + 1) : 1 = 14.11 : 3.76 : 1\tag{17}$$

This simple model will certainly help the preliminary design of a multi-stage rocket, though it is only of first-order accuracy, in the sense that we have totally neglected the aerodynamic forces, in the development of the theory behind it. In the analysis, we have also assumed that the I_{sp} of the propellants used in all the stages of a multi-stage rocket, has the same value.

However, in reality, we use a combination of solid and liquid propellants in various stages of a multi-stage rocket. A little more general model can be easily developed and is left as an exercise (Example 9.2) to the student.

EXERCISES

9.1 A liquid oxygen rocket has an exhaust speed of 2440 m/s. How far, a single-stage rocket burning liquid oxygen will travel from the Earth, if its fuel mass to total mass ratio is 2/3, and the whole fuel is burnt in 150 seconds? Assume that g is constant and neglect the payload mass.

9.2 A rocket consists of a satellite of fixed mass M_p, propelled by a three-stage launch vehicle of masses $M_i (i = 1, 2, 3)$ with structural factors $(1 - \varepsilon_i, \; i = 1, 2, 3)$. The exhaust speed of each stage is v_e. If the satellite requires a final speed V_r, show that the minimum total rocket mass required, that is, $M + M_p (M = \Sigma_{i=1}^{3} M_i)$ to attain this final speed is given by

$$M + M_p = \frac{\varepsilon_1 \varepsilon_2 \varepsilon_3 M_p \exp(V_\tau/v_e)}{[1 - b^{1/3} \exp(V_\tau/3v_e)]^3},$$

where

$$b = \prod_{i=1}^{3} (1 - \varepsilon_i).$$

Given, $V_\tau = 9.0 \times 10^3$ m/s and $v_e = 3.0 \times 10^3$ m/s, compare the values of M/M_p in the following cases

(i) $\varepsilon_1 = 0.9, \varepsilon_2 = 0.8, \varepsilon_3 = 0.7$
(ii) $\varepsilon_1 = \varepsilon_2 = \varepsilon_3 = 0.8$.

10

The Three- and *n*-Body Problems

All truths are easy to understand once they are discovered, the point is to discover them.
—Galileo

10.1 INTRODUCTION

It was Newton who first mathematically formulated the *n*-body problem. The statement of the problem is: At any time, given the positions and velocities of several massive bodies, each moving under the influence of others gravitational fields, to compute the position and velocity of each body at future times. That is, to predict the individual motion. These bodies are supposed to have spherical symmetry so that they can be regarded as point masses. It was also assumed that no other forces act on these bodies. This is also called *classical n-body problem*. This problem becomes much more complex, when the shapes of the bodies and their individual spins are also considered. Spin, in particular may not affect the gross motion of each body, while their shapes do affect. Such formulations are very useful to study the motion of planets in the solar system. The position of the planets changes during their orbital motion around the Sun, hence the gravitational forces acting on any particular planet also change. However, in the solar system, the Sun is the dominant centre of force. Thus, the resulting planetary motion closely approximates the motions observed, if the Sun and each planet comprise a pure two-body problem. Deviations from the two-body problem due to the gravitational forces of other planets are called *perturbations*.

To some extent, the motion of an Earth-bound satellite can be treated as a two-body problem. Though the satellite is of small mass, moving round the Earth, the dominant force centre is the Earth itself. However, due to the presence of the Sun and other planets, the oblate shape of the Earth, there will be a deviation from the two-body motion of a satellite. In this chapter, we shall derive the mathematical formulation of an *n*-body problem, of course, without considering their spin and shapes, and attempt to bring out some of the principal features of the *n*-body and three-body problems in that order.

10.2 MATHEMATICAL FORMULATION OF n-BODY PROBLEM

Let us consider an inertial frame of reference OXYZ and a system of n massive bodies of masses $m_i (i = 1, 2, ..., n)$, whose position vectors are $\vec{R}_i$, with reference to the origin O, as shown in Fig. 10.1, while their separation distances are given by $\vec{r}_{ij}$, where

$$\vec{r}_{ij} = \vec{R}_j - \vec{R}_i. \tag{10.1}$$

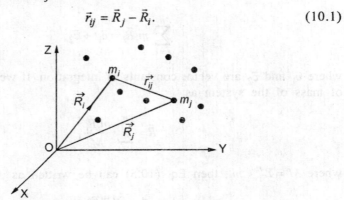

Figure 10.1 Vector relations in n-body problem.

Then, according to Newton's universal law of gravitation the equation of motion of the ith body relative to the inertial frame can be written as

$$m_i \ddot{\vec{R}}_i = G \sum_{j=1, j \neq i}^{n} \frac{m_i m_j}{r_{ij}^3} \vec{r}_{ij}, \quad i = 1, 2, ..., n. \tag{10.2}$$

Here, G is the constant of gravitation and vector $\vec{r}_{ij}$ means the vector joining m_i with m_j and directed from m_i to m_j. Thus,

$$\vec{r}_{ij} = -\vec{r}_{ji}. \tag{10.3}$$

In order to determine the motion of each body, we have to integrate Eq. (10.2) numerically. If a system involves hundreds of bodies, it is highly impractical to integrate and in such a case, we may have to rely on statistical techniques. In general, it is not possible to find a closed-form solution to an n-body formulation. However, through analytical means, only some general features of the dynamical behaviour of the bodies can be obtained, which we call integrals of motion and they are presented in the following section.

10.3 INTEGRALS OF MOTION

Summing all the equations of the set (10.2) and making use of the fact described in Eq. (10.3), we immediately get

$$\sum_{i=1}^{n} m_i \ddot{\vec{R}}_i = 0.$$

Integrating this result twice, we find

$$\sum_{i=1}^{n} m_i \dot{\vec{R}}_i = \vec{c}_1 \tag{10.4}$$

and

$$\sum_{i=1}^{n} m_i \vec{R}_i = \vec{c}_1 t + \vec{c}_2, \tag{10.5}$$

where $\vec{c}_1$ and $\vec{c}_2$ are vector constants of integration. If we define the centre of mass of the system as

$$\vec{R} = \sum_{i=1}^{n} \frac{m_i \vec{R}_i}{M}, \tag{10.6}$$

where $M = \Sigma_{i=1}^{n} m_i$, then Eq. (10.5) can be written as

$$\vec{R} = \frac{\vec{c}_1 t + \vec{c}_2}{M} \tag{10.7}$$

from which we get

$$\dot{\vec{R}} = \frac{\vec{c}_1}{M} = \text{constant}. \tag{10.8}$$

Equations (10.7) and (10.8) mean that the centre of mass of the system moves through space with constant speed in a straight line.

Taking the vector product on both sides of Eq. (10.2) with $\vec{R}_i$ and summing over i, we get

$$\sum_{i=1}^{n} m_i \vec{R}_i \times \ddot{\vec{R}}_i = G \sum_{i=1}^{n} \sum_{j=1}^{n} \frac{m_i m_j}{r_{ij}^3} \vec{R}_i \times \vec{r}_{ij}, \quad i \neq j. \tag{10.9}$$

But,

$$\vec{R}_i \times \vec{r}_{ij} = \vec{R}_i \times (\vec{R}_j - \vec{R}_i) = \vec{R}_i \times \vec{R}_j$$

and

$$\vec{R}_j \times \vec{r}_{ji} = \vec{R}_j \times (\vec{R}_i - \vec{R}_j) = -\vec{R}_i \times \vec{R}_j.$$

Therefore, the right-hand side of Eq. (10.9) reduces in pairs to zero and we are left with

$$\sum_{i=1}^{n} m_i \vec{R}_i \times \ddot{\vec{R}}_i = 0.$$

On integration, we have

$$\sum_{i=1}^{n} m_i \vec{R}_i \times \dot{\vec{R}}_i = \vec{h}, \qquad (10.10)$$

where $\vec{h}$ is a constant vector. The plane through the centre of mass and perpendicular to $\vec{h}$ is called an *invariable plane* of the system. Equation (10.10) also means that the sum of the moments of momenta or angular momenta of the system is a constant. It has been suggested to use this invariable plane in the planetary system as the fundamental reference plane, instead of the ecliptic plane. At present, the invariable plane is inclined at about 1°35′ to the ecliptic plane and lies between the planes of Jupiter and Saturn, the two most massive bodies among the planets. Of course, we should be careful before applying this idea to a physical system. The reason being that in our study of the n-body problem, we have assumed that all the bodies are rigid and spherical, then the orbital angular momentum remains constant. In reality all these conditions are not true, especially the shapes of the planets. Hence, the total angular momentum and the invariable plane, as defined in Eq. (10.10), may not be constant. However, the conditions are very nearly true, as such, we are justified in principle, to speak of invariable plane of our solar system. Its orientation is given by $\Omega = 107°$ and $i = 1°35′$ approximately.

If the expression (10.10) is resolved with respect to the inertial frame through O, the three integrals of area in Cartesian form can be written as

$$\sum_{i=1}^{n} m_i (y_i \dot{z}_i - z_i \dot{y}_i) = a_1$$

$$\sum_{i=1}^{n} m_i (z_i \dot{x}_i - x_i \dot{z}_i) = a_2$$

$$\sum_{i=1}^{n} m_i (x_i \dot{y}_i - y_i \dot{x}_i) = a_3,$$

where

$$\vec{h} = a_1 \hat{i} + a_2 \hat{j} + a_3 \hat{k}$$

and hence

$$|\vec{h}|^2 = a_1^2 + a_2^2 + a_3^2.$$

Thus, giving three constants of integration.

To obtain another constant, we may rewrite the equation of motion (10.2) in Cartesian coordinates with the help of Eq. (10.1) as

$$m_i \ddot{x}_i = -\sum_{j=1, j \neq i}^{n} \frac{Gm_i m_j}{r_{ij}^3} (x_i - x_j), \qquad i = 1, 2, \ldots, n$$

$$m_i \ddot{y}_i = -\sum_{j=1, j \neq i}^{n} \frac{Gm_i m_j}{r_{ij}^3} (y_i - y_j), \qquad i = 1, 2, \ldots, n \qquad (10.11)$$

$$m_i \ddot{z}_i = -\sum_{j=1, j \neq i}^{n} \frac{Gm_i m_j}{r_{ij}^3} (z_i - z_j), \qquad i = 1, 2, \ldots, n,$$

where,

$$r_{ij} = [(x_i - x_j)^2 + (y_i - y_j)^2 + (z_i - z_j)^2]^{1/2}. \qquad (10.12)$$

Now, of the n bodies, the ith body being at coordinates (x_i, y_i, z_i) and of mass m_i, the potential in which this body finds itself is given by the relation

$$U_i = -G \sum_{j=1}^{n} \frac{m_i m_j}{r_{ij}}, \qquad j \neq i. \qquad (10.13)$$

With the help of Eqs. (10.12) and (10.13), we can recast Eq. (10.11) in the following form:

$$m_i \ddot{x}_i = -\frac{\partial U_i}{\partial x_i}, \qquad m_i \ddot{y}_i = -\frac{\partial U_i}{\partial y_i}, \qquad m_i \ddot{z}_i = -\frac{\partial U_i}{\partial z_i}. \qquad (10.14)$$

Multiplying the first, second and third of Eqs. (10.14) respectively by $\dot{x}_i, \dot{y}_i, \dot{z}_i$ and summing over i, we have

$$\sum_{i=1}^{n} m_i (\dot{x}_i \ddot{x}_i + \dot{y}_i \ddot{y}_i + \dot{z}_i \ddot{z}_i) = -\sum_{i=1}^{n} \left(\frac{\partial U_i}{\partial x_i} \frac{\partial x_i}{\partial t} + \frac{\partial U_i}{\partial y_i} \frac{\partial y_i}{\partial t} + \frac{\partial U_i}{\partial z_i} \frac{\partial z_i}{\partial t} \right)$$

or

$$\sum_{i=1}^{n} \frac{m_i}{2} \frac{d}{dt} (\dot{x}_i^2 + \dot{y}_i^2 + \dot{z}_i^2) = -\sum_{i=1}^{n} \frac{dU_i}{dt}. \qquad (10.15)$$

Here, U_i is an implicit function of time. On integration, we obtain

$$\frac{1}{2} \sum_{i=1}^{n} m_i \vec{V}_i^2 + \sum_{i=1}^{n} U_i = \text{constant} = E \text{ (say)}$$

or

$$T + U = E, \qquad (10.16)$$

which is called *energy integral*. Thus, while neither kinetic energy nor potential energy is constant, the total energy E of the system is constant. In fact, there is a continuous exchange of energy between the bodies and the total energy for the n-body system remains constant and invariant with time and therefore conservative.

10.4 THE VIRIAL THEOREM

Consider the motion of an n-body system with respect to an inertial frame having origin at its centre of mass. Let us now introduce the idea of polar moment of inertia, I, of this system defined as

$$I = \sum_{i=1}^{n} m_i \vec{R}_i \cdot \vec{R}_i. \qquad (10.17)$$

Differentiating twice with respect to time, we find

$$\dot{I} = 2 \sum_{i=1}^{n} m_i \vec{R}_i \cdot \dot{\vec{R}}_i$$

$$\ddot{I} = 2 \sum_{i=1}^{n} m_i \dot{\vec{R}}_i \cdot \dot{\vec{R}}_i + 2 \sum m_i \vec{R}_i \cdot \ddot{\vec{R}}_i.$$

Now, taking the aid of Eq. (10.2), the above equation becomes

$$\ddot{I} = 4 \left(\sum_{i=1}^{n} \frac{1}{2} m_i \vec{V}_i^2 \right) + 2G \sum_{j=1, j \neq i}^{n} \sum_{i=1}^{n} \frac{m_i m_j}{r_{ij}^3} \vec{R}_i \cdot \vec{r}_{ij}.$$

Now, introducing U_i from Eq. (10.13), the above equation assumes the following form:

$$\ddot{I} = 4T + 2 \sum_{i=1}^{n} \vec{R}_i \cdot \nabla_i U. \qquad (10.18)$$

But,

$$\sum_{i=1}^{n} \vec{R}_i \cdot \nabla_i U = \sum_{i=1}^{n} \left(x_i \frac{\partial U}{\partial x_i} + y_i \frac{\partial U}{\partial y_i} + z_i \frac{\partial U}{\partial z_i} \right)$$

and U is a homogeneous function of all the coordinates. At this point, we may use Euler's theorem from calculus and write

$$\sum_{i=1}^{n} \vec{R}_i \cdot \nabla_i U = U.$$

Hence, Eq. (10.18) simply reduces to

$$\ddot{I} = 4T + 2U. \qquad (10.19)$$

However, using energy integral, that is, Eq. (10.16), we can write

$$\ddot{I} = -2U + 4E = 2T + 2E. \qquad (10.20)$$

It may be noted that Eqs. (10.19) and (10.20) are two forms of Lagrange–Jacobi identity. In Eq. (10.20), if E is such that $-2U + 4E$ or $2T + 2E$ are made positive, then $\ddot{I}$ is positive and therefore I increases with time without limit. This means, that the distance of at least one body from the origin ultimately approaches infinity. In other words, we say that such a body escapes from the system. Consequently, for a system of n bodies to be stable, that is, for a system, where all bodies remain in a restricted volume around the centre of mass, $\ddot{I}$ must be negative. This is of course only a necessary condition but not sufficient for the stability of the system under discussion.

Suppose, we consider the average values of $\ddot{I}, T$ and U over a certain interval of time $\Delta t = t_1 - t_0$, Eq. (10.19) gives

$$\frac{1}{\Delta t} \int_{t_0}^{t_1} \ddot{I}\, dt = \frac{4}{\Delta t} \int_{t_0}^{t_1} T\, dt + \frac{2}{\Delta t} \int_{t_0}^{t_1} U\, dt$$

or

$$\frac{1}{\Delta t}[\dot{I}]_{t_0}^{t_1} = \frac{2}{\Delta t}\left[\sum m_i \vec{R}_i \cdot \dot{\vec{R}}_i\right]_{t_0}^{t_1} = \frac{4}{\Delta t} \int_{t_0}^{t_1} T\, dt + \frac{2}{\Delta t} \int_{t_0}^{t_1} U\, dt$$

that is,

$$\frac{1}{\Delta t}\left(\sum m_i \vec{R}_i \cdot \frac{d\vec{R}_i}{dt}\right) = 2\bar{T} + \bar{U}, \qquad (10.21)$$

where $\bar{T}$ and $\bar{U}$ stand for the time-averaged values of T and U, respectively. For a stable system, $\vec{R}$, the distance of the body and $d\vec{R}/dt$, the velocity of that body are bounded and the left-hand side of Eq. (10.21), in fact tends to zero for large Δt. That is,

$$2\bar{T} + \bar{U} \to 0 \qquad \text{if } \Delta t \to \infty. \qquad (10.22)$$

But,

$$\bar{T} + \bar{U} = E. \qquad (10.23)$$

Thus, for a stable system, using relations (10.22) and (10.23) we have at once

$$\bar{T} = -E = -\frac{\bar{U}}{2}. \qquad (10.24)$$

This relation was first derived by Clausius in his investigations on kinetic theory of gases. This result is known as *virial theorem* or *Clausius theorem*, which states that for an n-body system, which is bounded in size and velocity, the average value of the kinetic energy over a large time interval equals minus half times the average value of the potential energy of the system and is also equal to minus times the total energy of the n-body system.

This theorem has wide applications in Stellar dynamics and forms the basis for statistical calculations of stellar clusters.

10.5 THE EQUATIONS OF RELATIVE MOTION

In the inertial frame, we have seen the formulation describing the motion of an n-body problem. In fact the motion of the ith body is given by the expression

$$m_i \ddot{\vec{R}}_i = G \sum_{j=1, j \neq i}^{n} \frac{m_i m_j}{r_{ij}^3} \vec{r}_{ij}, \qquad i = 1, 2, \ldots, n, \qquad (10.25)$$

where

$$\vec{r}_{ij} = \vec{R}_j - \vec{R}_i = -\vec{r}_{ji}.$$

Now, let the origin be shifted to the centre of one of the bodies, say the Sun, which is the dominant mass of the solar system of mass m_1 (say). Then, its equation of motion can be written as (see Fig. 10.2)

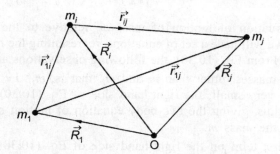

Figure 10.2 Vector relation in relative motion.

$$\ddot{\vec{R}}_1 = G \sum_{j=2}^{n} \frac{m_j}{r_{1j}^3} \vec{r}_{1j}. \qquad (10.26)$$

At the same time, the equation of motion of the ith body of mass m_i is

$$\ddot{\vec{R}}_i = G \sum_{j=1}^{n} \frac{m_j}{r_{ij}^3} \vec{r}_{ij}, \qquad i \neq 1, \ i = 2, 3, \ldots, n. \qquad (10.27)$$

Now, subtracting Eq. (10.26) from Eq. (10.27), we have

$$\ddot{\vec{R}}_i - \ddot{\vec{R}}_1 = \ddot{\vec{r}}_{1i} = G \left[-\left(m_i \frac{\vec{r}_{1i}}{r_{1i}^3} \right) + \frac{m_1}{r_{i1}^3} \vec{r}_{i1} + \sum_{j=2}^{n} \frac{m_j}{r_{ij}^3} \vec{r}_{ij} - \sum_{j=2}^{n} \frac{m_j \vec{r}_{1j}}{r_{1j}^3} \right], \qquad j \neq i$$

or

$$\ddot{\vec{r}}_{1i} = G\left[-(m_i + m_1)\frac{\vec{r}_{1i}}{r_{1i}^3} + \sum_{j=2, j\neq i}^{n} \frac{m_j \vec{r}_{ij}}{r_{ij}^3} - \sum_{j=2}^{n} \frac{m_j \vec{r}_{ij}}{r_{1j}^3}\right]. \quad (10.28)$$

But

$$\vec{r}_{ij} = \vec{r}_{1j} - \vec{r}_{1i}.$$

Therefore,

$$r_{ij}^3 = (\vec{r}_{ij} \cdot \vec{r}_{ij})^{3/2} = [(\vec{r}_{1j} - \vec{r}_{1i}) \cdot (\vec{r}_{1j} - \vec{r}_{1i})]^{3/2}.$$

Thus, Eq. (10.28) can be rewritten as

$$\ddot{\vec{r}}_{1i} + G(m_1 + m_i)\frac{\vec{r}_{1i}}{r_{1i}^3} = G\sum_{j=2, j\neq i}^{n} m_j \left(\frac{\vec{r}_{1j} - \vec{r}_{1i}}{r_{ij}^3} - \frac{\vec{r}_{1j}}{r_{1j}^3}\right). \quad (10.29)$$

Dropping the suffix 1, we have in general

$$\ddot{\vec{r}}_i + G(m + m_i)\frac{\vec{r}_i}{r_i^3} = G\sum_{j=2, j\neq i}^{n} m_j \left(\frac{\vec{r}_j - \vec{r}_i}{r_{ij}^3} - \frac{\vec{r}_j}{r_j^3}\right). \quad (10.30)$$

This is the equation of motion of mass m_i relative to the mass m. For $i = 2, 3, \ldots, n$, we will have a set of equations, representing the relative motion of the system. From Eq. (10.30), the following observations are to be noted:

(a) If the masses, other than m and m_i, that is, m_j, $j \neq i$ do not exist or are very small, the right-hand side of Eq. (10.30) can be made zero, thus, giving the two-body equation of motion of the mass m_i about the mass m.

(b) The first term on the right-hand side of Eq. (10.30) indicates the acceleration on mass m_i due to mass m_j, $j \neq i$.

(c) The second term on the right-hand side of Eq. (10.30) is the negative of the acceleration on the mass m due to the mass m_j, $j \neq i$.

Hence, the right-hand side of Eq. (10.30) consists of the perturbations by the masses m_j, $j \neq i$, on the orbit of m_i about m.

Suppose, we consider the three-body system consisting of the Sun, Earth and Moon, with the Earth as origin, the Moon as mass m_i and the Sun having mass m_j and also noting that

$$m_j \approx 330000 (m + m_i),$$

we arrive at the following paradox: It is known that the Sun's force on the Moon is much greater than the Earth's force on the moon, yet, the Moon revolves round the Earth, which is of a paradoxical behaviour. How to explain this paradox?

If we examine Eq. (10.30), we notice that it is the difference of the attractive force of the Sun on the Earth and on the Moon that appears on the right-hand side of Eq. (10.30). Both the Moon and the Earth are almost at the same distance from the Sun. This difference is small compared with the term due to the Earth itself and this can be treated as a perturbation of a two-body orbit of the Moon about the Earth.

In the place of Moon, if we consider an artificial satellite, the main perturbing force is due to non-spherical components of the Earth's gravitational field and also due to the atmospheric drag surrounding the Earth.*

10.6 THE GENERAL THREE-BODY PROBLEM

The general three-body problem is concerned with the motion of three bodies, moving under their mutual gravitational attractions, while no other external force is acting on them. As a special case, we can define a restricted three-body problem, wherein we consider one body to have a small mass when compared with the masses of the other two. That is, the motion of the other two massive bodies will not be influenced by the attraction of the third body. As an example, we may consider the motion of an artificial satellite or a spacecraft, within the Earth–Moon space, where the motion of a spacecraft becomes a restricted three-body problem in which the Earth and Moon represent two massive bodies.

10.6.1 Mathematical Formulation of General Three-body Problem

Let us consider three bodies which are spherically symmetric, so that, they attract one another like point masses. Let their masses be m_1, m_2 and m_3. Suppose, we imagine an inertial coordinate system OXYZ, with respect to which the position vectors of the three bodies be represented by $\vec{r}_1, \vec{r}_2, \vec{r}_3$ respectively as depicted in Fig. 10.3. Then, the equation of motion of each body, under their mutual gravitational attraction can be written as

$$m_1 \ddot{\vec{r}}_1 = G\left(\frac{m_1 m_2}{r_{12}^3} \vec{r}_{12} + \frac{m_1 m_3}{r_{13}^3} \vec{r}_{13} \right), \tag{10.31}$$

$$m_2 \ddot{\vec{r}}_2 = G\left(\frac{m_2 m_3}{r_{23}^3} \vec{r}_{23} + \frac{m_2 m_1}{r_{21}^3} \vec{r}_{21} \right), \tag{10.32}$$

$$m_3 \ddot{\vec{r}}_3 = G\left(\frac{m_3 m_1}{r_{31}^3} \vec{r}_{31} + \frac{m_3 m_2}{r_{32}^3} \vec{r}_{32} \right). \tag{10.33}$$

---

*For more details of these aspects, the interested reader may consult Escobal (1965).

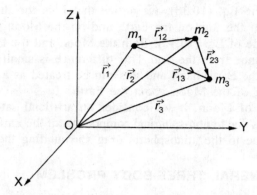

Figure 10.3 Vector relations in a three-body problem.

Adding Eqs. (10.31)–(10.33), we arrive at the relation

$$m_1\ddot{\vec{r}}_1 + m_2\ddot{\vec{r}}_2 + m_3\ddot{\vec{r}}_3 = 0,$$

which on integrating twice, we get

$$m_1\vec{r}_1 + m_2\vec{r}_2 + m_3\vec{r}_3 = \vec{c}_1 t + \vec{c}_2, \qquad (10.34)$$

where $\vec{c}_1$ and $\vec{c}_2$ are vector constants of integration. Introducing the definition of centre of mass of three bodies given by

$$\vec{R} = \frac{m_1\vec{r}_1 + m_2\vec{r}_2 + m_3\vec{r}_3}{m_1 + m_2 + m_3}.$$

Equation (10.34) becomes

$$\vec{R} = \frac{\vec{c}_1}{M}t + \frac{\vec{c}_2}{M}, \qquad (10.35)$$

where $M = m_1 + m_2 + m_3$. Equation (10.35) means that the centre of mass either remains at rest or moves uniformly in space, along a straight line.

10.6.2 Equations of Relative Motion of Three Bodies

To establish an inertial coordinate system in space, to which the motion of celestial bodies can be referred to is rather impossible. Hence, we consider the motion of $(n-1)$ bodies relative to the nth body. In the case of solar system, we choose the Sun as origin and describe the motion of other two bodies relative to the Sun, if we are considering a three-body problem. Let m_1 be the mass of the Sun, which we choose as the origin, m_2 and m_3 be the masses of other two bodies. In Fig. 10.4, let m_1 be chosen as the origin. Then, subtracting Eq. (10.31) from Eqs. (10.32) and (10.33), we get after some vector algebraic manipulations,

$$\ddot{\vec{r}}_{12} = -G\frac{m_1+m_2}{r_{12}^3}\vec{r}_{12} + Gm_3\left(\frac{\vec{r}_{23}}{r_{23}^3} - \frac{\vec{r}_{13}}{r_{13}^3}\right) \qquad (10.36)$$

and

$$\ddot{\vec{r}}_{13} = -G\frac{m_1 + m_3}{r_{13}^3}\vec{r}_{13} - Gm_2\left(\frac{\vec{r}_{23}}{r_{23}^3} + \frac{\vec{r}_{12}}{r_{12}^3}\right). \quad (10.37)$$

Hence, Eq. (10.36) represents the motion of m_2 relative to m_1, while Eq. (10.37) represents the motion of m_3 relative to m_1.

In our three-body system, let m_1, which is placed at the origin, be the dominant mass. The motions of the other two members of the system which could be either planets, comets or satellites, can be described relative to the Sun, the dominant mass. Let m_2 denote the body whose motion is to be investigated and let m_3 be the mass which disturbs the motion of m_2 about m_1. Then, the equation of motion m_2 can be written as (refer Fig. 10.4)

$$\ddot{\vec{r}} = -G\frac{m_1 + m_2}{r^3}\vec{r} + Gm_3\left[\frac{\vec{\rho}}{\rho^3} - \frac{\vec{r}'}{(r')^3}\right]. \quad (10.38)$$

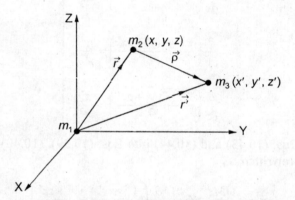

Figure 10.4 Relative motion of a three-body system.

In component form, we may write as

$$\ddot{x} = -\frac{GMx}{r^3} + Gm_3\left(\frac{x'-x}{\rho^3} - \frac{x'}{r'^3}\right), \quad (10.39)$$

$$\ddot{y} = -\frac{GMy}{r^3} + Gm_3\left(\frac{y'-y}{\rho^3} - \frac{y'}{r'^3}\right), \quad (10.40)$$

$$\ddot{z} = -\frac{GMz}{r^3} + Gm_3\left(\frac{z'-z}{\rho^3} - \frac{z'}{r'^3}\right), \quad (10.41)$$

where,

$$M = (m_1 + m_2), \qquad \vec{\rho} = \vec{r}' - \vec{r}. \quad (10.42)$$

Since,
$$\rho^2 = (x'-x)^2 + (y'-y)^2 + (z'-z)^2,$$
we have
$$\rho^{-1} = \frac{1}{\sqrt{(x'-x)^2 + (y'-y)^2 + (z'-z)^2}}$$
and
$$\frac{\partial \rho^{-1}}{\partial x} = \frac{x'-x}{\rho^3}, \quad \frac{\partial \rho^{-1}}{\partial y} = \frac{y'-y}{\rho^3}, \quad \frac{\partial \rho^{-1}}{\partial z} = \frac{z'-z}{\rho^3}. \quad (10.43)$$

Further, x, y and z are independent of x', y' and z'. Therefore, we can write at once

$$\left. \begin{array}{l} \dfrac{\partial}{\partial x}\left(\dfrac{xx' + yy' + zz'}{r'^3}\right) = \dfrac{x'}{r'^3} \\[2ex] \dfrac{\partial}{\partial y}\left(\dfrac{xx' + yy' + zz'}{r'^3}\right) = \dfrac{y'}{r'^3} \\[2ex] \dfrac{\partial}{\partial z}\left(\dfrac{xx' + yy' + zz'}{r'^3}\right) = \dfrac{z'}{r'^3}. \end{array} \right\} \quad (10.44)$$

Substituting Eqs. (10.43) and (10.44) into Eqs. (10.39), (10.40) and (10.41), they can be rewritten as

$$\left. \begin{array}{l} \ddot{x} = -\dfrac{GMx}{r^3} + Gm_3 \dfrac{\partial}{\partial x}\left(\dfrac{1}{\rho} - \dfrac{xx' + yy' + zz'}{r'^3}\right) \\[2ex] \ddot{y} = -\dfrac{GMy}{r^3} + Gm_3 \dfrac{\partial}{\partial y}\left(\dfrac{1}{\rho} - \dfrac{xx' + yy' + zz'}{r'^3}\right) \\[2ex] \ddot{z} = -\dfrac{GMz}{r^3} + Gm_3 \dfrac{\partial}{\partial z}\left(\dfrac{1}{\rho} - \dfrac{xx' + yy' + zz'}{r'^3}\right) \end{array} \right\} \quad (10.45)$$

Now, let us define a function R as

$$R = Gm_3 \left(\frac{1}{\rho} - \frac{xx' + yy' + zz'}{r'^3}\right). \quad (10.46)$$

Then, the set of Eqs. (10.45) can be recast as

$$\ddot{x} = -\frac{GMx}{r^3} + \frac{\partial R}{\partial x} \tag{10.47}$$

$$\ddot{y} = -\frac{GMy}{r^3} + \frac{\partial R}{\partial y} \tag{10.48}$$

$$\ddot{z} = -\frac{GMz}{r^3} + \frac{\partial R}{\partial z}. \tag{10.49}$$

Here, the function R is called the *disturbing function* or *perturbative function*. By setting $R = 0$, we can at once recognize that the above set of equations simplifies to the simple set of equations describing the motion of a two-body problem.

10.6.3 Stationary Solutions of a Three-body Problem

In the year 1772, Lagrange obtained two special solutions for a three-body problem with the following constraints:

(a) The geometric configurations of the three masses remain invariant with respect to time.
(b) The motion of the masses is such that their mutual distances from each other remain unaltered.
(c) The three-body configuration simply rotates in its own plane around their centre of mass.
(d) The three masses are initially projected in the same plane.

Further, it has been assumed that the three masses revolve in coplanar circular orbits around their centre of mass with constant angular speed. Let the position vectors of the masses relative to the centre of mass, C, be denoted by $\vec{r}_1 = r_1\hat{e}_1, \vec{r}_2 = r_2\hat{e}_2, \vec{r}_3 = r_3\hat{e}_3$ (see Fig. 10.5), where $\hat{e}_1, \hat{e}_2$ and $\hat{e}_3$ are unit vectors in the directions of $\vec{r}_1, \vec{r}_2$ and $\vec{r}_3$ respectively.

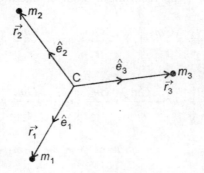

Figure 10.5 Orientation of a three-body problem relative to centre of mass.

For circular motion, it is known that, r_1, r_2 and r_3 remain constant and that the total coplanar acceleration of any one of the mass points is given by

$$\vec{a} = (\ddot{r} - r\dot\theta^2)\hat{e}_r + (2\dot{r}\dot\theta + r\ddot\theta)\hat{e}_\theta, \qquad (10.50)$$

where $\hat{e}_r$ and $\hat{e}_\theta$ are the unit vectors in the radial and transverse directions. Thus, in this special case with so many constraints, we will have

$$\dot{r} = \ddot{r} = 0$$

$$\dot\theta = n \text{ (constant angular speed about C)}$$

$$\ddot\theta = 0$$

Thus, the acceleration becomes

$$\vec{a}_i = -r_i n^2 \hat{e}_i \qquad (i = 1, 2, 3). \qquad (10.51)$$

Using these values into Eqs. (10.31) to (10.33), we obtain

$$-n^2 r_1 \hat{e}_1 = G\left[\frac{m_2}{r_{12}^3}(r_2\hat{e}_2 - r_1\hat{e}_1) + \frac{m_3}{r_{13}^3}(r_3\hat{e}_3 - r_1\hat{e}_1)\right] \qquad (10.52)$$

$$-n^2 r_2 \hat{e}_2 = G\left[\frac{m_3}{r_{23}^3}(r_3\hat{e}_3 - r_2\hat{e}_2) - \frac{m_1}{r_{12}^3}(r_2\hat{e}_2 - r_1\hat{e}_1)\right] \qquad (10.53)$$

$$-n^2 r_3 \hat{e}_3 = G\left[-\frac{m_2}{r_{23}^3}(r_3\hat{e}_3 - r_2\hat{e}_2) - \frac{m_1}{r_{13}^3}(r_3\hat{e}_3 - r_1\hat{e}_1)\right]. \qquad (10.54)$$

Since the origin is chosen at the centre of mass, we have

$$m_1 r_1 \hat{e}_1 + m_2 r_2 \hat{e}_2 + m_3 r_3 \hat{e}_3 = 0. \qquad (10.55)$$

Now, multiplying Eq. (10.52) by m_1 and Eq. (10.53) by m_2 and adding, we get an equation, in which if we use Eq. (10.55), the result is found to be Eq. (10.54). Therefore, the resulting system of equations to be solved are

$$\left(-n^2 + \frac{Gm_2}{r_{12}^3} + \frac{Gm_3}{r_{13}^3}\right) r_1\hat{e}_1 - \frac{Gm_2}{r_{12}^3} r_2\hat{e}_2 - \frac{Gm_3}{r_{13}^3} r_3\hat{e}_3 = 0 \qquad (10.56)$$

$$-\frac{Gm_1}{r_{12}^3} r_1\hat{e}_1 + \left(-n^2 + \frac{Gm_3}{r_{23}^3} + \frac{Gm_1}{r_{12}^3}\right) r_2\hat{e}_2 - \frac{Gm_3}{r_{23}^3} r_3\hat{e}_3 = 0 \qquad (10.57)$$

$$m_1 r_1 \hat{e}_1 + m_2 r_2 \hat{e}_2 + m_3 r_3 \hat{e}_3 = 0 \qquad (10.58)$$

These conditions must be satisfied, if the three mass points are to move, in plane circular orbits about their centre of mass with uniform angular speed of, n, radians/second.

Let us now consider a rectangular coordinate system with origin at C (refer Fig. 10.5), which is rotating with angular speed n radians per second in the anticlockwise direction. In this coordinate system, let the unit vectors $\hat{e}_1, \hat{e}_2$ and $\hat{e}_3$ be fixed in position, and their orientation relative to this rectangular coordinate system be given by

$$\hat{e}_s = \cos\theta_s \hat{i} + \sin\theta_s \hat{j}, \qquad s = 1, 2, 3. \tag{10.59}$$

Now, taking the scalar product of Eqs. (10.56) to (10.58) with $\hat{i}$ and using Eq. (10.59), we have on simplification

$$-n^2 x_1 + \frac{Gm_2}{r_{12}^3}(x_1 - x_2) + \frac{Gm_3}{r_{13}^3}(x_1 - x_3) = 0, \tag{10.60}$$

$$-n^2 x_2 + \frac{Gm_1}{r_{12}^3}(x_2 - x_1) + \frac{Gm_3}{r_{23}^3}(x_2 - x_3) = 0, \tag{10.61}$$

$$m_1 x_1 + m_2 x_2 + m_3 x_3 = 0. \tag{10.62}$$

Similarly, taking the scalar product of Eqs. (10.56) to (10.58) with $\hat{j}$ and using Eqs. (10.59) and after simplification, we obtain

$$-n^2 y_1 + \frac{Gm_2}{r_{12}^3}(y_1 - y_2) + \frac{Gm_3}{r_{13}^3}(y_1 - y_3) = 0 \tag{10.63}$$

$$-n^2 y_2 + \frac{Gm_1}{r_{12}^3}(y_2 - y_1) + \frac{Gm_3}{r_{23}^3}(y_2 - y_3) = 0 \tag{10.64}$$

$$m_1 y_1 + m_2 y_2 + m_3 y_3 = 0, \tag{10.65}$$

where

$$x_i = r_i \cos\theta_i, \qquad y_i = r_i \sin\theta_i, \qquad i = 1, 2, 3 \tag{10.66}$$

Thus, Eqs. (10.60)–(10.65) constitute six simultaneous equations in unknowns $(x_i, y_i), i = 1, 2, 3$. Now, we shall attempt to give two distinct solutions as derived by Lagrange.

In the first solution, we assume that the masses are located at the vertices of an equilateral triangle. That is, let $r_{12} = r_{13} = r_{23} = \rho$ for all times. Then, Eqs. (10.60) – (10.65) can be written as

$$\left.\begin{aligned}
(-n^2 + Gm_2/\rho^3 + Gm_3/\rho^3)x_1 - (Gm_2/\rho^3)x_2 - (Gm_3/\rho^3)x_3 &= 0 \\
-(Gm_1/\rho^3)x_1 + [-n^2 + G(m_1 + m_3)/\rho^3]x_2 - (Gm_3/\rho^3)x_3 &= 0 \\
m_1 x_1 + m_2 x_2 + m_3 x_3 &= 0 \\
(-n^2 + Gm_2/\rho^3 + Gm_3/\rho^3)y_1 - (Gm_2/\rho^3)y_2 - (Gm_3/\rho^3)y_3 &= 0 \\
-(Gm_1/\rho^3)y_1 + [-n^2 + G(m_1 + m_3)/\rho^3]y_2 - (Gm_3/\rho^3)y_3 &= 0 \\
m_1 y_1 + m_2 y_2 + m_3 y_3 &= 0
\end{aligned}\right\}. \quad (10.67)$$

This is a homogeneous system, which will have a non-trivial solution, if the determinant of its coefficient matrix is zero. That is, if

$$\begin{vmatrix}
(-n^2 + Gm_2/\rho^3 + Gm_3/\rho^3) & -Gm_2/\rho^3 & -Gm_3/\rho^3 \\
-Gm_1/\rho^3 & (-n^2 + Gm_1/\rho^3 + Gm_3/\rho^3) & -Gm_3/\rho^3 \\
m_1 & m_2 & m_3 \\
0 & 0 & 0 \\
0 & 0 & 0 \\
0 & 0 & 0 \\
0 & 0 & 0 \\
0 & 0 & 0 \\
0 & 0 & 0 \\
(-n^2 + Gm_2/\rho^3 + Gm_3/\rho^3) & -Gm_2/\rho^3 & -Gm_3/\rho^3 \\
-Gm_1/\rho^3 & (-n^2 + Gm_1/\rho^3 + Gm_3/\rho^3) & -Gm_3/\rho^3 \\
m_1 & m_2 & m_3
\end{vmatrix} = 0.$$

(10.68)

Its value, when evaluated using row, column manipulations, is found to be equal to

$$m_3^2 \left(\frac{GM}{\rho^3} - n^2 \right)^4, \quad (10.69)$$

where $M = m_1 + m_2 + m_3$, the total mass of the system we are considering. Since m_3 cannot be zero, we at once get

$$n^2 = \frac{GM}{\rho^3}. \qquad (10.70)$$

One can easily verify that this equation is dimensionally consistent and indicates that n is in radians per unit time. With $n^2 = GM$ in Eqs. (10.67), we can choose any two pairs of coordinates (x_i, y_i) arbitrarily, adjust the scale, and determine the third pair in such a way that $\rho = 1$. Thus, the first problem described by Eqs. (10.67) has a solution.

The second solution to Eqs. (10.60)–(10.65) is found by assuming that $y_1 = y_2 = y_3 = 0$. It means that all mass points are on the X-axis. Then, Eqs. (10.63)–(10.65) are satisfied. Suppose that the masses are arranged on the X-axis, as shown in Fig. 10.6.

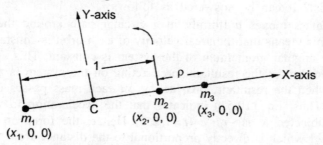

Figure 10.6 Straight line solution.

Let us denote the distance $(x_3 - x_2)$ by ρ and choose the scale in such a way that $(x_2 - x_1) = 1$. Then, the following inequality

$$x_1 < x_2 < x_3$$

is obvious and Eqs. (10.60)–(10.62) reduce to

$$\left. \begin{array}{r} -n^2 x_1 + Gm_2 + Gm_3 (1+\rho)^{-2} = 0 \\ n^2(1+x_1) - Gm_1 + Gm_3 \rho^{-2} = 0 \\ m_1 x_1 + m_2 (1+x_1) + m_3 (1+x_1+\rho) = 0 \end{array} \right\}. \qquad (10.71)$$

Eliminating n^2 from the first-two equations of the set (10.71), we obtain

$$x_1 = \frac{[m_2 (1+\rho)^2 + m_3]\rho^2}{(1+2\rho)m_3 - (m_1 + m_2)(1+\rho)^2 \rho^2}.$$

Substituting this value of x_1 into the third equation of (10.71) and after simplification, we arrive at a fifth degree polynomial in ρ as

$$(m_1 + m_2)\rho^5 + (3m_1 + 2m_2)\rho^4 + (3m_1 + m_2)\rho^3 - (m_2 + 3m_3)\rho^2$$
$$- (2m_2 + 3m_3)\rho - (m_2 + m_3) = 0. \qquad (10.72)$$

By Descarte's rule of signs, we observe that there is only one real positive root, as there is only one change of sign in the coefficients of the polynomial

Eq. (10.72). Thus, if the three masses are arranged along X-axis, as shown in Fig. 10.6, we get only one positive solution for ρ. When ρ is found, we can locate m_3 uniquely as m_1 and m_2 are already located to set the scale of distance $x_2 - x_1 = 1$. Even if we change the order of location of masses, we get a similar solution for each arrangement. The solution of Eq. (10.72) can be easily obtained using any iterative method such as Newton–Raphson method on a digital computer.*

Thus, the three-body problem possesses essentially two distinct solutions as found by Lagrange, the equilateral triangle solution and the straight line solution. These solutions are valid for any masses moving in coplanar, circular orbits around their centre of mass with uniform angular speed. At this point, it is natural to ask a question—under what conditions will such a dynamical system exist? It can be answered as follows:

Each mass moves uniformly in a circular path around the centre of mass C. This means that the areal velocity of each orbit is constant. In other words, the angular momentum of the system is constant. This again means that the moment of the resultant force acting on the system is zero, which happens when the resultant force acting on each mass passes through the origin C. Thus, Eq. (10.51) indicates that the acceleration to which each mass is subjected to is $-r_i n^2 \hat{e}_i$ ($i = 1, 2, 3$). Hence, the force on each mass is $-m_i r_i n^2 \hat{e}_i$, which is directly proportional to the distance from their centre of mass C. The negative sign means, it is acting towards C.

Thus, if two of the masses, say m_1 and m_2, are taken as points of reference, then there are five possible locations, where the third body can be placed, as shown in Fig. 10.7. These points L_1, L_2 L_3, L_4 and L_5 are called *Lagrange points* or *Libration points* or *equilibrium points*.

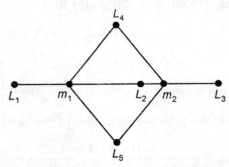

Figure 10.7 Lagrange points in a three-body problem.

Initially both the equilateral triangle solution and straight line solution are thought to be only of academic interest and such formations are unlikely to exist. Surprisingly, both types of solutions are realized in the solar system.

In the Sun–Jupiter system, there exists in fact some dozen asteroids oscillating about the points L_4 and L_5. They are called Trojans, each one with Sun and Jupiter, providing an example of the equilateral triangle solution.

*See Sankara Rao, 2004

It is observed that a Trojan wanders 20° or more as measured from the Sun from the points L_4 and L_5. Recently, in the year 1960, with increased interest in space explorations, the question of existence of such points with respect to other primaries especially for the Earth–Moon system arose quite naturally. The points L_4 and L_5 are found to be occupied by meteoric particles. With respect to straight line solution, in the Sun–Earth system, a visible light is observed after sunset in the plane of the ecliptic in a direction opposite to that of the Sun, which is attributed to the Sun's illumination of meteoric particles at the Lagrange point L_3.

To gain further insight into the motion of the three-body problem, Poincare and Hill have studied the restricted three-body problem and obtained another particular solution. They assumed that one of the three bodies has a very small mass (say an artificial satellite) when compared with the other two massive bodies, so that the motion of the two massive bodies will not be influenced by the attraction of the third body, while the satellite moves under the combined gravitational attraction of the two massive bodies.*

Example 10.1 Masses $m_1 = 1$, $m_2 = 3$, $m_3 = 5$ are situated along the X-axis in such a way that $x_1 < x_2 < x_3$ as in the straight line solution of the three-body problem with $x_2 - x_1 = 1$. Find the position of m_3.

Solution If the masses are arranged on the X-axis, as shown in Fig. 10.6, and denoting the distance of m_3 from m_2 by ρ and choosing the scale $(x_2 - x_1) = 1$. Then, ρ is obtained from the polynomial Eq. (10.72). That is,

$$(m_1 + m_2)\rho^5 + (3m_1 + 2m_2)\rho^4 + (3m_1 + m_2)\rho^3 - (m_2 + 3m_3)\rho^2$$
$$- (2m_2 + 3m_3)\rho - (m_2 + m_3) = 0. \tag{1}$$

Using the given data: $m_1 = 1, m_2 = 3, m_3 = 5$, Eq. (1) reduces to

$$f(\rho) = 4\rho^5 + 9\rho^4 + 6\rho^3 - 18\rho^2 - 21\rho - 8 = 0. \tag{2}$$

By Descarte's rule of signs, we observe that there is only one real positive root. To find that positive root of Eq. (2), we use Newton–Raphson iterative method, as follows: We note that $f(1) =$ negative (–28), $f(2) =$ positive (198). Therefore, the root lies in the interval (1, 2). Starting with $\rho_0 = 2$, we find iteratively**

$$\rho_1 = \rho_0 - \frac{f(\rho_0)}{f'(\rho_0)} = 2 - \frac{198}{587} = 1.6627$$

$$\rho_2 = \rho_1 - \frac{f(\rho_1)}{f'(\rho_1)} = 1.6627 - \frac{26.9167}{286.2448} = 1.5687$$

---

*For detailed discussion and the stability of motion about Lagrange points, the reader may refer McCuskey (1963) and the Astronomical Journal for recent works on this topic.
**Refer Sankara Rao, 2004.

$$\rho_3 = \rho_2 - \frac{f(\rho_2)}{f'(\rho_2)} = 1.5687 - \frac{33.2821}{244.058} = 1.4323$$

$$\rho_4 = \rho_3 - \frac{f(\rho_3)}{f'(\rho_3)} = 1.4328 - \frac{4.6141}{154.3186} = 1.4024$$

Thus, after fourth iteration, we may take the root as $\rho = 1.4$ units. Hence m_3 is located at a distance of 1.4 units from m_2.

EXERCISES

10.1 What is a three-body problem? Derive the equations of motion for a three-body problem, stating the assumptions made. Hence, explain the concept of a perturbative function after rewriting the equations of motion in an appropriate manner.

10.2 The mass of the Moon is (1/81) that of the Earth and the distance from the Earth to Moon is 384400 km. Assuming that the Moon's orbit is circular, where on the line joining the Earth and Moon would one locate a space platform, so that it would always remain between the two?

Multiple-Choice Questions

Choose the correct answer to the following questions:

1. For the Hamiltonian $H = (1/2)(q^{-2} + p^2 q^4)$, the equation of motion for q can be written as $f(q, \dot{q}, \ddot{q}) = 0$, where $f(q, \dot{q}, \ddot{q})$ is given by
   (a) $\ddot{q} - q\dot{q}$
   (b) $\ddot{q} - q/\dot{q}$
   (c) $\ddot{q} - 2\dot{q}^2/q - q$
   (d) $\ddot{q} - 2\dot{q}^2 - q$.

2. A particle moves in the xy-plane under the influence of a central force depending only on its distance from the origin. Then, the Hamiltonian for such a system is
   (a) $\frac{1}{2}m(\dot{r}^2) - V(r)$
   (b) $p_r^2/(2m) + p_\theta^2/(2mr^2) + V(r)$
   (c) $(1/2)m(\dot{r}^2 + r^2\dot{\theta}^2) - V(r)$
   (d) $p_r^2/(2m) + p_\theta^2/(2m) + V(r)$.

3. A particle moves in the xy-plane under the influence of a central force which varies inversely to its distance from the centre of force. If T denotes the kinetic energy, V the potential energy, L the Lagrangian and H the Hamiltonian of this system, then,
   (a) $L = H - V$
   (b) $H = T - V$
   (c) $H = T + V$
   (d) None of these.

4. In the motion of a two-particle system, if two particles are connected by a rigid weightless rod of constant length, then the number of degrees of freedom of the system is
   (a) 2
   (b) 3
   (c) 5
   (d) 6.

5. Euler's equations of motion of a rigid body about a fixed point O, in the absence of external forces are:

$$I_{xx}\dot{\omega}_x - (I_{yy} - I_{zz})\omega_y\omega_z = 0,$$

$$I_{yy}\dot{\omega}_y - (I_{zz} - I_{xx})\omega_z\omega_x = 0,$$

$$I_{zz}\dot{\omega}_z - (I_{xx} - I_{yy})\omega_x\omega_y = 0,$$

where $\omega_x, \omega_y, \omega_z$ are the components of the angular velocity of a rigid body in the direction of principal axes of inertia at O; I_{xx}, I_{yy}, I_{zz}

are the principal moments of inertia at O. Let T denotes the kinetic energy during the motion and h, the magnitude of the angular momentum of the body about O. Then, during the motion

(a) Both T and h vary

(b) T varies but h remains constant

(c) T remains constant but h varies

(d) Both T and h remain constant.

6. If the total kinetic energy of a system of particles about the origin is equal to its kinetic energy about the centre of mass, then the centre of mass is

(a) at rest (b) moving along a circle
(c) moving on a straight line (d) moving along an ellipse.

7. The number of generalized coordinates required to describe the motion of a rigid body with one of its points fixed is

(a) 9 (b) 6
(c) 3 (d) 1.

8. The number of degrees of freedom for a rigid body which has two fixed ends is

(a) 6 (b) 3
(c) 2 (d) 1.

9. If Euler's equations of motion for a rigid body about a fixed point O, in the absence of external forces are:

$$I_{xx}\dot\omega_x - (I_{yy} - I_{zz})\omega_y\omega_z = 0,$$

$$I_{yy}\dot\omega_y - (I_{zz} - I_{xx})\omega_z\omega_x = 0,$$

$$I_{zz}\dot\omega_z - (I_{xx} - I_{yy})\omega_x\omega_y = 0,$$

then, the magnitude of the angular velocity

(a) is constant (b) varies
(c) is zero (d) is unity.

10. The moment of inertia of a rectangular plate of length $2a$ and width $2b$ about an axis through its centre O and perpendicular to its plane (if its mass is M) is

(a) $(1/2)M(a^2 + b^2)$ (b) $M(a^2 + b^2)$
(c) $(1/3)M(a^2 + b^2)$ (d) Ma^2b^2.

11. If M is the mass of a sphere, the moment of inertia of a solid sphere about a diameter is

(a) $(1/2)Ma^2$ (b) Ma^2
(c) $(1/3)Ma^2$ (d) $(2/5)Ma^2$.

12. The angular momentum of a particle has the dimensions of
    (a) MLT^{-1}
    (b) ML^2T^{-1}
    (c) ML^2T
    (d) MLT.

13. In the Newton's gravitational law
    $$\vec{F} = \frac{GMm}{r^2},$$
    GM has the dimensions of
    (a) L^3/T^2
    (b) L^2/T^2
    (c) L^3/T^3
    (d) L^2/T^3

14. The value of GM in Newton's universal gravitational law (taking $g = 9.806$, $r_E = 6378$ km) when the attracting force is Earth, is
    (a) 3.989×10^{14} m^3/s^2
    (b) 3.989×10^{11} m^3/s^2
    (c) 9.806×10^6 m^3/s^2
    (d) 9.806×10^{14} m^3/s^2.

15. The magnitude of the escape velocity of an object from the surface of the Earth is
    (a) 11.2 km/s
    (b) 5.9 km/s
    (c) 7.9 km/s
    (d) 10.4 km/s.

16. A curve with given boundary points whose rotation about the x-axis generates a surface of minimum area is a
    (a) cycloid
    (b) parabola
    (c) ellipse
    (d) catenary.

17. The shape of a solid body moving in a flow of fluid with least resistance is
    (a) spherical
    (b) cylindrical
    (c) parabola of degree 3/4
    (d) parabola of degree 3/2.

18. The velocity of a particle when its kinetic energy equals its rest energy is
    (a) $2c$
    (b) $\sqrt{3}c/2$
    (c) c
    (d) $c/2$.

19. A force with components $(-7, 4, -5)$ acts at the point $(2, 4, -3)$. Then the torque with respect to the origin exerted by this force is
    (a) $-8\hat{i} + 31\hat{j} + 36\hat{k}$
    (b) $-8\hat{i} - 31\hat{j} - 36\hat{k}$
    (c) null vector
    (d) $\hat{i} + \hat{j} + \hat{k}$.

20. If three particles of masses 1, 2 and 4 units move under a force field such that their position vectors at any time t are respectively given by
    $$\vec{r}_1 = 2\hat{i} + 4t^2\hat{k},$$
    $$\vec{r}_2 = 4t\hat{i} - \hat{k},$$
    $$\vec{r}_3 = (\cos \pi t)\hat{i} + (\sin \pi t)\hat{j}.$$

then, the angular momentum of the system about the origin at $t=1/2$ is given by

(a) zero vector
(b) $4(-4\hat{j}+\pi\hat{k})$
(c) $-4(-4\hat{j}+\pi\hat{k})$
(d) $-4(\pi\hat{j}+\hat{k})$.

21. For a particle of unit mass moving under gravitational field, along the cycloid $x = \phi - \sin\phi$, $y = 1 + \cos\phi$, the Lagrangian for the motion is given by

(a) $\dot{\phi}^2(1+\cos\phi)-g(1-\cos\phi)$
(b) $\dot{\phi}^2(1-\cos\phi)+g(1+\cos\phi)$
(c) $\dot{\phi}^2(1-\cos\phi)-g(1+\cos\phi)$
(d) $2\dot{\phi}^2(1-\cos\phi)-g(1+\cos\phi)$.

22. A particle of mass m is moving under gravity, on a smooth conical surface. The cone has semi-vertical angle α and is held with its axis vertical and apex pointing down, as shown below. Then, the Lagrangian for the motion is

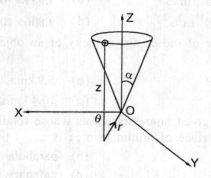

(a) $(m/2)(\dot{r}^2 \operatorname{cosec}^2\alpha + r^2\dot{\theta}^2) - mgr \cot\alpha$
(b) $(m/2)(\dot{r}^2 \cot^2\alpha + r^2\dot{\theta}^2) - mgr \cot\alpha$
(c) $(1/2)(\operatorname{cosec}^2\alpha + r^2) + mgr \cos\alpha$
(d) $(1/2)(r^2 \operatorname{cosec}^2\alpha + r^2) - mgr \cos\alpha$.

23. A cylinder having its axis vertical floats in a liquid of density ρ. It is pushed down slightly and released. If the equilibrium position of the cylinder is at a distance z from the liquid surface at any time, the period of oscillation of the cylinder, whose weight is W with cross-sectional area A, is given by

(a) $2\pi/\rho$
(b) $2\pi/\sqrt{A\rho}$
(c) $2\pi/\sqrt{W\rho}$
(d) $2\pi/\sqrt{W/(g^2 A\rho)}$.

24. If the motion of a particle on the circumference of a unit circle under a central force directed towards the origin given by $r = 2\cos\theta$, then the law of force is

(a) an inverse square power
(b) an inverse fifth power
(c) an inverse cube power
(d) an inverse seventh power.

25. On the surface of the Earth, that is, when $r = R$, where R is the radius of the Earth, the force of attraction of the Earth on an object of mass m is equal to its weight mg of the object. Thus, if M is the mass of the Earth, then
    (a) $GM = gR^2$
    (b) $GM = gR$
    (c) $GM = \sqrt{gR}$
    (d) $GM = \sqrt{g/R}$.

26. Three particles of masses $3g$, $5g$ and $2g$ are located respectively at $(1, 0, -1)$, $(-2, 1, 3)$ and $(3, -1, 1)$. Then, their centre of mass is located at
    (a) $(0, 0, 0)$
    (b) $(1, 1, 1)$
    (c) $(3/10, 5/10, 2/10)$
    (d) $(-1/10, 3/10, 7/5)$.

27. A particle of mass 4 units moving along the x-axis is attracted towards the origin by a force whose magnitude is $8x$. If it is initially at rest at $x = 10$, then, the frequency of the particle is
    (a) $10\sqrt{2}\pi$
    (b) π
    (c) $10/(\pi\sqrt{2})$
    (d) $1/(\sqrt{2}\pi)$.

28. A simple pendulum is taken inside a deep mine. Relative to the period of oscillation on the surface, the time period inside the mine
    (a) remains the same
    (b) decreases
    (c) increases
    (d) becomes infinite.

29. Which one of the following particles experiences a coriolis force?
    (a) A particle at rest with respect to Earth at Bhopal
    (b) A particle thrown vertically upwards at Bhopal
    (c) A particle thrown vertically upwards at the North Pole
    (d) A particle moving horizontally along the north–south direction.

30. In the frame of reference of a rotating turntable, an insect of mass m is moving radially outwards ($+x$ direction) with a speed v. If the turntable is rotating with a constant angular velocity $\omega\hat{k}$ in the vertically upward direction, the net pseudo force on the insect, when it is at a distance r from the axis of the turntable, is
    (a) $m\omega^2 r\hat{i} + 2m\omega v\hat{j}$
    (b) $m\omega^2 r\hat{i} - 2m\omega v\hat{j}$
    (c) $-m\omega^2 r\hat{i} + 2m\omega v\hat{j}$
    (d) $m\omega^2 r\hat{i}$.

31. The moment of inertia of a disk of radius R and mass M rotating about an axis, as shown in the figure below, given by is

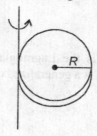

(a) $(1/2)MR^2$
(b) MR^2
(c) $(3/2)MR^2$
(d) $2MR^2$.

32. A cube of edge s and mass M is suspended vertically from one of its corners, then the length of the equivalent simple pendulum is
    (a) $(4/3)s$
    (b) $(2/3)s$
    (c) $(2\sqrt{2}/3)s$
    (d) $2\sqrt{2}s$.

33. A particle is placed on the top of a sphere in a gravitational field and allowed to slide without friction. Then, the motion has
    (a) no constraint
    (b) a holonomic constraint
    (c) a non-holonomic constraint
    (d) a rheonomic constraint.

34. If the escape velocity from the surface of a spherical planet of mass M is given by
$$\sqrt{\frac{GM}{2R}},$$
then, the radius of the planet is
    (a) $R/2$
    (b) R
    (c) $2R$
    (d) $4R$.

35. When a planet moves around the Sun, then
    (a) its areal velocity is constant
    (b) its areal velocity depends on its position
    (c) its linear velocity is constant
    (d) its angular velocity is constant.

36. A planet moves around the Sun in an elliptical orbit with semi-major axis a and time period τ. Then, τ is proportional to
    (a) a^2
    (b) $a^{1/2}$
    (c) $a^{3/2}$
    (d) a^3.

37. A particle moves in a circular orbit about the origin under the action of a central force $\vec{F} = -k\hat{r}/r^3$. If the potential energy is zero at infinity, the total energy of the particle is
    (a) $-k/r^2$
    (b) $-k/2r^2$
    (c) zero
    (d) k/r^2.

38. The Lagrangian of the Sun–Earth system is
    (a) $\frac{1}{2}m\dot{r}^2 + \frac{1}{2}mr^2\dot{\theta}^2 - \frac{GMm}{r}$
    (b) $\frac{1}{2}m\dot{r}^2 + \frac{1}{2}mr^2\dot{\theta}^2 + \frac{GMm}{r}$
    (c) $\frac{1}{2}m\dot{r}^2 - \frac{GMm}{r}$
    (d) $\frac{1}{2}mr^2\dot{\theta}^2 + \frac{GMm}{r}$

39. If $\partial L/\partial q_n = 0$, where L is the Lagrangian for a conservative system, without constraint and q_n is a generalized coordinate, then the generalized momenta is

(a) a cyclic coordinate (b) a constant of motion
(c) equal to $\dfrac{d}{dt}\left(\dfrac{\partial L}{\partial q_n}\right)$ (d) undefined.

40. The Lagrangian for the Kepler's problem is given by
$$L = \frac{1}{2}(\dot{r}^2 + r^2\dot{\theta}^2) + \frac{\mu}{r}, \qquad (\mu > 0)$$
where (r, θ) denote the polar coordinates and the mass of the particle is unity. Then,
(a) $p_\theta = 2r^2\dot{\theta}$ (b) $p_r = 2\dot{r}$
(c) The angular momentum of the particle about the centre of attraction is a constant
(d) The total energy of the particle is time-dependent.

41. A particle of mass m is constrained to move on the plane curve $xy = c\,(c > 0)$ under gravity (y-axis vertical). Then, the Lagrangian of the particle is
(a) $\dfrac{m}{2}\dot{x}^2\left(1 + \dfrac{c^2}{x^4}\right) + \dfrac{mgc}{x}$ (b) $\dfrac{m}{2}\dot{x}^2\left(1 + \dfrac{c^2}{x^4}\right) - \dfrac{mgc}{x}$
(c) $\dfrac{m\dot{x}^2}{2}\left(1 + \dfrac{c^2}{x^2}\right) + \dfrac{mgc}{x}$ (d) $\dfrac{m}{2}\dot{x}^2\left(1 + \dfrac{c^2}{x^2}\right) - \dfrac{mgc}{x}$.

42. The potential energy of a classical particle moving in one dimension is kx^4, where k is a constant. If the particle moves from a point x_1 at time t_1 to a point x_2 at time t_2, the actual path followed by a particle is that which makes the following integral extremum. Select the correct integral:
(a) $\int_{t_1}^{t_2}\left(\dfrac{1}{2}mv^2 + kx^4\right)dt$ (b) $\int_{t_1}^{t_2}\left(\dfrac{1}{2}mv^2 - kx^4\right)dt$
(c) $\int_{t_1}^{t_2}\left(\dfrac{1}{2}mv^2 + 4kx^3\right)dt$ (d) $\int_{t_1}^{t_2}\left(\dfrac{1}{2}mv^2 - 4kx^3\right)dt$.

43. A linear transformation of a generalized coordinate q and the corresponding momenta p to Q and P given by
$$Q = q + p, \qquad P = q + \alpha p$$
is canonical if the value of the constant α is
(a) -1 (b) 0
(c) 1 (d) 2.

44. A particle has rest mass m_0 and momentum $m_0 c$, where c is the velocity of light. Then, the total energy and the velocity of the particle are respectively given as

(a) $\sqrt{2}\, m_0c^2$ and c (b) $2m_0c^2$ and $c/\sqrt{2}$
(c) $\sqrt{2}\, m_0c^2$ and $c/2$ (d) $2m_0c^2$ and c.

45. The kinetic energy of a relativistic particle of rest mass m moving with speed v is

   (a) $\dfrac{1}{2}mv^2$

   (b) $\dfrac{mc^2}{\sqrt{1-\dfrac{v^2}{c^2}}}$

   (c) $\dfrac{mc^2}{\sqrt{1-\dfrac{v^2}{c^2}}} - mc^2$

   (d) $\dfrac{1}{2}m(v^2 - c^2)$

46. The momentum of an electron of mass m which has the same kinetic energy as its rest mass energy (if c is the velocity of light) is

   (a) $\sqrt{3}\, mc$ (b) $\sqrt{2}\, mc$
   (c) mc (d) $mc/\sqrt{2}$.

47. A circle of radius 5 m lies at rest in the xy-plane of the laboratory. For an observer moving with a uniform velocity v along the y-direction, the circle appears to be an ellipse with an equation

   $$\dfrac{x^2}{25} + \dfrac{y^2}{9} = 1.$$

   Then the speed of the observer in terms of the velocity of light is

   (a) $9c/25$ (b) $3c/5$
   (c) $4c/5$ (d) $16c/25$.

48. Introducing the new coordinate $Q = q \sin \omega t$ in the Lagrangian

   $$L = \dfrac{m}{2}(\dot{q}^2 \sin^2 \omega t + \dot{q} q \omega \sin 2\omega t + q^2 \omega^2)$$

   the Hamiltonian H equals (p is conjugate momentum)

   (a) $\dfrac{m}{2}\left(\dfrac{p^2}{m^2} - Q^2 \omega^2\right)$

   (b) $\dfrac{m}{2}\left(\dfrac{p^2}{m^2} + Q^2 \omega^2\right)$

   (c) $\dfrac{m}{2}\left(\dfrac{p^2}{m^2} - Q^2 \omega^2 + 2\omega^2 Q \,\text{cosec}\, \omega t\right)$

   (d) $\dfrac{m}{2}\left(\dfrac{p^2}{m^2} + Q^2 \omega^2 - 2\omega^2 Q \,\text{cosec}\, \omega t\right)$.

49. A satellite is in a circular orbit which is concentric and coplanar with the Earth's equator. At what radius of the orbit will the satellite appear to remain stationary when viewed by observers on the Earth? Assume that

the sense of rotation of the orbit is same as that of the Earth (take $G = 6.67 \times 10^{-11}$ N-m²/kg, mass of the Earth, $M_e = 5.98 \times 10^{24}$ kg, angular velocity of the Earth, $\omega_e = 7.3 \times 10^{-5}$ rad/s.)

(a) 4.2×10^4 km (b) 4.2×10^6 km
(c) 6.38×10^4 km (d) 6.38×10^6 km.

50. Which of the following equations is relativistically invariant? (α, β, γ and δ are constants of suitable dimensions.)

(a) $\dfrac{\partial \phi(x,t)}{\partial t} = \alpha \dfrac{\partial^2 \phi(x,t)}{\partial x^2}$ (b) $\dfrac{\partial^2 \phi(x,t)}{\partial t^2} = \beta \dfrac{\partial^2 \phi(x,t)}{\partial x^2}$

(b) $\dfrac{\partial^2 \phi(x,t)}{\partial t^2} = \gamma \dfrac{\partial \phi(x,t)}{\partial x}$ (d) $\dfrac{\partial \phi(x,t)}{\partial t} = \delta \dfrac{\partial^3 \phi(x,t)}{\partial x^3}$.

51. A person has a mass of 100 kg on the Earth. When he is on the spacecraft, an observer on the Earth finds his mass as 101 kg. Then, the speed of the spacecraft is

(a) 4.2×10^7 m/s (b) 4.2×10^8 m/s
(c) 4.2×10^6 m/s (d) 4.2×10^5 m/s.

52. If a body is dropped from rest at a height of 200 m above the surface of the Earth at a latitude of 45°, then, the magnitude of deflection is

(a) 4.379 cm (b) 6 cm
(c) 2.5 cm (d) 1.5 m.

Bibliography

Arnold, V.I., *Mathematical Methods of Classical Mechanics*, Springer-Verlag, New York, 1978.

Ball, K.J. and G.F. Osborne, *Space Vehicle Dynamics*, Oxford University Press, 1967.

Barger, V.D. and M.G. Olsson, *Classical Mechanics—A Modern Perspective*, McGraw-Hill, 1995.

Bate, R.R., D.D. Mueller, and J.E. White, *Fundamentals of Astrodynamics*, Dover Publications, New York, 1971.

Desloge, E.A., *Classical Mechanics*, Vols. 1 and 2, John Wiley & Sons, New York, 1982.

Escobal, P.R., *Methods of Orbit Determination*, John Wiley & Sons, New York, 1965.

French, A.P., *Special Relativity*, Introductory Physics Series, MIT Press, Nelson, 1969.

Gerlach, O.H., Attitude stabilization and control of Earth satellites, *Space Science Reviews*, Reidel Publishing Company, Dordrecht, No. 4, pp. 541–582, 1965.

Goldstein, Herbert, *Classical Mechanics*, Addison-Wesley, World Student Series, 1980.

Greenwood, Donald T., *Principles of Dynamics*, 2nd ed., Prentice-Hall of India, New Delhi, 1988.

Hauser, Walter, *Introduction to the Principles of Mechanics*, Addison-Wesley, Massachusetts, 1965.

McCuskey, S.W., *Introduction to Celestial Mechanics*, Addison-Wesley, Massachusetts, 1963.

Meriam, J.L. and L.G. Kraige, *Engineering Mechanics*, Vol. 2: *Dynamics*, John Wiley & Sons, Singapore, 1999.

Roy, Archie E., *The Foundations of Astrodynamics*, MacMillan, New York, 1965.

Sankara Rao, K., Computation of time in Sunlight for an elliptic orbit,' *S.S.T.C-AMD-TR-2-73* V.S.S.C, Trivandrum, 1973.

Sankara Rao, K., *Numerical Methods for Scientists and Engineers*, 2nd ed., Prentice-Hall of India, New Delhi, 2004.

Synge, J.L. and B.A. Griffith, *Principles of Mechanics*, McGraw-Hill, 1984.

Tolman, R.C., *Relativity, Thermodynamics and Cosmology*, Oxford University Press, 1958.

Answers to Exercises

Chapter 1

1.1 The dimensions of $GM = L^3/T^2$.

1.2 *Hint:* If the attracting body is the Earth and if the attracted body of mass m is at the surface of the Earth, then, $\vec{F} = mg$, $r = r_e$ = mean radius of the Earth = 6378 kilometres. Substituting these values along with $g = 9.806$ m/s² into the universal law of gravitation

$$mg = \frac{GMm}{r_e^2} \quad \text{or} \quad GM = r_e^2 g.$$

Therefore,

$$GM = (6378 \times 1000)^2 (9.806) = 3.979 \times 10^{14} \text{ m}^3/\text{s}^2.$$

1.3 If the net force acting on the particle is zero, the linear momentum is constant. If the moment of the force acting on the particle is zero, then the angular momentum is constant.

1.4 $\vec{r} = 4t^3 \hat{i} + 4(t^4 + 1)\hat{j} + \left(3 + \dfrac{1}{\pi} - \dfrac{1}{\pi} \cos \pi t\right) \hat{k}$ m.

Chapter 2

2.1 $\vec{R} = -\left(\dfrac{a}{14}\right)\hat{i}, \quad \vec{V} = b\left(\dfrac{2}{7}\hat{i} + \hat{j}\right)$ m/s,

$\vec{a} = -\dfrac{1}{7} b\hat{i} + \dfrac{1}{7}(4b - 6a)\hat{j}$ m/s²,

$\vec{H} = -mb(12.25a + 0.5b)\hat{k}$ kg-m²/s

2.2 $\vec{R} = 0, \vec{V} = \dfrac{1}{3}(2\hat{i} + \hat{j})$ m/s, $\vec{a} = 0$,

$\vec{H} = 48\hat{i} - 60\hat{j} + 12\hat{k}$ kg-m²/s

2.3 $\vec{R} = \hat{i} + 1.2\hat{j} + 2\hat{k}$

$\sum m_i \vec{v}_i = 15.5\hat{i} - 6\hat{j} + 2.5\hat{k}$ kg-m/s

$\vec{H} = 18\hat{i} + 29\hat{j} - 14\hat{k}$ kg-m²/s

2.4 A circle of radius R with centre at P, the path of the second particle given by

$$x = R\cos\theta + \rho_x, \quad y = R\sin\theta + \rho_y,$$

where $\vec{\rho} = \rho_x \hat{i} + \rho_y \hat{j}$ such that

$$(x - \rho_x)^2 + (y - \rho_y)^2 = R^2$$

2.5 Refer Fig. 2.10 and note that $\phi = \theta + 30° = \theta + \pi/6$ radians

$$a_Q = \ddot{\vec{r}} = \ddot{\vec{R}} + \ddot{\vec{p}} = -\Omega^2(\vec{R} + \vec{p}) = -\Omega^2 \vec{r}$$

2.6 Apply the principle of angular momentum. Choose the origin at the centre of mass of the ring:

$$\vec{G} = \dot{\vec{H}} = \frac{d}{dt}(rm\vec{V}) = \frac{d}{dt}(r^2 m\vec{\omega})$$

Therefore,

$$\dot{\omega} = \frac{G}{r^2 m} = \frac{300 \times 2}{4 \times 100} = 1.5 \text{ rad/s}^2.$$

$$a_c = \frac{300}{100} = 3 \text{ m/s}^2.$$

2.7 In the inertial frame, the desired force is the centripetal force $-mr\omega^2$. Therefore, the force on the stone is

$$-m\omega^2 r = -1(2 \times 3.1415 \times 5)^2 \times 2 \approx -1974 \text{ N}.$$

In the non-inertial frame rotating with the stone, the acceleration of the stone is zero. Thus, the total force in this frame is zero.

2.9 $\frac{1}{3}\sqrt{\frac{8h^3}{g}} \, \omega \cos 40° = 0.017$ m

Chapter 3

3.1 $I_{xx} = I_{yy} = I_{zz} = (8/3)Ma^2$, $I_{yz} = I_{zx} = I_{xy} = Ma^2$, where M is the mass of the solid cube.

3.3 $I_{xx} = (M/3)(b^2 + c^2)$, $I_{yy} = (M/3)(c^2 + a^2)$, $I_{zz} = (M/3)(a^2 + b^2)$, where M is the mass of the block.

3.5 $I^*_{xx} = 3$, $I^*_{yy} = 2$, $I^*_{zz} = 4$

$\hat{e}_1 = i - 2j - 2k$, $\hat{e}_2 = -2i + j - 2k$, $\hat{e}_3 = -2i - 2j + k$.

3.7 Let the radii of the disks be R_1 and R_2, respectively. Since the disks have the same mass and thickness, we have at once $\rho_1 R_1^2 = \rho_2 R_2^2$. Equivalently

$$\frac{R_1^2}{R_2^2} = \frac{\rho_2}{\rho_1}.$$

The M.I. of the disks are

$$I_1 = \frac{MR_1^2}{2}, \quad I_2 = \frac{MR_2^2}{2}.$$

Therefore,

$$\frac{I_1}{I_2} = \frac{R_1^2}{R_2^2} = \frac{\rho_2}{\rho_1}.$$

Since $\rho_1 < \rho_2$, $I_1 > I_2$. Hence the disk 1 has larger M.I.

3.8 Let P be the force applied, r the radius of the disk. Then $P - mg = ma_y$, $0 = ma_x$,

Therefore,

$$a_x = 0, \quad a_y = \frac{200 - 10 \times 9.806}{10} = 10.194 \text{ m/s}^2.$$

Also from the angular momentum principle, $\vec{G} = \dot{\vec{H}} = I\dot{\vec{\omega}}$, we have

$$-Pr = I_c \dot{\vec{\omega}} = \frac{mr^2}{2} \dot{\vec{\omega}},$$

which gives

$$\dot{\vec{\omega}} = \frac{-200 \times 0.5 \times 2}{10 \times (0.5)^2} = -80 \text{ rad/s}^2.$$

indicating that the angular acceleration is in the clockwise direction.

3.9 (Similar to solved Example 3.19) angular acceleration $\alpha = 2.4$ rad/s^2.

$$A_x = 6 \text{ N} \leftarrow, \quad A_y = 19.612 \text{ N} \uparrow$$

Chapter 4

4.4 $I_{zz}^2 s^2 \geq 4 I_{xx} Mgl \cos\theta$

Hint: Refer to Eq. (4.36) and note that for steady precession $A^2 = (s + \dot{\psi} \cos\theta)^2 = s^2$ etc.

4.5 The discussion is on similar lines to that of the rotation of the Earth when it is free from torque, as in Section 4.3.

4.6 $\omega_x = \omega_0 \cos(\alpha t + \varepsilon)$, $\omega_y = \omega_0 \sin(\alpha t + \varepsilon)$, $\omega_z = $ constant

Hint: Set $I_{xx} = I_{yy} = I$ in Euler's equations and get $\omega_z = $ constant $= \Omega$ (say) and $\dot{\omega}_x = -\dfrac{I_{zz} - I}{I} \Omega \omega_y$,

$\dot{\omega}_y = \dfrac{I_{zz} - I}{I} \Omega \omega_x$. Differentiate and get

$$\ddot{\omega}_x = -\left(\dfrac{I_{zz} - I}{I} \Omega\right)^2 \omega_x \text{ and } \ddot{\omega}_y = -\left(\dfrac{I_{zz} - I}{I} \Omega\right)^2 \omega_y, \text{ etc}\ldots$$

4.7 (a) 6, (b) 3, (c) 1.

4.8 In the solved Example 4.5, if we set $a = b$ which corresponds to a square plate, we find $G_z = 0$.

4.9 For a 'sleeping top', we must have $\theta = 0$, $\dot{\theta} = 0$, since the axis of the top should be vertical and no rotation takes place. Thus, Eq. (4.21), which represents the total energy, becomes

$$\dfrac{I_{zz}}{2} A^2 + Mgl = E$$

or

$$I_{zz} A^2 = 2(E - Mgl). \tag{1}$$

Using this result, Eq. (4.30), can be re-written as

$$\left. \begin{array}{l} \alpha = \dfrac{2Mgl}{I_{xx}}, \ \beta = \dfrac{2Mgl}{I_{xx}}, \\[2mm] \gamma = \dfrac{I_{zz} A}{I_{xx}}, \ \delta = \dfrac{I_{zz} A}{I_{xx}}, \end{array} \right\} \tag{2}$$

from which, we observe that $\alpha = \beta$ and $\gamma = \delta$.
Now, Eq. (4.31) can be recast as

$$f(u) = (1 - u)^2 [\alpha(1 + u) - \gamma^2],$$

where $u = \cos\theta$. We observe from the above equation that $f(u) = 0$ has a double root at $u = 1$, while the third root is found to be

$$u = \dfrac{\gamma^2}{\alpha} - 1 = \dfrac{I_{zz}^2 A^2}{2 Mgl I_{xx}} - 1. \quad \text{[Follows from (2)]}$$

Then, the top will sleep, if and only if this root is greater than or at most equal to unity. Therefore, we have the condition as

$$A^2 \geq \dfrac{4 Mgl I_{xx}}{I_{zz}^2}.$$

Chapter 5

5.8 Apogee of the elliptic orbit is 7961.59 km. Hence the maximum altitude of the satellite from the surface of the Earth $= 7961.59 - 6378 \approx 1584$ km.

5.9 The semi-major axis of the elliptic orbit $= 6927.94$ km; period $= 95.645$ min; eccentricity, $e = 0.036$.

5.10 The orbit is an ellipse, using

$$v^2 = GM\left(\frac{2}{r} - \frac{1}{a}\right)$$

At perigee, $r = q = a(1-e)$. At apogee, $r = a(1+e)$.

$$v_p = 10.28 \text{ km/s}, \qquad v_a = 1.42 \text{ km/s}.$$

5.11 $F(r) = \alpha r^n$, $\left(\dfrac{du}{d\theta}\right)^2 + u^2 = \dfrac{\beta}{n+1} u^{-(n+1)} + c$

At apse, $\dot{r} = 0$ implying $\dfrac{du}{d\theta} = 0$. Therefore,

$$u^2 = c + \frac{\beta}{n+1}\left(\frac{1}{u^{n+1}}\right), \text{ i.e. } \frac{1}{r^2} = c + \frac{\beta}{n+1} r^{n+1} \text{ and}$$

$$r^{n+3} + \frac{n+1}{\beta} cr^2 - \frac{n+1}{\beta} = 0, \text{ which gives the apsidal distances.}$$

5.12 $F \propto \dfrac{1}{r^4}$.

5.13 Kepler's equation can be written as $e \sin E = E - M = f(M)$. Since $\sin E$ is a periodic function possessing Fourier series,

$$e \sin E = 2 \sum_{k=1}^{\infty} (a_k \cos kM + b_k \sin kM),$$

$\sin E$ being an odd function, $a_k = 0$ for $k = 1, 2, \ldots, \infty$. Hence

$$e \sin E = 2 \sum_{k=1}^{\infty} b_k \sin kM,$$

where

$$b_k = \frac{1}{2\pi} \int_0^{2\pi} (e \sin E) \sin kM \, dM = \frac{1}{k} J_k(ke)$$

etc.

5.14 $E_2 = 67°.2851$, $M_2 = 61°.9999$.

5.15 $E_2 = 54°.3064$, $M_2 = 45°.0001$.

5.16 $(r, v) = (3.61637 \text{ A.U.}, 127°.50733)$

5.17 $\dot{r} = \sqrt{\dfrac{GMe^2}{a(1-e^2)}}.$

5.19 $a = 1.33$ A.U., $e = 0.2383$
$i = 26°.8$, $\Omega = 45°$

5.20 From the solved Example 5.1, we see that $GM_E = g_E R_E^2$. Similarly, $GM_m = g_m R_m^2$. Their division yields

$$\frac{g_m}{g_E} = \frac{M_m}{M_E}\left(\frac{R_E}{R_m}\right)^2 = \frac{(3.66)^2}{81.6} = 0.1641617 \approx \frac{1}{6}.$$

v_{escape}(moon) $= 2.37$ km/s.

Chapter 6

6.1 Given $r_i = r_i(\vec{q}, t)$,

$$\dot{\vec{r}}_i = \sum_{\rho=1}^{n} \frac{\partial \vec{r}_i}{\partial q_\rho} \dot{q}_\rho + \frac{\partial \vec{r}_i}{\partial t},$$

The Kinetic energy,

$$T = \frac{1}{2}\sum_{i=1}^{n} m_i \dot{\vec{r}}_i \cdot \dot{\vec{r}}_i.$$

That is,

$$T = \frac{1}{2}\sum_{i=1}^{n} m_i \left(\sum_{j=1}^{n}\sum_{k=1}^{n} \frac{\partial \vec{r}_i}{\partial q_j}\cdot\frac{\partial \vec{r}_i}{\partial q_k} \dot{q}_j \dot{q}_k + 2\frac{\partial \vec{r}_i}{\partial t}\cdot \sum_{j=1}^{n}\frac{\partial \vec{r}_i}{\partial q_j}\dot{q}_j + \frac{\partial \vec{r}_i}{\partial t}\cdot\frac{\partial \vec{r}_i}{\partial t} \right) \quad (1)$$

Introduce

$$\alpha_{jk} = \sum_{i=1}^{n} m_i \frac{\partial \vec{r}_i}{\partial q_j}\cdot\frac{\partial \vec{r}_i}{\partial q_k},$$

$$\beta_j = \sum_{i=1}^{n} m_i \frac{\partial \vec{r}_i}{\partial t}\cdot\frac{\partial \vec{r}_i}{\partial q_j},$$

$$\gamma_j = \frac{1}{2}\sum_{i=1}^{n} m_i \frac{\partial \vec{r}_i}{\partial t}\cdot\frac{\partial \vec{r}_i}{\partial t}$$

Then, Eq. (1) becomes

$$T = \frac{1}{2}\left(\sum_{j=1}^{n}\sum_{k=1}^{n}\alpha_{jk}\dot{q}_j\dot{q}_k + 2\sum_{j=1}^{n}\beta_j\dot{q}_j + \gamma\right)$$

$= \frac{1}{2}$ (a homogeneous quadratic function in generalized velocities

+ a linear homogeneous function of generalized velocities

+ a function of generalized coordinates and time).

6.2 Since $V = 0$, $L = T$. In terms of Euler angles, $\omega_x = \dot{\theta}$, $\omega_y = \dot{\psi}\sin\theta$, $\omega_z = \dot{\psi}\cos\theta + \dot{\phi}$

$$L = \frac{1}{2}[I_{xx}(\dot{\theta})^2 + I_{yy}(\dot{\psi}\sin\theta)^2 + I_{zz}(\dot{\psi}\cos\theta + \dot{\phi})^2].$$

Then, the Lagrange equations are

$$\frac{d}{dt}\left(\frac{\partial L}{\partial \dot{\theta}}\right) - \frac{\partial L}{\partial \theta} = 0, \quad \frac{d}{dt}\left(\frac{\partial L}{\partial \dot{\phi}}\right) - \frac{\partial L}{\partial \phi} = 0, \quad \frac{d}{dt}\left(\frac{\partial L}{\partial \dot{\psi}}\right) - \frac{\partial L}{\partial \psi} = 0,$$

which gives

$$I_{xx}\ddot{\theta} - I_{yy}\dot{\psi}^2\sin\theta\cos\theta + I_{zz}\dot{\psi}^2\cos\theta\sin\theta + \dot{\psi}\dot{\phi}\sin\theta = 0,$$

$$\frac{d}{dt}[I_{zz}(\dot{\phi} + \dot{\psi}\cos\theta)] = 0,$$

$$\frac{d}{dt}(I_{yy}\dot{\psi}\sin^2\theta + I_{zz}\dot{\psi}\cos^2\theta + \dot{\phi}\cos\theta) = 0.$$

6.3 $L = \frac{1}{2}l^2(m_1 + m_2)\dot{\theta}^2 + \frac{1}{2}l^2 m_2\dot{\phi}^2 + l^2 m_2\dot{\theta}\dot{\phi}\cos(\theta - \phi)$

$+ (m_1 + m_2)gl\cos\theta + m_2 gl\cos\phi$

The Lagranges equations of motion

$$\ddot{\theta} + \frac{m_2}{m_1 + m_2}[\dot{\phi}^2\sin(\theta - \phi) + \ddot{\phi}\cos(\theta - \phi)] + \frac{g}{l}\sin\theta = 0$$

$$\ddot{\phi} + [-\dot{\theta}^2\sin(\theta - \phi) + \ddot{\theta}\cos(\theta - \phi)] + \frac{g}{l}\sin\phi = 0$$

6.4 $p_\phi = \frac{\partial L}{\partial \dot{\phi}} = I_{zz}(\dot{\psi}\cos\theta + \dot{\phi})$

$p_\psi = \frac{\partial L}{\partial \dot{\psi}} = I_{xx}\dot{\psi}\sin^2\theta + I_{zz}(\dot{\psi}\cos\theta + \dot{\phi})\cos\theta$

$H = I_{xx}\frac{\dot{\theta}^2}{2} - \frac{(p_\psi - p_\phi\cos\theta)^2}{2I_{xx}\sin^2\theta} - \frac{p_\phi}{2I_{zz}} + mgl\cos\theta$

6.11 $T = \frac{1}{2}m(\dot{r}^2 + r^2\dot{\theta}^2)$, $V = mgr\cos\theta$

$L = T - V = \frac{1}{2}mr^2(\dot{\theta}^2 + \omega^2\sin^2\theta) - mgr\cos\theta$

The Lagrange's equation of motion for the particle is

$$\ddot{\theta} - \omega^2\sin\theta\cos\theta - \frac{g\sin\theta}{r} = 0$$

6.12 The degrees of freedom of the particle is two.

6.13 (a) The Lagrangian

$$L = \frac{1}{2}m(\dot{r}^2 + r^2\dot{\theta}^2 + r^2\sin^2\theta\,\dot{\phi}^2)$$

(b) The Lagrange's equations of motion of the satellite are:

$$\frac{d}{dt}\left(\frac{\partial L}{\partial \dot{r}}\right) - \frac{\partial L}{\partial r} = m\ddot{r} - mr\dot{\theta}^2 - mr\sin^2\theta\,\dot{\phi}^2 = 0$$

$$\frac{d}{dt}\left(\frac{\partial L}{\partial \dot{\theta}}\right) - \frac{\partial L}{\partial \theta} = mr^2\ddot{\theta} - mr^2\dot{\phi}^2\sin\theta\cos\theta = 0$$

$$\frac{d}{dt}\left(\frac{\partial L}{\partial \dot{\phi}}\right) - \frac{\partial L}{\partial \phi} = mr^2\sin^2\theta\,\ddot{\phi} = 0$$

(c) $p_r = \frac{\partial L}{\partial \dot{r}} = m\dot{r}$, $p_\theta = \frac{\partial L}{\partial \dot{\theta}} = mr^2\dot{\theta}$, $p_\phi = \frac{\partial L}{\partial \dot{\phi}} = mr^2\sin^2\theta\,\dot{\phi}$

Chapter 7

7.2 $F = (1/2)pq$

7.3 $F = pq + p\log p$

7.4 $F(q, p, t_0) = -(p+1)e^p t_0 + F(q_0, p_0, t_0)$

7.12 The phase space is six-dimensional. The configuration space is three-dimensional.

Chapter 8

8.4 $m = \dfrac{m_0}{\sqrt{1 - \dfrac{v^2}{c^2}}} = \dfrac{1.6725 \times 10^{-27}}{\sqrt{1 - \left(\dfrac{2.7 \times 10^8}{3 \times 10^8}\right)^2}} = 3.841 \times 10^{-27}$ kg

Momentum, $p = mv = 1.04 \times 10^{-18}$ kg-m/s

If the velocity of the coalesced particle is V, then using the principle of conservation of momentum

$$p = (m_0 + M)V, \quad \text{which gives } V = 3.9 \times 10^7 \text{ m/s}$$

8.5 2.598×10^8 m/s

8.6 Using Lorentz transformation,

$$dx^2 + dy^2 + dz^2 - c^2 dt^2 = \left(\frac{dx' + vdt'}{\sqrt{1 - \frac{v^2}{c^2}}}\right)^2 + dy'^2 + dz'^2 - c^2 \left(\frac{dt' + vdx'}{\sqrt{1 - \frac{v^2}{c^2}}}\right)^2$$

$$= dx'^2 + dy'^2 + dz'^2 - c^2 dt'^2$$

8.7 $F = \dfrac{d}{dt}(mv) = \dfrac{d}{dt}\left[\dfrac{m_0 v}{\sqrt{1 - \dfrac{v^2}{c^2}}}\right] = m_0 \dfrac{d}{dt}\left[\dfrac{v}{\left(1 - \dfrac{v^2}{c^2}\right)^{1/2}}\right]$. Now differentiate using the quotient rule, etc...

Chapter 9

9.1 Given: $v_e = 2440$ m/s, $M_f/M_0 = 2/3$, $g = 9.8$ m/s^2 $t_{b.0} = \tau = 150$. Recall Eq. (5) of solved Example 9.1 and note $\tau = \varepsilon M_0/k$, and $M_f/M_0 = \varepsilon = 2/3$, then we get

$$h = -v_e \tau \log(1-\varepsilon) + v_e \tau + \left(v_e \frac{\tau}{\varepsilon}\right)\log(1-\varepsilon) - \frac{1}{2}g\tau^2 = 56.8 \text{ km}.$$

(i) Thus, the height attained while the fuel is burning = 56.8 km.

(ii) In free flight, the height increase is obtained using

$$v^2 = u^2 - 2gh' = [-v_e \log(1-\varepsilon) - g\tau]^2 - 2gh' = 0$$

which gives $h' = 74.775$ km.

(iii) The total height attained by the rocket $= h + h'$, etc...

9.2 Case (i) $\dfrac{M}{M_p} = \dfrac{0.9 \times 0.8 \times 0.7 \times \exp(3)}{[1 - (0.1 \times 0.2 \times 0.3)^{1/3} \exp(1)]^3} - 1 \approx 77$

Case (ii) $\dfrac{M}{M_p} = \dfrac{(0.8)^3 \exp(3)}{[1 - 0.2 \exp(1)]^3} - 1 \approx 107$.

Chapter 10

10.2 Let m_1 be the mass of the earth and m_2 that of moon. Suppose the space platform of mass m is located at a distance ρ_1 from Earth and ρ_2 from Moon of course in between Earth and Moon. Then, using the universal law of gravitation.

$$\frac{Gmm_1}{\rho_1^2} = \frac{Gmm_2}{\rho_2^2} = \frac{Gm(m_1/81)}{\rho_2^2}$$

Giving

$$\frac{1}{\rho_1^2} = \frac{1}{81\rho_2^2}.$$

Let $\rho_1 + \rho_2 = 1$ unit. Then,

$$\frac{1}{(1-\rho_2)^2} = \frac{1}{81\rho_2^2} \quad \text{or} \quad \frac{\rho_2}{1-\rho_2} = \frac{1}{9}$$

That is, $\rho_2 = 0.1$ units. Hence, the space platform should be located at a distance of 38,440 km from the Moon.

Answers to Multiple-Choice Questions

1. The Lagrangian and Hamiltonian of a system is related by the expression
$$L = \dot{q}p - H \quad \text{and} \quad \dot{q} = \frac{\partial H}{\partial p}.$$
We are given $H = (1/2)(q^{-2} + p^2 q^4)$. Therefore,
$$\dot{q} = \frac{\partial H}{\partial p} = pq^4 \qquad (1)$$

and
$$L = p^2 q^4 - \frac{1}{2}(q^{-2} + p^2 q^4)$$

or
$$L = \frac{\dot{q}^2}{2q^4} - \frac{1}{2q^2} \quad \text{[using (1)]}.$$

Now, the Lagrange's equation of motion is
$$\frac{d}{dt}\left(\frac{\partial L}{\partial \dot{q}}\right) - \frac{\partial L}{\partial q} = 0.$$

That is,
$$\frac{d}{dt}\left(\frac{\dot{q}}{q^4}\right) + \frac{2\dot{q}^2}{q^5} - \frac{1}{q^3} = 0 = \ddot{q} - \frac{2\dot{q}^2}{q} - q.$$

Thus, the required equation of motion is
$$f(q, \dot{q}, \ddot{q}) = \ddot{q} - 2\frac{\dot{q}^2}{q} - q.$$

Hence, the correct choice is (c).

2. Assuming that the particle is located by its polar coordinates (r, θ), the potential due to the given central force is denoted by $V(r)$. The kinetic energy of the particle is given by $T = (1/2) m(\dot{r}^2 + r^2 \dot{\theta}^2)$. Then, the Lagrangian of the system is
$$L = T - V = \frac{1}{2}m(\dot{r}^2 + r^2\dot{\theta}^2) - V(r). \qquad (1)$$

387

The generalized momenta is

$$p_r = \frac{\partial L}{\partial \dot{r}} = m\dot{r}, \qquad p_\theta = \frac{\partial L}{\partial \dot{\theta}} = mr^2\dot{\theta}. \qquad (2)$$

Then,

$$\dot{r} = \frac{p_r}{m}, \qquad \dot{\theta} = \frac{p_\theta}{mr^2}. \qquad (3)$$

Now, the Hamiltonian is given by

$$H = \sum_{\rho=1}^{n} p_\rho \dot{q}_\rho - L$$

$$= p_r \dot{r} + p_\theta \dot{\theta} - \left[\frac{1}{2}m(\dot{r}^2 + r^2\dot{\theta}^2) - V(r)\right]$$

$$= p_r \left(\frac{p_r}{m}\right) + p_\theta \left(\frac{p_\theta}{mr^2}\right) - \left[\frac{1}{2}m\left(\frac{p_r^2}{m^2} + \frac{p_\theta^2}{m^2 r^2}\right) - V(r)\right]$$

or

$$H = \frac{p_r^2}{2m} + \frac{p_\theta^2}{2mr^2} + V(r)$$

Therefore, the correct choice is (b).

3. The correct choice is (c).
4. The correct choice is (d).
5. The correct choice is (b).
6. The correct choice is (a).
7. To describe the motion of a rigid body, the number of generalized coordinates required are six (three for translational motion and three for rotational motion). Since the rigid body is fixed with one of its points, there is no translational motion. Hence the correct choice is (c).
8. Since the rigid body is fixed at two points, the body can rotate about an axis, joining the two fixed points. Hence the number of degrees of freedom is one. Therefore, the correct choice is (d).
9. If $I_{xx} = I_{yy}$, we get $\dot{\omega}_z = 0$, implying that ω_z is constant. Mutliplying the first equation by ω_x, the second equation by ω_y and adding, we get

$$I_{xx}(\dot{\omega}_x \omega_x + \dot{\omega}_y \omega_y) = 0.$$

As, $I_{xx} \neq 0$, we have

$$\dot{\omega}_x \omega_x + \dot{\omega}_y \omega_y = \frac{1}{2}\frac{d}{dt}(\omega_x^2 + \omega_y^2) = 0$$

implying, $\omega_x^2 + \omega_y^2$ = constant. Therfore,
$$|\vec{\omega}| = (\omega_x^2 + \omega_y^2 + \omega_z^2)^{1/2} = \text{constant.}$$
Hence, the correct choice is (a).

10. The correct choice is (c) (see Example 3.1).
11. The correct choice is (d) (see Example 3.5).
12. The correct choice is (b).
13. The correct choice is (a).
14. The correct choice is (a).
15. $v_e = \sqrt{2gR}$. Taking $g = 9.806$ m/s^2 and radius of the earth $R = 6378$ km, we find that the escape velocity $v_e = 11.184$ km/s (also refer to Eq. 5.46). Therefore, the correct choice is (a).
16. The correct choice is (d). (see Example 7.9).
17. The correct choice is (c).
18. The kinetic energy of a particle of rest mass m_0 is given by
$$T = E - m_0 c^2 = m_0 c^2 \left(\frac{1}{\sqrt{1 - \frac{v^2}{c^2}}} - 1 \right)$$

As this equals $m_0 c^2$, we have at once
$$\frac{1}{\sqrt{1 - \frac{v^2}{c^2}}} = 2.$$

Therefore,
$$1 - \frac{v^2}{c^2} = \frac{1}{4}.$$

That is, $v = \sqrt{3}c/2$. Hence, the correct choice is (b).

19. Torque $\vec{G} = \vec{r} \times \vec{F} = \begin{vmatrix} \hat{i} & \hat{j} & \hat{k} \\ 2 & 4 & -3 \\ -7 & 4 & -5 \end{vmatrix} = -8\hat{i} + 31\hat{j} + 36\hat{k}$

Hence, the correct choice is (a).

20. The angular momentum of the system is
$$\vec{h} = \sum \vec{r}_i \times m_i \dot{\vec{r}}_i$$

$= (2\hat{i} + 4t^2 \hat{k}) \times 8t\hat{k} + 2(4t\hat{i} - \hat{k}) \times 4\hat{i} + 4[(\cos \pi t)\hat{i} + (\sin \pi t)\hat{j}] \times$
$[\pi(-\sin \pi t)\hat{i} + \pi(\cos \pi t)\hat{j}]$

$= -16t\hat{j} + 8\hat{j} + 4\pi(\cos^2 \pi t \hat{k} + \sin^2 \pi t \hat{k})$

Then,
$$\vec{h}\Big|_{t=1/2} = 4\pi \hat{k}.$$

Hence none of the given choices is correct.

21. The kinetic energy of the system is given by
$$T = \frac{1}{2}(\dot{x}^2 + \dot{y}^2)$$
$$= \frac{1}{2}\{(\dot{\phi} - \dot{\phi}\cos\phi)^2 + [(-\sin\phi)\dot{\phi}]^2\}$$
$$= \dot{\phi}^2(1 - \cos\phi).$$

The potential energy of the system is
$$V = mgh = gy = g(1 + \cos\phi).$$

Now, the Lagrangian of the system is: $T - V = \dot{\phi}^2(1 - \cos\phi) - g(1 + \cos\phi)$.
Hence the correct choice is (c).

22. Taking (r, θ) as generalized coordinates, let (x, y, z) be the Cartesian coordinates of the particle at a given instant, then we have $x = r\cos\theta$, $y = r\sin\theta$, $z = r\cot\alpha$. Therefore,
$$\dot{x} = \dot{r}\cos\theta - r\dot{\theta}\sin\theta$$
$$\dot{y} = \dot{r}\sin\theta + r\dot{\theta}\cos\theta$$
$$\dot{z} = \dot{r}\cot\alpha.$$

Hence the kinetic energy, T, of the particle is
$$T = \frac{1}{2}m(\dot{x}^2 + \dot{y}^2 + \dot{z}^2) = \frac{m}{2}(\dot{r}^2\csc^2\alpha + r^2\dot{\theta}^2).$$

The potential energy of the particle is given by
$$V = mgz = mgr\cot\alpha.$$

Thus, the Lagrangian for the motion is
$$L = T - V = \frac{m}{2}(\dot{r}^2\csc^2\alpha + r^2\dot{\theta}^2) - mgr\cot\alpha.$$

Therefore, the correct choice is (a).

23. Using Archimede's principle, the buoyant force on the cylinder is $(Az)\rho g$. Then, by Newton's second law, we have
$$\frac{W}{g}\frac{d^2z}{dt^2} = -Az\rho g$$

or
$$\frac{d^2z}{dt^2} + \frac{g^2 A\rho}{W}z = 0$$

The period of oscillation is thus found to be $2\pi\sqrt{W/(g^2 A\rho)}$. Hence the correct choice is (d).

24. Refer Example 5.10. The correct choice is (b).
25. Given,
$$\frac{GMm}{R^2} = mg \quad \text{or} \quad GM = gR^2$$
Therefore, the correct choice is (a).
26. The position vectors of the particles are respectively given by
$$\vec{r}_1 = \hat{i} - \hat{k}, \qquad \vec{r}_2 = -2\hat{i} + \hat{j} + 3\hat{k}, \qquad \vec{r}_3 = 3\hat{i} - \hat{j} + \hat{k}.$$
Then, the centre of mass of the system is given by
$$\vec{r} = \frac{3(\hat{i}-\hat{k}) + 5(-2\hat{i}+\hat{j}+\hat{k}) + 2(3\hat{i}-\hat{j}+\hat{k})}{3+5+2} = -\frac{1}{10}\hat{i} + \frac{3}{10}\hat{j} + \frac{7}{5}\hat{k}.$$
Hence the correct choice is (d).
27. Let $\vec{r} = x\hat{i}$ be the initial position vector of the particle P(say). Then, the acceleration of P is $(d^2x/dt^2)\hat{i}$. By Newton's second law, we have
$$4\frac{d^2x}{dt^2}\hat{i} = -8x\hat{i} \quad \text{or} \quad \frac{d^2x}{dt^2} + 2x = 0. \tag{1}$$

This is the governing equation of motion of the particle. The period of oscillation of the particle is $2\pi/\sqrt{2} = \sqrt{2}\pi$. Hence its frequency is $1/\sqrt{2}\pi$. Therefore, the correct choice is (d).
28. The time period τ of a simple pendulum is given by the formula
$$\tau = 2\pi\sqrt{\frac{l}{g}}.$$
It is known that g decreases as we go inside the mine, and, therefore, we observe that τ increases. Hence the correct choice is (c).
29. Recall Eq. (2.34), the Coriolis force is given by
$$F_{\text{cor}} = -2m\vec{\omega} \times \vec{V},$$
where $\vec{\omega}$ is the angular velocity of the rotating frame, and $\vec{V}$ is the velocity of the body in the rotating frame. If a body is thrown vertically upwards, $\vec{\omega}$ and $\vec{V}$ become parallel and therefore $F_{\text{cor}} = 0$ in the cases (b), (c) and (d).
Hence the right choice is (a).
30. Equation (2.33) gives the net fictitious force acting on the insect as
$$-m\vec{\omega}\times(\vec{\omega}\times\vec{r}) - 2m(\vec{\omega}\times\vec{v}) = -m[\omega\hat{k}\cdot r\hat{i})\omega\hat{k} - (\omega\hat{k}\cdot\omega\hat{k})r\hat{i}] - 2m\omega\hat{k}\times v\hat{i}$$
$$= (m\omega^2 r)\hat{i} - (2m\omega v)\hat{j}.$$
Hence the correct choice is (b).

31. Using the parallel axis theorem, we have $I = I_0 + MR^2$, where I_0 is the M.I. of the disc about an axis passing through its centre of mass and perpendicular to it (see Example 3.2). Thus,

$$I = \frac{1}{2}MR^2 + MR^2 = \frac{3}{2}MR^2.$$

Hence the correct choice is (c).

32. The M.I. of a cube about an edge is same as that of a square plate about a side, as in the following figure:

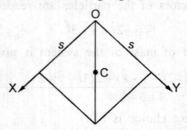

The M.I. of the square plate about OX is $(1/3)Ms^2$ (see Example 3.1). Similarly, the M.I. of the square plate about OY is also $(1/3)Ms^2$. Now, using the perpendicular axis theorem, the M.I. of the square plate about OZ, through O and perpendicular to the plane of the plate is

$$I_0 = \frac{M}{3}(s^2 + s^2) = \frac{2}{3}Ms^2. \qquad (1)$$

Since the cube is oscillating about the OZ-axis, it is a compound pendulum. The length of the equivalent simple pendulum is given by

$$l = \frac{I_0}{Ma} \qquad (2)$$

In this example,

$$a = OC = \frac{1}{2}\sqrt{2}s \qquad (3)$$

Now, substituting Eqs. (1) and (3) into Eq. (2), we get

$$l = \frac{(2/3)Ms^2}{(M/\sqrt{2})s} = \frac{2\sqrt{2}}{3}s.$$

Hence the correct choice is (c).

33. The correct choice is (b).
34. The correct choice is (d).
35. In view of Kepler's second law, the law of areas, the correct choice is (a).

36. As a consequence of Kepler's third law, the harmonic law, the correct choice is (c).

37. Since the particle is moving in a circular orbit, we have

$$\frac{mv^2}{r} = \frac{k}{r^3}.$$

Therefore,

$$\text{Kinetic energy} = \frac{1}{2}mv^2 = \frac{k}{2r^2}.$$

Also, from Eq. (5.17),

$$\text{Potential energy} = k \int_\infty^r \frac{1}{r^3} dr = -\frac{k}{2r^2}.$$

Thus, the total energy, that is, the sum of the K.E. and P.E. is zero. Hence, the correct choice is (c).

38. Observe that

$$L = T - V = \frac{1}{2}m\dot{r}^2 + \frac{1}{2}mr^2\dot{\theta}^2 - \left(-\frac{GMm}{r}\right).$$

That is,

$$L = \frac{1}{2}m\dot{r}^2 + \frac{1}{2}mr^2\dot{\theta}^2 + \frac{GMm}{r}.$$

Hence the correct choice is (b).

39. The Lagrange's equation of motion of the system is

$$\frac{\partial L}{\partial q_n} - \frac{d}{dt}\left(\frac{\partial L}{\partial \dot{q}_n}\right) = 0, \quad \text{implying} \quad \frac{d}{dt}\left(\frac{\partial L}{\partial \dot{q}_n}\right) = 0,$$

which means $\partial L/\partial \dot{q}_n$ = constant. In other words, the generalized momenta is constant. Hence the correct choice is (b).

40. The correct choice is (c).

41. The kinetic energy

$$T = \frac{1}{2}m(\dot{x}^2 + \dot{y}^2) \tag{1}$$

given $xy = c$ or $y = c/x$ and therefore,

$$\dot{y} = -\frac{c\dot{x}}{x^2}.$$

Thus,

$$T = \frac{1}{2}m\dot{x}^2\left(1 + \frac{c^2}{x^4}\right) \quad \text{and} \quad V = mgy = \frac{mgc}{x}$$

and
$$L = T - V = \frac{1}{2}m\dot{x}^2\left(1 + \frac{c^2}{x^4}\right) - \frac{mgc}{x}$$

Hence the correct choice is (b).

42. The correct choice is (b). This concept is known as *Hamilton variational principle*.

43. The given transformation, that is,
$$Q = q + p, \qquad P = q + \alpha p$$
is canonical if and only if the Lagrange brackets satisfy:
$$[q, q] = [p, p] = 0 \quad \text{and} \quad [q, p] = 1.$$
But
$$[q, p] = \frac{\partial Q}{\partial q}\frac{\partial P}{\partial p} - \frac{\partial Q}{\partial p}\frac{\partial P}{\partial q} = (1)(\alpha) - (1)(1) = 1,$$
which yields $\alpha = 2$. Hence the correct choice is (d).

44. We know that the total energy of a relativistic particle is $E^2 = p^2c^2 + m_0^2c^4$. But, given that $p = m_0 c$,
$$E^2 = 2m_0^2 c^4 \quad \text{or} \quad E = \sqrt{2}\, m_0 c^2.$$
Then, the momentum of a relativistic particle is found to be
$$p = m_0 c = \frac{m_0 v}{\sqrt{1 - \frac{v^2}{c^2}}} \quad \text{or} \quad v = \frac{c}{\sqrt{2}}.$$

Hence the correct choice is (c).

45. Let us recall Eq. (8.47) which states that the kinetic energy of a moving particle = total energy − rest mass energy. Therefore,
$$T = E - mc^2 = m'c^2 - mc^2 = \frac{mc^2}{\sqrt{1 - \frac{v^2}{c^2}}} - mc^2.$$

Hence the correct choice is (c).

46. We know that $E^2 = p^2c^2 + m^2c^4$. Now, using the given data, we find
$$(2mc^2)^2 = p^2c^2 + m^2c^4 \quad \text{or} \quad p^2 = 3m^2c^2.$$

Therefore, $p = \sqrt{3}\, mc$. Hence the correct choice is (a).

47. It is known that the Fitzgerald length contraction formula is given by
$$L' = L\sqrt{1 - \frac{v^2}{c^2}}.$$

Using the given data, we have at once

$$3 = 5\sqrt{1 - \frac{v^2}{c^2}} \quad \text{or} \quad 9 = 25\left(1 - \frac{v^2}{c^2}\right)$$

which simplifies to $v = 4c/5$. Hence the correct choice is (c).

48. Given

$$L = \frac{m}{2}(\dot{q}^2 \sin^2 \omega t + \dot{q}q\omega \sin 2\omega t + q^2\omega^2).$$

The transformation is $Q = q \sin \omega t$, therefore,

$$\dot{Q} = \dot{q} \sin \omega t + q\omega \cos \omega t$$

and

$$\dot{Q}^2 = \dot{q}^2 \sin^2 \omega t + q^2\omega^2 \cos^2 \omega t + q\dot{q}\omega \sin 2\omega t.$$

Now, using this result, L becomes

$$L = \frac{m}{2}(\dot{Q}^2 + q^2\omega^2 \sin^2 \omega t) = \frac{m}{2}(\dot{Q}^2 + Q^2\omega^2)$$

From the definition of H, we have

$$H = \frac{\partial L}{\partial \dot{Q}}\dot{Q} - L = \frac{m}{2}(\dot{Q}^2 - Q^2\omega^2) \tag{1}$$

But,

$$P = \frac{\partial L}{\partial \dot{Q}} = m\dot{Q},$$

Using this result, Eq. (1) becomes

$$H = \frac{m}{2}\left(\frac{P^2}{m^2} - \omega^2 Q^2\right).$$

Hence the correct choice is (b).

49. In a circular orbit, the gravitational attraction is equal and opposite to the centrifugal force. Therefore,

$$\frac{GM_e M_s}{r^2} = M_s \omega^2 r$$

or

$$r^3 = \frac{GM_e}{\omega^2} = \frac{GM_e \tau^2}{(2\pi)^2}, \tag{1}$$

where τ is the period. We want ω for the satellite orbit to be equal to the angular frequency ω_e of the Earth about its axis. The angular frequency of the Earth $\omega_e = 2\pi/(8.64 \times 10^4) = 7.3 \times 10^{-5}$ rad/s.

Taking $\omega = \omega_e$, Eq. (1) gives

$$r^3 = \frac{(6.67 \times 10^{-11})(5.98 \times 10^{24})}{(7.3 \times 10^{-5})^2}$$

$$\approx 0.75 \times 10^{23}$$

$$= 75 \times 10^{21} \text{ cubic metres}$$

$$= 75 \times 10^{27} \text{ cubic cm}$$

Therefore,

$$r = 4.2 \times 10^9 \text{ cm} = 4.2 \times 10^4 \text{ km}$$

Hence the correct choice is (a).

50. We know that the wave equation is invariant under Lorentz transformation. Hence, the correct choice is (b).

51. We know,

$$m = \frac{m_0}{\sqrt{1 - \frac{v^2}{c^2}}}.$$

Therefore,

$$1 - \frac{v^2}{c^2} = \left(\frac{m_0}{m}\right)^2 = \left(\frac{100}{101}\right)^2 = 0.9803$$

Thus,

$$v = 0.14c = 0.14 \times 3 \times 10^8 = 4.2 \times 10^7 \text{ m/s}$$

Hence the correct choice is (a).

52. Note from Eq. (12) of Example 2.3, the deflection is given by

$$d = \frac{\omega g}{3} \left(\frac{2h}{g}\right)^{3/2} \cos \lambda$$

$$= \frac{\omega}{3} \left(\frac{8h^3}{g}\right)^{1/2} \cos \lambda$$

$$= \frac{7.3 \times 10^{-5}}{3} \left[\frac{8 \times (200)^3}{9.8}\right]^{1/2} \cos 45°$$

$$= 0.04379 \text{ m}$$

$$= 4.379 \text{ cm}.$$

Hence the correct choice is (a).

Index

Action along a curve, 249
Addition law of velocities, 304
Angular momentum principle, 7, 31, 32, 120
Angular velocity, 35, 37, 106
Apogee/Aphelion, 151
Apses, 172
Apsidal quadratic, 162
Areal velocity, 148
Attitudinal stability of satellite, 139
Autumnal equinox, 152
Azimuth, 154

Bakers equation, 203
Booster, 331
Brachistachrone, 292
Burn-out time, 322

Calculus of variation, 283
Canonical transformations, 273
    group property of, 279
Catenoids, 290
Celestial sphere, 150
    equator, 151
    horizon, 151
    latitude, 152
    longitude, 152
Central force motion, 155
Centre of mass, 28, 29, 61, 180
Centrifugal force, 44
Centripetal acceleration, 10, 37
Clausius theorem, 350
Communication satellites, 169
Compound pendulum, 109

Conservation of energy, 7, 19
Conservative system, 210
Constraints, 207
Contact transformation, 248
Coordinate systems
    cylindrical, 12
    radial and transverse, 10
    spherical, 13
Coriolis force, 43, 45
Cycloid, 290

D'Alemberts principle, 225
Declination, 153
Degrees of freedom, 208
Delta variation, 284
Double pendulum, 245
Dynamics of rigid body, 59

Earth
    equator, 152
    radius, 164
Eccentric anomaly, 189, 191
Ecliptic, 152
Einstein's mass–energy
    relation, 310
Elevation, 154
Elliptic
    integral, 25
    orbit, 164
Euler angles, 103
Euler–Lagranges equations, 285
Euler's equations for a rigid body motion, 113, 115
Exhaust velocity, single-stage rocket, 321

398 Index

Fitzgerald contraction, 320
Functionals, 283

Galilean transformation, 296
Generalized coordinates, 208
  momenta, 231, 238
Geo-stationary
  launch vehicle, 319
  satellite, 169
Gravitational constant, 149
Great circle, 150
Gyroscopic motion, 118, 143

Hamiltonian function, 231
Hamilton–Jacobi equation, 260
Hamilton's canonical equations, 232
Hamilton's characteristic function, 286
Hamilton's principle, 248, 253, 256
Harmonic oscillator, 26, 214, 233, 265
Holonomic system, 211, 245
Hyperbolic orbit, 166, 203

Ignorable coordinates, 238
Impulsive force, 56
Inclination of the orbit, 187
Inertia tensor, 92
Inertial frames, 3, 40
Integrals of energy, 156
Invariable plane, 347
Inverse square force, 159

Jacobi's identity, 283, 294
Jacobi theorem, 261

Kepler's equation, 189
Kepler's laws, 147–149
Kepler's problem, 215, 238, 242, 263
Kinetic energy, 8, 87, 89
Konig theorem, 89

Lagrange brackets, 274
Lagrange multiplier, 338
Lagrange points, 362
Lagrange's equations, 211

Latitude and longitude of a place, 153
Line of nodes, 105
Linear momentum principle, 28
Liquid propellants, 325
Liouville's theorem, 267
Lorentz transformation, 298
Launch-site, selection of, 339

Meridian, 153
Michelson–Morley experiment, 297
Momental ellipsoid, 80
Moment of inertia, 59, 64–78
Momentum
  angular 6, 7
  linear, 2, 5
Multi-statge rocket, 335

n-body problem, 344, 345
Natural motion, 248
Newtonian relativity, 41
Newton's law of gravitation, 149
Newton's laws of motion, 1
Non-inertial frame, 40
Nose cone, single stage rocket, 321
Nutational motion, 123

Orbital elements, 186
Orbital motion, 147
Optimization of multi-stage rocket, 335

Parabolic orbit, 163
Parallel-axis theorem, 62
Payload, 321
Perigee, 151
Perpendicular axis theorem, 63
Perturbations, 344
Purtabitive function, 357
Phase space, 267
Point transformation, 271
Poisson brackets, 281
Polar equation of orbit, 158
Potential energy, 16
Precessional motion, 122
Principal axes, 84
Principal moment of inertia, 85
Principle of least action, 257

Principle of virtual work, 225
Product of inertia, 60
Propellant, 321

Radius of gyration, 64
Rate of
   growth, 40
   transport, 40
Relative motion, 181
Rest mass, 308
Rheonomic system, 210
Right ascention, 153
Rigid body, 34, 87
Rocket dynamics, 319
Rotating frames, 38
Routhian function, 240

Satellite, 150
   launch vehicle, 319
   orbits, 169, 174, 175
Scleronomic system, 210
Simple pendulum, 22
Single-stage rocket, 321
Solar system, 150
Space of events, 249
Space–time equivalence, 298
Special theory of relativity, 295
Specific impulse, 325
Spherical pendulum, 217, 239
Spinning top, 221
Structural factor, 322
Symmetrical top, 90, 243

Three-body problem, 353
Thrust of rocket motor, 324
Thrust–time curve, 324
Time dialation, 303
Top
   sleeping condition, 146
   steady precession of, 128, 129
Torque, 6
Torque-free motion of earth, 132
Translational motion, 34
Two-body problem, 180
Two-stage rocket, 332

Units and dimensions, 4

V-2 rocket, 319
Variation
   of action, 250
   of mass with velocity, 308
Vernal equinox, 152
Virial theorem, 349
Virtual work, 225
von Braun, 319

Work–energy theorem, 16

Zenith, 152
Zenith angle, 13